Understanding Modern Transistors and Diodes

Written in a concise, easy-to-read style, this text for senior undergraduate and graduate courses covers all key topics thoroughly. It is also a useful self-study guide for practising engineers who need a complete, up-to-date review of the subject.

Key features:

- Rigorous theoretical treatment combined with practical detail
- A theoretical framework built up systematically from the Schrödinger Wave Equation and the Boltzmann Transport Equation
- Covers MOSFETS, HBTs, HJFETS, solar cells and LEDs.
- Uses the PSP model for MOSFETS
- Describes the operation of modern, high-performance transistors and diodes
- Evaluates the suitability of various transistor types and diodes for specific modern applications
- Examines solar cells and LEDs for their potential impact on energy generation and reduction
- Includes a chapter on nanotransistors to prepare students and professionals for the future
- Rigorous treatment of device capacitance
- Provides results of detailed numerical simulations to compare with analytical solutions
- End-of-chapter exercises to aid understanding
- Online availability of sets of lecture slides for undergraduate and graduate courses

David L. Pulfrey is a Professor in the Department of Electrical and Computer Engineering at the University of British Columbia, Canada, where he has been since receiving his Ph.D. in 1968 from the University of Manchester, UK. He has won teaching awards at the university-, provincial- and international-levels. Most recently he won the 2009 IEEE Electron Devices Society Education Award "for contributions to the teaching of electron devices at both the undergraduate and graduate levels". He has received recognition for his research work on a wide range of semiconductor devices by being elected Fellow of the IEEE in 2000, and Fellow of the Canadian Academy of Engineering in 2003.

Understanding Modern Transistors and Diodes

DAVID L. PULFREY

Department of Electrical and Computer Engineering
University of British Columbia
Vancouver, BC V6T1Z4
Canada

CAMBRIDGE UNIVERSITY PRESS
Cambridge, New York, Melbourne, Madrid, Cape Town, Singapore,
São Paulo, Delhi, Dubai, Tokyo

Cambridge University Press
The Edinburgh Building, Cambridge CB2 8RU, UK

Published in the United States of America by Cambridge University Press, New York

www.cambridge.org
Information on this title: www.cambridge.org/9780521514606

First published 2010

Printed in the United Kingdom at the University Press, Cambridge

A catalogue record for this publication is available from the British Library

ISBN 978-0-521-51460-6 Hardback

Additional resources for this publication at www.cambridge.org/9780521514606

To Eileen

Contents

Preface

Understanding Modern Transistors and Diodes is a textbook on semiconductor devices with three objectives: (i) to provide a rigorous, yet readable, account of the theoretical basis of the subject of semiconductor devices; (ii) to apply this theory to contemporary transistors and diodes so that their design and operation can be thoroughly understood; (iii) to leave readers with a sense of confidence that they are well equipped to appreciate the workings of tomorrow's devices, and to participate in their development.

There are many books on semiconductor devices, often with similar objectives, and it is reasonable to ask: why write another one? The answer is two-fold: firstly, after teaching and researching in the area for 40 years, I have a strong personal viewpoint on how the subject can best be presented to students; secondly, we are at a particularly interesting point in the development of the subject – we are at the micro/nano boundary for high-performance transistors, and we are on the threshold of seeing optoelectronic diodes make a contribution to our planet's sustainability.

These circumstances are new, and are quite different from those of 20 years ago when I was last moved to write a book on semiconductor devices. At that time the major development was the incorporation of thousands of transistors into monolithic integrated circuits. To design and analyse such circuits, the transistors were represented by a set of model parameters. One could use these parameters to design a circuit without understanding how they related to the physical properties of the actual transistors comprising the circuit. To address this deficiency I co-authored a book with Garry Tarr in 1989 that specifically linked circuit-model parameters to the physical properties of transistors and diodes.[1]

Today, after a further 20 years of teaching and researching in the area of solid-state devices, I find myself lecturing on, and needing to know more about: the effect of miniaturization on the performance of silicon field-effect transistors, as used in increasingly dense integrated circuits and memories; the displacement of the silicon bipolar transistor from its traditional areas of strength (high-frequency, high-power, low-noise) by heterostructural devices based on compound semiconductors; how device engineers and physicists can address sustainability issues in their domain, particularly the generation of electricity from a renewable source via more cost-effective solar cells, and the reduction of electricity usage for lighting via high-brightness light-emitting diodes. Sometimes I feel as though the trends in semiconductor devices are creating

---

[1] D.L. Pulfrey and N.G. Tarr, *Introduction to Microelectronic Devices*, Prentice-Hall, 1989.

an impossible situation: the need for greater depth of knowledge in a wider variety of devices.

The solution to this dilemma comes back to the first objective of this book: provide a rigorous and digestible theoretical basis, from which the understanding of devices of the modern era, and of the near future, follows naturally. This is how *Understanding Modern Transistors and Diodes* meets the challenge of covering a wide breadth of topics in the depth they warrant, while managing to limit the material to that which can be covered in one or two one-term courses. The requisite physics is treated properly once and is then approximated, and seen to be approximated, where justifiable, when being applied to various devices. The physics has to be quantum mechanical for several reasons: band structure is important for all the devices we discuss, particularly for heterostructural diodes and transistors of both field-effect and bipolar varieties; electron-photon interactions are obviously relevant in solar cells and light-emitting diodes; tunnelling is an important leakage-current mechanism in field-effect transistors; future one-dimensional transistors may be so short that ballistic, rather than dissipative, transport will be operative. Even in 'classical' devices transport must be treated rigorously in view of the trends towards miniaturization: the Drift-Diffusion Equation cannot be blindly applied, but must be justified after a proper treatment of its parent, the Boltzmann Transport Equation. One intermediate solution to this equation, the charge-density continuity equation, provides the basis for our rigorous and formal description of capacitance. This device property is crucially important to the transistors presented in the application-specific chapters in the book on digital switching, high-frequency performance and semiconductor memories. As a final emphasis on the rigour of this book, the traditional SPICE-related model for the MOS field-effect transistor is put in its rightful place, i.e., as a computationally expedient approximation to the 'surface-potential' model. If SPICE has helped design circuits that have enabled higher performance computers, then that has been its downfall, because those computers can now permit the more rigorous surface-potential model to be used for the more accurate simulation of integrated circuits!

Understanding Modern Transistors and Diodes is intended for students at the graduate or senior-undergraduate level who are studying electronics, microelectronics or nanoelectronics, within the disciplines of electrical and computer engineering, engineering physics or physics. However, there is sufficient material on basic semiconductor theory and elementary device physics for the book to be appropriate also for a junior-level course on solid-state electronic devices. Additionally, the inclusion in the book of specific chapters on the application of the foundation material to modern, high-performance transistors and diodes, and a glimpse into the future of true nanotransistors, should make the book of interest to practitioners and managers in the semiconductor industry, particularly those who have not had the opportunity to keep up with recent developments in the field. It is my hope that the depth and breadth of this book might make it a 'one-stop shop' for several levels of courses on semiconductor devices, and for device-practitioner neophytes and veterans alike. The material in this book, in various stages of development, has been used by me for senior-level undergraduate courses and for graduate-level courses on semiconductor devices at UBC, for short courses to engineers at PMC-Sierra in Vancouver, and to graduate students at the University of Pisa and at the Technical

University of Vienna. I thank all those students of these courses who have commented on the material and have sought to improve it.

As an undergraduate I focused on 'heavy-current electrical engineering', and never benefited from a course on semiconductor devices. I am basically 'self-taught' in the area, and I think that this has attuned me particularly well to the nature of the difficulties many students face in trying to master this profound subject. Hopefully this book circumvents most of these obstacles to the understanding of how semiconductor devices work. If it does, then thanks are due to many people who have enlightened me over my 40 years of working in the subject area, both as a professor at the University of British Columbia, and as a visiting research engineer at various industry, government, and university laboratories around the world. I particularly want to mention Lawrence Young, who hired me as a postdoc in 1968, and thereby started my transformation to a 'light-current electrical engineer'. I owe a great debt of gratitude to my graduate students, with whom I have worked collegially, learning with them, and sharing the work 'in the trenches' as much as possible. One of the great pleasures of writing this book has been to call on some of them, and on some former undergraduates too, to make sure that the material in some of the device-specific chapters in the book is truly modern. Particularly, I wish to thank: Alvin Loke (AMD, Colorado) for his enthusiastic support, his insights into the finer points of modern, high-performance CMOS devices and his arrangement of the procurement of the cover photograph from AMD's Dresden laboratory; Tony St. Denis (Triquint, Portland) for provision of material on high-frequency and low-noise heterojunction field-effect transistors; Mani Vaidyanathan (University of Alberta) for his insights into high-frequency devices, and for his encouragement; Leonardo Castro (Qimonda, Munich) for helpful details on DRAMs; David John (NXP, Eindhoven) for useful information on silicon power transistors, and for alerting me to Philips' version of the MOSFET surface-potential model; Shawn Searles (AMD, Austin) for sharing his thoughts on where Si CMOS is heading; Gary Tarr (Carleton University) for commenting on the solar cell chapter. I also wish to thank Ivan Pesic of Silvaco Data Systems for making a copy of his company's excellent simulation software, Atlas, available to me during 2008. At Cambridge University Press, England, I thank Julie Lancashire for her encouragement of this project, and Sarah Matthews, Caroline Brown and Richard Marston for their assistance in bringing it to fruition.

Most 'part-time' authors of technical books comment on the interruptions to their family life that writing a textbook entails, and I am no exception. My children, their spouses and my grandchildren are my friends, and I am conscious of the time I have missed spending with them. I hope they will think that this book has been worth it. The writing of it has been sustained by the encouragement, support and understanding of my wife, Eileen, to whom I give my deepest thanks.

David Pulfrey
Vancouver

1   Introduction

It is highly probable that you will use a laptop computer when doing the exercises in this book. If so, you may be interested to know that the central processing unit of your computer resides in a thin sliver of silicon, about 1 square centimetre in area. This small chip contains over 100,000,000 Si MOSFETs,[1] each about a thousand times smaller than the diameter of a human hair! The slender computer that you nonchalantly stuff into your backpack has more computing power than the vacuum-tube computers that occupied an entire room when I was a student over 40 years ago.

When you are reading this book, you may be distracted by an incoming call on your cell 'phone. That may get you wondering what's inside your sleek 'mobile'. If you opened it up, and knew where to look, you'd find some GaAs HBTs.[2] These transistors can operate at the high frequencies required for local-area-network telecommunications, and they can deliver the power necessary for the transmission of signals.

Of course, a cell 'phone nowadays is no longer just a replacement for those clunking, tethered, hand-sets of not so long ago: it is also a camera and a juke box. The immense storage requirements of these applications are met by Flash memory, comprising more millions of Si MOSFETs.

Your cell 'phone is really a PDA,[3] and probably also allows internet access, in which case you may wonder how signals from around the globe find their way into your machine. Somewhere in the communications chain there's probably a low-noise amplifier to receive tiny signals and not add undue noise to them. GaAs HBTs are good for this, but even better are InP HEMTs.[4] If satellites are involved, then the base station will employ high-power transistors, possibly lateral-diffused Si MOSFETs, or maybe GaN HJFETs.[5]

So, without straying very far from where you are sitting as you read this, you have tangible evidence of the dramatic influence electronics has on the way many of us conduct our business and recreation. All the different transistors mentioned above are described in this book, and are grouped according to their ability to perform: in high-speed digital logic; at high frequencies; with low noise; at high output power; in semiconductor memory.

---

[1] Metal-Oxide-Semiconductor Field-Effect Transistors.
[2] Heterojunction Bipolar Transistors.
[3] Personal Digital Assistant.
[4] High Electron Mobility Transistors.
[5] Heterojunction Field-Effect Transistors.

Of course, our electronics-oriented activities would not be possible if the supply of electricity were curtailed. This could happen, either by the exhaustion of the Earth's store of fossil fuels, or by the threat to our habitable environment that the extraction and use of them entails. Alternate, and renewable, forms of electrical energy generation are desirable; photovoltaics, using semiconductor diodes as solar cells, is an attractive proposition. How solar cells work is described in this book. We look at traditional Si cells, and at both thin-film cells and tandem cells for possible implementation in the future.

You may know that about 20% of the world's energy consumption goes into producing light. Glance up at the incandescent light bulb that is illuminating your room: it's so inefficient that if you had a few of them in use, then you probably wouldn't need to heat your study in winter! Again, some alternative is needed; LEDs[6] using diodes made from compound semiconductors are beginning to make an impact in this area. We describe how high-brightness LEDs work, and look at ways of producing white light.

To understand the operation of all these transistors and diodes, and to provide the knowledge base that will enable you to understand new devices as they appear, and to design better devices yourself, a solid, physical understanding of semiconductors must be attained. The first part of this book is devoted to this. The emphasis is on Quantum Mechanics, as this branch of physics is needed increasingly to understand transistors as they move from the micro- to the nano-realm, and also, of course, to understand interactions between electrons and holes and photons in optoelectronic diodes.

The book ends with a brief look at cylindrical nanotransistors, the future development of which may perhaps involve you?

Enjoy the book!

---

[6] Light-Emitting Diodes.

2   Energy band basics

Louis de Broglie, in his Ph.D. thesis of 1924, postulated that every object that has
momentum p also has a wavelength λ:

$$p = \frac{h}{\lambda}, \tag{2.1}$$

where h is Planck's constant. Macroscopic objects of our everyday experience have
extremely short wavelengths, so they are invariably viewed as particles, with a point
mass and an observable trajectory. Contrarily, microscopic objects can have much longer
wavelengths, and may do wave-like things, such as diffract around other microscopic
objects. Electrons and atoms are microscopic objects, so when we need to consider them
both together we must take a quantum-mechanical, rather than a classical, approach.
This is what we do in this chapter. Our initial goal is to develop the concept of energy
bands, representing ranges of permissible energies for electrons within a solid. We then
seek to provide an understanding of related concepts that are used throughout this book:
electron states, crystal momentum, band structure, holes, effective mass, energy band
diagrams. These objectives are most directly arrived at from a consideration of the
periodic nature of the potential through which the electrons would move in a perfectly
crystalline material.

2.1   Periodic structures

Crystalline structures are based on a matrix of points called a **Bravais lattice**. For the
Group IV semiconductors and most of the III-V semiconductors that are considered in
this book, the Bravais lattice is the face-centred cubic lattice. To this underlying structure
are added the actual atoms that constitute the **basis** of a particular material. The basis
for Si, Ge, GaAs, InP, for example, comprises two atoms, which are shown as any
neighbouring pair of shaded and unshaded atoms in Fig. 2.1. Each atom occupies a site
on a face-centred cubic lattice, so the actual structure comprises two, interpenetrating,
face-centred cubic lattices. When the two atoms are the same, as in the elemental
semiconductors Si and Ge, the structure is called **diamond**. When the two atoms are
different, e.g., Ga and As, the structure is referred to as **sphalerite** or **zinc blende**. The
bonding of atoms in these structures is tetragonal, as shown by the linkages in Fig. 2.1.

Instead of trying to deal with the countless numbers of atoms that comprise an actual
piece of crystalline material, it is often convenient to capture the structural essence of a

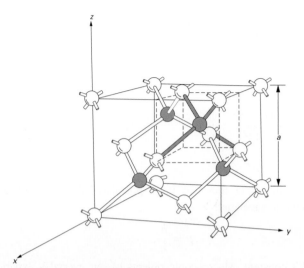

Figure 2.1 The diamond and sphalerite crystal structure. There are two, interpenetrating, face-centred cubic (FCC) lattices, one comprising the shaded atoms and the other comprising the unshaded atoms. The corresponding points in each FCC lattice are displaced by $\frac{a}{4}(\hat{x} + \hat{y} + \hat{z})$, where a, the **lattice constant**, is the length of the side of the cube. Adapted from Sze [1], © John Wiley & Sons, Inc. 1985, reproduced with permission.

crystal in its primitive unit cell, or, simply, **primitive cell**. This is a volume, containing precisely one lattice point, from which, by appropriate rotations and translations, the space of the Bravais lattice can be exactly filled. There is no unique primitive unit cell for a given Bravais lattice, and one of them is shown by the dashed lines in Fig. 2.1. Another primitive unit cell is the **Wigner-Seitz primitive cell**, the construction of which is illustrated in Fig. 2.2 for a simple face-centred rectangular matrix of unshaded atoms. The primitive unit cell in this case is a hexagon, which also contains one of the shaded atoms from an identical matrix of atoms. Thus, this particular crystal structure has a basis of two. For a real 3-D crystal the lines between nearest-neighbour atoms are bisected by planes; and for the face-centred cubic lattice the Wigner-Seitz cell is a rhombic dodecahedron [2, Fig. 1.8b].

2.2   Periodic potential

To illustrate the relationship between energy and momentum in a crystalline material, we consider a 'toy' structure comprising a one-dimensional array of primitive cells, with each cell having a basis of unity, and the atom being monovalent (see Fig. 2.3a). The potential energy of a single electron due to Coulombic interaction with the ion cores of the monovalent atoms is shown in Fig. 2.3b. However, we are not interested here in the precise form of the potential energy: we are only concerned with its periodicity. Therefore, we reduce the potential-energy profile to the delta-function representation shown in Fig. 2.3d. Don't be alarmed that the last profile might not be very realistic:

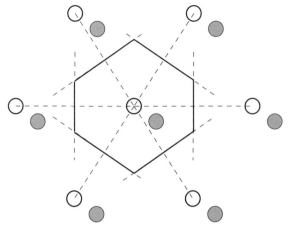

Figure 2.2 Example of a 2-D crystal comprising simple face-centred rectangular arrays of unshaded and shaded atoms. The Wigner-Seitz primitive unit cell is shown by the solid lines. These lines connect the perpendicular bisectors of the lines joining one unshaded atom to each neighbouring unshaded atom. One atom from the shaded array falls within the primitive unit cell; thus, this crystal structure has a basis of two atoms.

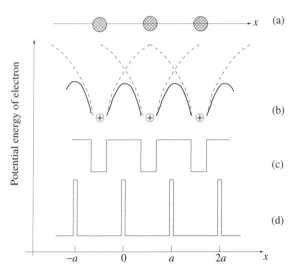

Figure 2.3 (a) 1-D periodic array of primitive cells, each cell containing one monovalent atom. (b) 1-D Coulombic potential energy for an electron in the 1-D array. Dashed lines are the potential energies due to a single ion core. Solid lines are the total potential energy. (c) 1-D square well representation of (b). (d) 1-D delta-function representation of (c).

even Fig. 2.3b is inaccurate, as it omits effects such as: the potential energy of an electron due to the proximity of other electrons; the different spacing between atoms in different directions of the real (3-D) crystal; and the possible presence of dissimilar elements in the crystal, e.g., as in compound semiconductors, such as GaAs. The important fact is that any periodic potential leads to the revelation of energy bands, and, therefore, will do for our present purpose.[1] The profile in Fig. 2.3d, comprising N delta-function potential barriers spaced a apart, can be expressed as

$$U(x) = \beta \sum_{l=0}^{N-1} \delta(x - la),$$ (2.2)

where $\delta(x)$ is the Dirac delta function and β is some constant.[2]

2.3   Schrödinger's equation

When considering the fine details of an electron's motion in a solid, we need to consider its wave-like nature. The appropriate equation is the Schrödinger Wave Equation, which was originally postulated in 1925 to provide a formal description of the experimentally observed, discrete frequencies of light emission from an excited hydrogen atom. You can have confidence in the equation because, in the intervening 80+ years, no experiments have been reported that give results contrary to the predictions of the equation. The form of the equation of interest to us here is the **time-independent Schrödinger Wave Equation**, i.e., in one dimension,

$$-\frac{\hbar^2}{2m_0} \frac{d^2\psi(x)}{dx^2} + U(x)\psi(x) = E\psi(x),$$ (2.3)

where m_0 is the electron rest mass, $\psi(x)$ is the position-dependent part of the **electron wavefunction** $\Psi(x, t)$, U is the potential energy and E is the total energy.[3]

Thinking in terms of conservation of energy, it can be seen that the first term in (2.3) must relate to kinetic energy. Often, the first two terms are grouped together and described as the **Hamiltonian** of the system

$$\mathcal{H}\psi = E\psi,$$ (2.4)

where the Hamiltonian $\mathcal{H}$ operates on the wavefunction to describe the total energy of the system.

Niels Bohr's statistical interpretation of the wavefunction is particularly helpful in getting a feeling for what Ψ really is: $\Psi \Psi^* dx \equiv |\Psi(x, t)|^2\, dx$ is the probability of finding the electron between x and $(x + dx)$ at time t.[4] If the electron is somewhere in x (1-D case), then it follows that $\int_{-\infty}^{+\infty} |\Psi(x, t)|^2\, dx = 1$. Equivalently, $\int_{-\infty}^{+\infty} |\psi(x)|^2\, dx = 1$. Thus, $\Psi(x, t)$ and $\psi(x)$ enable us to compute the probability of finding an electron

---

[1] If you insist on giving some physical significance to the potential profile in Fig. 2.3d, then you may wish to view the electron as being largely confined to the vicinity of an atom, but having some probability of tunnelling to a neighbouring, identical, region through a thin potential barrier.

[2] The property of the delta function that is relevant here is: $\delta(y) = 0$ if $y \neq 0$, and $\delta(y) = \infty$ if $y = 0$.

[3] This equation follows from the full, time-dependent Schrödinger Wave Equation, which describes the full wavefunction, i.e., in the 1-D case, $\Psi(x, t)$. In all our work we will take the potential energy to be independent of time. This allows the full equation to be solved by the method of Separation of Variables, for which solutions are simply: $\Psi(x, t) = \psi(x) f(t)$, where $f(t) = \exp(-iEt/\hbar)$ and $E = \hbar\omega$. Thus, we can solve (2.3) for $\psi(x)$, and then always multiply by $f(t)$ to get the full time dependence if we need it.

[4] The superscript * denotes the complex conjugate.

somewhere in space at some time. This is how quantum mechanics works: it deals in probabilities. This is not an inadequacy of the theory; it is a description of how Nature appears to work at the level of very tiny entities.

2.4   Energy bands

Consider the periodic delta-function potential in Fig. 2.3d. Here, we use it to develop an understanding of energy bands, closely following the treatment of Griffiths [3]. In the region $0 < x < a$ the potential energy is zero, so, from (2.3)

$$\frac{d^2\psi}{dx^2} + g^2\psi(x) = 0\,, \tag{2.5}$$

where

$$g = \frac{\sqrt{2m_0 E}}{\hbar}\,. \tag{2.6}$$

The general solution is

$$\psi(x) = A\sin(gx) + B\cos(gx), \quad (0 < x < a)\,. \tag{2.7}$$

A and B are constants that need to be evaluated by considering the boundary conditions. The general rules are:

- ψ must be continuous at a boundary;
- $d\psi/dx$ must be continuous at a boundary, except when the potential energy goes to infinity.[5]

In our problem we have lots of boundaries, and at each one $U \to \infty$. Fortunately, because of the periodic nature of the potential, we can reach a solution quite easily by appealing to **Bloch's Theorem**, which states that for a periodic potential $U(x + a) = U(x)$, the solutions to Schrödinger's equation satisfy

$$\psi_k(x) = u_k(x)e^{ikx}\,, \tag{2.8}$$

where $u_k(x)$ has the periodicity of the lattice, and the subscript k indicates that $u(x)$ has different functional forms for different values of the **Bloch wavenumber** k. Note that if u is not periodic but is a constant, then the Bloch wave becomes a plane wave. Therefore, a Bloch wave, given by (2.8), is a plane wave modulated by a function that has the periodicity of the lattice. An alternative way of stating Bloch's Theorem follows from (2.8), namely

$$\psi_k(x + a) = e^{ika}\psi_k(x)\,. \tag{2.9}$$

---

[5] If there is a discontinuity in $d\psi/dx$, then the kinetic-energy term in (2.3) $\to \infty$, but the equation is still satisfied if $U \to \infty$. When we resort to the 'Effective-mass Schrödinger Wave Equation', the boundary condition for the derivative of ψ must also include what we shall call the effective mass, if this property changes across the boundary (see Section 2.11).

Note that this equation does not state that $\psi_k(x)$ is periodic, but it does lead to $|\psi_k(x)|^2$ being periodic. The latter is comforting because one would expect an electron to have an equal probability of being at any of the identical sites in the linear array. The periodicity breaks down at the edges of the crystal, but that shouldn't have a significant effect on the electrons deep within the crystal if the array is very long compared to the separation between atoms, i.e., if N, the number of primitive cells, is very large. Mathematically, we can impose complete periodicity by bending the array into a circle so that $x = -a$ follows $x = (N-2)a$ in Fig. 2.3d. We then have a convenient, so-called **periodic boundary condition**:

$$\psi_k(x + Na) = \psi_k(x).\tag{2.10}$$

Using this in (2.9), yields

$$e^{ikNa}\psi_k(x) = \psi_k(x),\tag{2.11}$$

from which it is clear that

$$k = \frac{2\pi n}{Na},\quad (n = 0, \pm1, \pm2, \pm3, \cdots),\tag{2.12}$$

where n is an integer. (2.9) can now be used to obtain the wavefunction in the region $-a < x < 0$ of Fig. 2.3d:

$$\psi_k(x) = e^{-ika}[A \sin g(x+a) + B \cos g(x+a)], \quad (-a < x < 0).\tag{2.13}$$

Now that we have expressions for the wavefunctions in adjoining regions we can use the matching conditions for ψ and $d\psi/dx$ to evaluate or eliminate the constants A and B. Matching the wavefunctions at $x = 0$ gives

$$B = e^{-ika}[A \sin(ga) + B \cos(ga)].\tag{2.14}$$

Because of the delta function, the derivative of ψ is not continuous at $x = 0$, so we need to find the discontinuity in order to get another expression linking A and B. For $U(x) = \beta\delta(x)$, which comes from (2.2), the discontinuity is

$$\Delta(\frac{d\psi}{dx}) = \frac{2m_0\beta}{\hbar^2}\psi(0).^6\tag{2.15}$$

Thus, it follows from the derivatives of ψ at $x = 0$ that

$$gA - e^{-ika}g[A \cos(ga) - B \sin(ga)] = \frac{2m_0\beta}{\hbar^2}B.\tag{2.16}$$

---

[6] To obtain this, integrate Schrödinger's equation over a tiny interval spanning $x = 0$. The integral of the $d^2\psi/dx^2$ term is precisely the discontinuity we seek. It is equal to the integrals over the $E\psi$ and $U\psi$ terms. In the former term E is a constant and ψ is finite, so integrating over an infinitesimal interval gives zero. The same would usually be true for the $U\psi$ term, but because $U = \infty$ at $x = 0$, the integral is finite and equals $\beta\psi(0)$, where we have used another property of the delta function: $\int_{-\infty}^{\infty} \delta(x)\,dx = 1$.

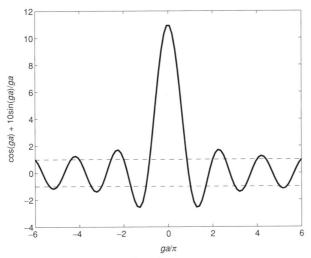

Figure 2.4 Plot of (2.17) for $\left[\frac{m_0\beta a}{\hbar^2}\right] = 10$, showing the allowed values of ga, i.e., those within the dashed lines. The forbidden values of ga lie in the areas outside the dashed lines.

From (2.14) and (2.16), after some manipulation, an expression devoid of A and B results:

$$\cos(ka) = \cos(ga) + \left[\frac{m_0\beta a}{\hbar^2}\right]\frac{\sin(ga)}{ga}. \tag{2.17}$$

This key equation unlocks the secret of bands: the right-hand side is a function of ga, and g is a function of the energy E from (2.6), but the left-hand side decrees that $f(ga)$ must be bounded by ± 1. Thus, values of E are only allowed when $-1 \le f(ga) \le 1$. This is illustrated by the plot of (2.17) in Fig. 2.4. Note that this figure is arbitrarily truncated at $g = 6\pi/a$, but, in reality, g could be extended indefinitely; thus, there are an infinite number of ranges of allowed energy, each one of which is called an **energy band**.

The energy bands corresponding to the allowed values of ga, and the forbidden regions (bandgaps) separating the bands, are usually displayed on a plot of energy E versus Bloch wavevector k. The version shown in Fig. 2.5 is known as an **extended-zone** plot. The first zone spans the range $-\pi/a < k < \pi/a$; the second zone is split into two: $-2\pi/a < k < -\pi/a$ and $\pi/a < k < 2\pi/a$; etc. Thus each zone extends over a range of $2\pi/a$ in k. From (2.12), it is seen that the corresponding range in n is N, the number of primitive calls. As the latter number will be usually very large in semiconductor devices, the separation of neighbouring k values ($=2\pi/Na$), is so small that the E-k relation appears continuous within a band.

An E-k plot is often interpreted as an energy-momentum relationship. This is because, from (2.1), momentum can be written as $\hbar k$, where $\hbar = h/2\pi$ is Dirac's constant, and $k = 2\pi/\lambda$ is the general relationship between wavelength and wavevector. For the specific case of a Bloch wavevector, $\hbar k$ is called the **crystal momentum**. The crystal

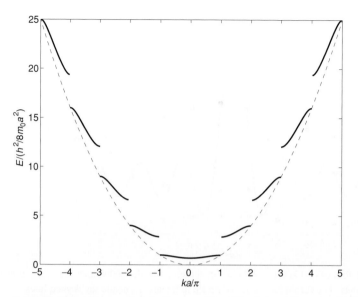

Figure 2.5 'Extended-zone' plot of energy (from Fig. 2.4 and Equation (2.6)) for the first five allowed energy bands. For example, the first band of ga runs from $ga = 0.83\pi$ to $ga = 1.00\pi$ (see Fig. 2.4). This range of ga values, and their negatives, are then used in (2.6) to obtain the first allowed band of energies. The corresponding ka range for the first band is $-\pi < ka < \pi$. The parabola shown by the dashed curve is the E-k relation for a free electron. Note how the allowed bands become closer to this parabola as the energy increases, indicating the increasing 'freedom' of the higher energy electrons.

momentum is not the actual mechanical momentum of the electron: it is the momentum of the electron due to the action of applied forces, as we show in Section 2.9.

2.5   Reduced-zone plot

An alternative way of displaying the E-k relationship is to compress all of its information into the first zone. This is achieved by horizontally shifting each of the curves from the higher order zones in the extended-zone plot by an appropriate multiple of $2\pi/a$. For example, consider the positive wavevectors in the 4th and 5th zones, i.e., $3\pi/a < k < 5\pi/a$. Now, write the wavevector as

$$k = \frac{4\pi}{a} + k',$$ (2.18)

where the new wavevector k' is constrained to $-\pi/a \le k' \le \pi/a$, i.e., to the first zone. The Bloch wavefunction from (2.8) then becomes

$$\psi_k(x) = u_k(x)e^{i4\pi x/a}e^{ik'x}$$

$$\equiv u'_k(x)e^{ik'x}$$

$$= \psi_{k'}(x).$$ (2.19)

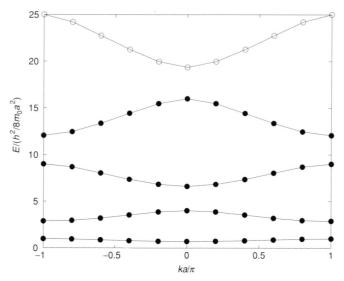

Figure 2.6 Reduced-zone plot of energy (from Fig. 2.4 and Equation (2.6)), for the case of $N = 10$. The ten crystal momentum states in each band are shown.

The terms $\exp(i4\pi x/a)$ and $u_k(x)$ have the same period a, so they have been amalgamated into a new periodic function $u'_k(x)$. The changes to u_k and to k are complementary in that they leave the wavefunction unchanged. In our example, the shift in k of $4\pi/a$ takes the band of the 4th zone (positive k) to the range $-\pi/a < k' < 0$, and the 5th band to $0 < k' < \pi/a$. The bands in the new scheme are completed by similar operations on the corresponding, negative-k portions of the 4th and 5th bands from the extended-zone plot. Similar actions, with translations of appropriate multiples of $2\pi/a$, bring all of the other bands into the first zone. The resulting plot is called a **reduced-zone** plot, an example of which is shown in Fig. 2.6. The first zone, which now contains all the bands, is called the **first Brillouin zone**, or often just *the* Brillouin zone. In the reduced-zone plot $\hbar k$ is properly called the **reduced crystal momentum**.

2.6   Origin of the bandgaps

We have seen how energy bandgaps arise from a mathematical treatment of a periodic structure. For a physical explanation, consider a beam of electrons of wavelength λ propagating through our 1-D lattice, and imagine that there is scattering of the beam from two neighbouring lattice sites. The two portions of the reflected beam would reinforce constructively if the **Bragg condition** for normal incidence were satisfied, i.e.,

$$2a = b\lambda, \tag{2.20}$$

where a is the spacing between lattice sites and $b = 1, 2, 3, \cdots$ is an integer. Further Bragg reflections would lead to our beam bouncing around in the crystal, being reflected back and forth, and taking-on the property of a standing wave, rather than that of a propagating wave. The wavevectors at which this would occur are

$$k = \pm\frac{2\pi}{\lambda} = \pm\frac{b\pi}{a} \equiv \pm\frac{1}{2}G_b. \tag{2.21}$$

Thus, energy bandgaps, within which there are no propagating waves, open up at the Brillouin-zone boundaries because of the strong Bragg reflection.

In (2.21), $G_b = b2\pi/a$ is a set of multiples of 2π and the reciprocal of the lattice spacing a. Collectively, the multiples are called **reciprocal lattice numbers**, and become vectors in 2-D and 3-D systems. The translation numbers used to obtain the reduced-zone plot from the extended-zone plot can now be seen to be reciprocal lattice numbers.

2.7   Quantum states and material classification

The reduced-zone plot of Fig. 2.6 has been drawn for the particular case of ten monovalent primitive unit cells ($N = 10$). Because we are considering a reduced-zone plot, $|k_{max}| = \pi/a$, so n is restricted, from (2.12), to $|n_{max}| = N/2$. The allowed values of ka/π (from (2.12)) are, therefore: $0, \pm0.2, \pm0.4, \pm0.6, \pm0.8, \pm1.0$. These are then used to solve (2.17) for the corresponding ga, from which the allowed energies follow from (2.6). Each circle on the plot of Fig. 2.6 corresponds to a particular value of n, the quantum number defining the allowed values of k in (2.12). Thus, n designates a **state** of reduced crystal momentum that can be occupied by an electron. Note that the end-values, $n = \pm N/2$ in (2.12), are one and the same point, so that the total number of distinct n numbers in the reduced-zone scheme is precisely equal to N, the number of points in our lattice of primitive unit cells.

In fact, each reduced-crystal-momentum state can be occupied by two electrons, providing that they have opposite spin. This is a manifestation of **Pauli's Exclusion Principle**, which observes that no two electrons can have the same quantum numbers. The quantum number for electron spin is $\pm\frac{1}{2}$ and, so far, we have one quantum number (n) for the crystal momentum. In the reduced-zone scheme, where n is restricted to values between $-N/2$ and $N/2$, we need another number to distinguish between states with the same value of reduced wavevector, but with different values of energy. This number is called the **band index**. In Fig. 2.6, the band index runs upwards from 1 to 5.

Thus, each band contains $2N$ states, where N is the number of primitive unit cells that form the real crystal lattice. For the case of a primitive cell containing a single atom that is monovalent, there will be N valence electrons. At temperature $T = 0\,\text{K}$ these electrons will occupy the bottom half of the first band. If there were 2 valence electrons per primitive cell, the entire first band would be occupied at $0\,\text{K}$. More generally, bands will be either completely filled or completely empty if there is an even number of electrons in the primitive unit cell. The highest fully occupied band at $0\,\text{K}$ is called the **valence band**, and the lowest unfilled band at $0\,\text{K}$ is called the **conduction band**. The energy gap between these bands is called the **bandgap**, and is designated E_g.

When a band is completely full, a filled state with crystal momentum $+\hbar k$ is matched by a filled state with crystal momentum $-\hbar k$. Thus, there is no net crystal momentum.[7] We have already alluded to the fact that crystal momentum is the electron momentum due to external forces, such as an applied electric field: therefore, there can be no net motion of charge carriers, i.e., no current, no matter how high the applied field is, provided the electrons stay in the full band.

Now, let us put some thermal energy into the system by increasing T. In the monovalent case the electrons can respond to this stimulus by moving into allowed states of higher energy and crystal momentum within the half-full first band. If an electric field were also applied, electrons could be accelerated into states of higher crystal momentum, and there would be a current. This is the case for most **metals**.

In the divalent case, the only possibility for getting a net gain in crystal momentum would be if some electrons could somehow acquire enough energy to cross the forbidden energy bandgap and then populate some of the states in the empty second band, in which they would then be 'free' to gain crystal momentum from an applied field. If this bandgap is very large, it is unlikely that electrons can be excited into it, and so we have an **insulator**. If the bandgap is not too large, some electrons can be excited into the conduction band, and we have a **semiconductor**. Typically, useful semiconductors have a bandgap in the range 0.5–3.5 eV.

For silicon, the dominant semiconductor material, the atoms are arranged in the diamond lattice structure, as shown in Fig. 2.1, and the primitive unit cell comprises 2 atoms, each of which has 4 valence electrons. Thus, in the entire material there are $8N$ valence electrons; at 0 K these would fill-up the first 4 bands. Therefore, in Si, the gap between the 4th and 5th bands is the bandgap: its value is $E_g = 1.12$ eV at 300 K.

2.8   Band structure of real semiconductors

In our simple 1-D example, a reciprocal lattice number G_b was introduced, and its magnitude was some multiple of 2π divided by the spacing between primitive unit cells in a linear array. Thus, the reciprocal lattice number can be envisaged as residing in **reciprocal space**, which, in this simple case, consists of a linear array of points separated by $2\pi/a$, where a is the spacing of primitive unit cells in the **direct lattice**, or in **real space**. In 3-D, the primitive unit cell in real space becomes a volume, and we have reciprocal lattice vectors which have a magnitude of some multiple of 2π divided by the spacing between planes of atoms. The direction of the reciprocal lattice vector in reciprocal space is perpendicular to that of the planes in real space [4].

The primitive unit cell in reciprocal space for the face-centred cubic lattice in real space is a truncated octahedron (see Fig. 2.7). The Cartesian axes refer to directions of the Bloch wavevector k. As stated above, these directions are perpendicular to planes in the direct lattice, so it is reasonable to give them the same designation as is used for

---

[7] Strictly speaking, if we are alluding to the reduced-zone plot, we should be talking about reduced crystal momentum, but, for brevity, we don't always make this distinction.

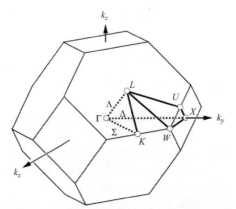

Figure 2.7 The Brillouin zone, or the primitive unit cell in reciprocal space, for the real-space face-centred-cubic lattice. Various symmetry points are labelled. Courtesy of John Davies, University of Glasgow.

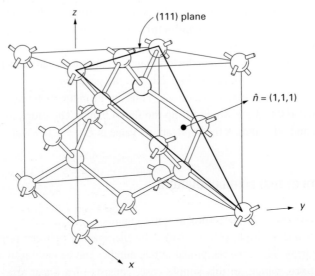

Figure 2.8 Real-space diamond structure, with the (111) plane highlighted. Reproduced from Pulfrey and Tarr [5].

the normals to crystal planes. The notation is that of **Miller indices**, and is illustrated in Fig. 2.8 for the diamond/sphalerite structure. For example, in the natural Cartesian coordinate system of the direct lattice, as illustrated in Fig. 2.1, the (100) plane intersects the x-,y-,z-axes at a, ∞, ∞, respectively. The latter set becomes (100) by taking the reciprocal of each intercept and reducing to integer values. The normal to this plane is specified by the same set of numbers, but with a different parenthesis, i.e., [100]. Because the labeling of the axes is arbitrary, surfaces such as $(-1,0,0)$ and $(0,1,0)$ should have exactly the same properties as (100) surfaces. Collectively, such surfaces are denoted {100}, and equivalent normal directions as $\langle 100 \rangle$. Thus, in reciprocal space,

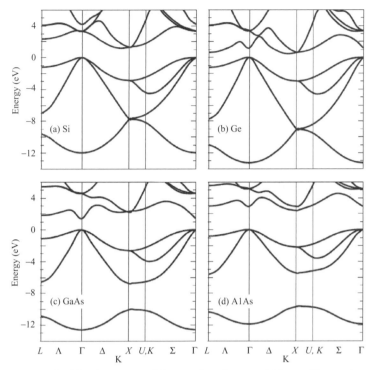

Figure 2.9 Real band structure of four semiconductors widely used in transistors and diodes: (a) Si, (b) Ge, (c) GaAs, (d) AlAs. From Davies [7, Fig. 2.16], © Cambridge University Press 1998, reproduced with permission. Original data from G.P. Srivastava, University of Exeter.

[100], for example, denotes the normal to the (k_x, k_y, k_z)-surface that has intercepts in reciprocal space of $(G_{1,x}, \infty, \infty)$. On Fig. 2.7 this direction is from the origin of **k-space** at the so-called Γ**-point** out through the centre of the square surface at the so-called X**-point**. These point symbols come from Group Theory. The other direction of interest to us for the transistors and diodes discussed in this book is the [111] direction, which passes from Γ to L at the centre of the hexagonal faces of the reciprocal lattice unit cell; in the direct lattice it is the normal to the plane that is highlighted in Fig. 2.8.

The band structure for the 3-D case is not as easily arrived at as in the 1-D case. Numerical calculations are necessary, and a clear example of one particular method is given by Datta [6]. Results of detailed calculations for some common semiconductor materials are shown in Fig. 2.9. Note, firstly, the similarity around the Γ-point of the lowest three bands for all of these materials. These are the valence bands and they are similar because they relate to the similar bonding coordination of the diamond and sphalerite structures. In crystals of Si and GaAs, for example, the orbitals of the valence electrons hybridize ($3sp^3$ in Si and $4sp^3$ in GaAs). The lowest band retains some of the symmetrical character of the atomic s-orbitals. The remaining valence bands are more directional, and derive more from the three atomic p-orbitals. Only two higher valence bands appear in Fig. 2.9 because the highest is actually two bands with the same

E-k relationship: they are said to be doubly **degenerate**. Thus there are, in fact, four valence bands, as required to accept the 8N electrons of the N primitive unit cells for these materials. When the magnetic moment of the spinning electrons and the angular momentum due to their orbital motion is taken into account, this **spin-orbit coupling** lowers the energy of one of the degenerate bands: the **splitting** is slight in Si (0.04 eV) and larger in GaAs (0.34 eV).

The conduction bands for Si and GaAs are noticeably different: the atomic-orbital-character of the electrons is lost because the wavefunctions of the conduction electrons are not required to yield a high probability density in the immediate locality of the atoms, i.e., conduction electrons are 'freer' than valence electrons.

To construct a conduction band in the [100] direction, for example, start at the Γ-point and move through the zone to the X-point. There will be an energy gap at this boundary of the zone, and a new conduction band will fold back into the zone, just as we determined in our 1-D example. However, the 3-D situation is complicated by the fact that bands in a particular zone can arise due to wavevectors arriving at the zone boundary by various routes. For example, with reference to Fig. 2.7, starting at the origin again, and moving to the K-point, proceeding to W and then to X would produce a state at X with a different energy to that of the state arising from the direct path of Γ-X. Starting at this new state and then proceeding directly to Γ produces another band in the [100] direction. Thus, the conduction-band structure is very complicated, with overlapping bands and some degeneracies. Fortunately, the region of greatest interest for the transistors and diodes considered in this book is centred around the bottom of the lowest conduction band. In GaAs, this occurs at the Γ-point, and is non-degenerate, so the conduction band is isotropic in k-space. As the valence band extrema occur at the same value of k, GaAs is said to have a **direct bandgap**. In Si, the lowest minimum of the conduction bands occurs at a point that is about 80% of $\frac{1}{2}G_{100}$, where G_{100} is the reciprocal lattice vector for the first band in the [100] direction. Recall that its length is 2π divided by the spacing between (100) planes, which is $a/2$ in the face-centred cubic structure, i.e., $|k_{100}|_{\max} = 2\pi/a$. The band minimum does not occur at the same value of crystal momentum as the extremum in the valence band: thus, Si is an example of an **indirect bandgap** material.

Finally, in Fig. 2.7, focus on the point on the k_x axis where the conduction-band minimum occurs in Si. Now, move away from this point in any perpendicular direction. The edges of the Brillouin zone are equidistant from the point in these perpendicular directions, but the zone lengths are different from that in the Γ-X direction. Thus, the bottom of the conduction band in Si is anisotropic in k-space.

2.9   Crystal momentum and effective mass

In Section 2.5 we hinted at a relationship between an electron's crystal momentum and an external force that may be acting on it. Here, we derive this relationship and, along the way, define the very useful concept of the effective mass of a mobile charge carrier.

Consider the 1-D case of an electron in either the conduction band or the partially filled valence band and subject to an external force $F_{x,\,\text{ext}}$, which could be due to an applied electric field $\mathcal{E}$, for example.[8] The electron gains energy from the field according to

$$\frac{dE}{dt} = F_{x,\,\text{ext}}\,\frac{dx}{dt} = F_{x,\,\text{ext}}\,v_x\,, \tag{2.22}$$

where v_x is the velocity in the x-direction.

The question arises: what is the appropriate velocity? So far we have described electrons in a crystal via Bloch wavefunctions. These tell us that the probability of finding an electron at some point in a primitive unit cell is the same for all of the primitive unit cells of the crystal. This is not too helpful if we wish to have a better idea of where the electron is in the actual semiconductor device. We would expect to need such information when considering the effect of external forces applied to an actual semiconductor device. For example, whether a photo-excited electron is in the quasi-neutral- or depletion-region of a solar cell is important to know regarding the likelihood of that electron contributing to the photocurrent.[9] While a single wavefunction doesn't give us precise spatial information about the electron, it does give us the electron's crystal momentum, via the wavevector k. By superposing waves of slightly different k, a **wavepacket** can be constructed: the wider the range of k's used, the more tightly constrained in space the wavepacket will be, and the more the electron will appear to have mass at a point, i.e., to be particle-like. The electron can then be treated classically, and be endowed with a trajectory, which is obviously helpful when following an electron through a device. Thus, the velocity to use in (2.22) is the velocity of the centre of the wavepacket: the **group velocity**.

Recall that, from general wave theory, $v_{\text{group}} = d\omega/dk$, where the angular frequency ω is related to the energy by $E = \hbar\omega$. Therefore, in our case, where we have a 1-D Bloch wavevector k in the x- direction,

$$v_x = \frac{1}{\hbar}\frac{dE}{dk_x}\,. \tag{2.23}$$

Substituting into (2.22) and using

$$\frac{dE}{dt} = \frac{dE}{dk_x}\frac{dk_x}{dt}\,, \tag{2.24}$$

we arrive at

$$F_{x,\,\text{ext}} = \frac{d(\hbar k_x)}{dt}\,. \tag{2.25}$$

This is an amazing result: it tells us that $\hbar k$ behaves as the momentum for external forces applied to an electron moving through a periodic structure! In other words, we don't have to know the actual, mechanical momentum of the electron, which will change periodically in response to the crystal field. Instead, the response to an external force can

---

[8] $F = -q\mathcal{E}$ for an electron.
[9] The terms 'quasi-neutral region' and 'depletion region' are explained in Chapter 6.

Table 2.1 Some band parameters for Si and GaAs. The effective masses are for $T = 4\,\mathrm{K}$, and are taken from Pierret [8]. The values for A, B, and C are from Reggiani [9].

Semiconductor	E_g (eV)	m_e^* (m_0)	m_l^* (m_0)	m_t^* (m_0)	m_{hh}^* (m_0)	m_{lh}^* (m_0)	A	B	C
Si	1.12		0.92	0.19	0.54	0.15	4.22	0.78	4.80
GaAs	1.42	0.067			0.51	0.08	7.65	4.82	7.71

be simply calculated from a consideration of only the time-dependence of the crystal momentum.

To make use of this remarkable fact, consider the acceleration of the electron (again using 1-D for simplicity)

$$a_x = \frac{dv_x}{dt} = \frac{1}{\hbar}\frac{d^2E}{dk_x^2}\frac{dk_x}{dt} = \frac{1}{\hbar^2}\frac{d^2E}{dk_x^2}\frac{d(\hbar k_x)}{dt}. \tag{2.26}$$

Using (2.25), leads to

$$a_x = \left[\frac{1}{\hbar^2}\frac{d^2E}{dk_x^2}\right]F_{x,\,\text{ext}}. \tag{2.27}$$

This equation has the familiar form of Newton's Second Law of Motion, allowing us to associate a mass, which is called the **effective mass** $m^*(E)$, with the bracketed term in (2.27), i.e., for our 1-D case

$$m_x^*(E) = \left[\frac{1}{\hbar^2}\frac{d^2E}{dk_x^2}\right]^{-1}. \tag{2.28}$$

Evidently, $m^*(E)$ depends on the direction, and in multi-dimension systems is, in fact, a tensor [2, p.66]. Also, because $m^*(E)$ depends on the band structure, which depends on the potential energy environment of the crystal, the effective mass is not expected to be equal to the free-electron mass m_0. Some values are given in Table 2.1 at the end of Section 2.10.

To emphasize the key point of this section: an electron moving under the combined influence of an externally applied force and the forces associated with the lattice ion cores, responds to the external force just as if it were a free particle, but with a mass that is determined by the band structure of the host material. This means that, once we know the band structure of a semiconductor, we can compute $m^*(E)$, and, thereafter, not concern ourselves about the internal details of how the potential varies according to the electron-ion core interactions.

2.9.1   Negative effective mass

From (2.28) we see that $m^*(E)$ is positive at the bottom of bands, i.e., where the E-k relation is concave upwards, and negative at the top of bands, where the E-k relation is convex upwards. Let us deal with the conduction band first.

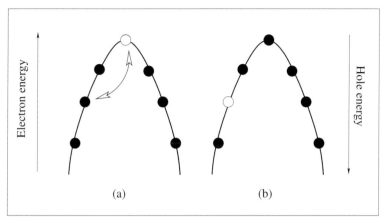

Figure 2.10 Hole energy. (a) Excitation of an electron to a higher energy state in the valence band. (b) After the excitation, the hole can be viewed as having gained energy, i.e., the hole energy increases downwards. The x-variable is k_x.

At the bottom of the conduction band $m^*(E)$ is positive, so a positive force causes a positive change in crystal momentum, i.e., the electron accelerates in the direction of the applied force. However, as the electron moves up the band, it passes through a crystal momentum state at which $m^*(E)$ becomes infinite and, thereafter, is negative. The transition from positive to negative effective mass marks the point where the acceleration due to the external force is overcome by the increasing Bragg reflection of the Bloch waves as the Brillouin-zone boundary is approached. In other words, the momentum transfer from the applied force to the electron becomes less than the momentum transfer from the lattice to the electron. As we point out in Section 4.1, the conduction-band electrons usually don't enter this part of the zone, so these electrons can usually be relied upon to stay near the bottom of the band, and to accelerate in the direction of the applied force.

In the valence band, however, it is the top of the band that is most important regarding the motion of charge carriers. In order for there to be a net change in crystal momentum of the elecrons in the valence band, there must be empty states in the band into which the electrons can move. How such empty states can arise is discussed in Chapter 3, but it can be appreciated that they will exist near the top of the band because the electrons will tend to gravitate to their lowest possible energy states (see Fig. 2.10a). The empty states near the top of the valence band are called **holes**. If an electron is somehow excited into one of these empty states, an empty state will appear lower down in the band (see Fig. 2.10b). This exchange can be thought of as giving energy to the hole, i.e., the hole energy increases in the downwards direction of the E-k diagram, which is invariably drawn from the perspective of electrons. This means that, from the hole point of view, $(d^2E/dk^2) > 0$, and the hole effective mass is positive near the top of the band. Thus, holes accelerate in the same direction as the applied external force, just like 'normal' objects. For this reason, and for the fact that it is easier to keep track of the movement of a relatively few number of holes, rather than of the large number of electrons in the valence band, we choose to designate charge conduction in the valence band as being due solely to holes. The situation then becomes analogous to that near the bottom of the conduction band, where there are positive-effective-mass electrons moving in a predominantly empty band.

From now on, when we talk of conduction by electrons we are implicity referring to the lower regions of the conduction band. The top of the valence band is the domain of holes, and we consider them to represent the carriers of current in this band.

2.9.2 Hole polarity

Although the two types of charge carrier discussed above have positive effective mass, they have a different polarity of charge. To see this, imagine that we have an intrinsic semiconductor with a full valence band and an empty conduction band: the material is neutral as the electron charge balances the charge of the atomic cores

$$\int_{\Omega} (-qn_{i,\mathrm{VB}} + qA)\, d\Omega = 0 \,, \tag{2.29}$$

where $q = 1.602 \times 10^{-19}$ C is the magnitude of the electronic charge, $n_{i,\mathrm{VB}}$ and A are the concentrations of electrons in the valence band and of atoms, respectively, and Ω is the volume of the material. Now, consider raising the temperature so that some electrons are excited from the valence band to the conduction band. The new valence-band electron concentration is $n'_{i,\mathrm{VB}}$, and the charge equation becomes

$$\int_{\Omega} (-qn'_{i\mathrm{VB}} - qn_i + qA)\, d\Omega = 0$$

$$\int_{\Omega} (q[A - n'_{i\mathrm{VB}}] - qn_i)\, d\Omega = 0$$

$$\int_{\Omega} (qp_i - qn_i)\, d\Omega = 0 \,, \tag{2.30}$$

where p_i is the concentration of holes in the valence band and n_i is the concentration of electrons in the conduction band. Thus, a positive charge is associated with the holes.

2.9.3 Parabolic-band approximation

Given that the regions of the *E-k* diagram near to the band extrema are of particular importance, we can anticipate that it would be useful for analytical purposes if the effective mass in these regions could be treated as a constant, rather than as being energy dependent. Inspection of (2.28) informs that a parabolic *E-k* relationship would yield such a constant effective mass. The *E-k* relationship for a *free* electron is truly parabolic, i.e., $E = (\hbar k)^2 / 2m_0$. By analogy, for electrons near the bottom of the conduction band, and for holes near the top of the valence band, we write the kinetic energy as

$$E - \hat{E} = \frac{\hbar^2 k^2}{2m^*} \,, \tag{2.31}$$

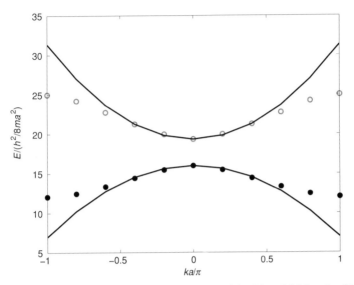

Figure 2.11 Fitting of parabolae to the extrema of the 4th and 5th bands of Fig. 2.6.

where $\hat{E} > 0$ is the energy of the extremum of the appropriate band, and m^* is a constant near the extremum, and is called the **parabolic-band effective mass**.

As an example, let us apply the parabolic-band approximation to the 4th and 5th bands from Fig. 2.6. The result is shown in Fig. 2.11, which gives an idea of the limited applicability of the approximation. However, the parabolic-band approximation is very useful because, in reality, it is in these regions that the charge carriers in which we are interested are found most often. For example, consider the upper band of Fig. 2.11, and imagine it to contain some electrons near the bottom of the band. If an electric field is now applied these electrons will gain crystal momentum and move 'up' the band. However, they will inevitably collide with the atoms of the lattice, thereby losing energy and momentum, and be returned to states near the bottom of the band.

One further implication of the constant effective-mass description is that crystal momentum states $\hbar k$ can be viewed as velocity states. The conduction band, for example, can be envisaged as comprising electrons of higher and higher velocity as the band is populated from the bottom. This picture will prove extremely helpful when considering the injection of carriers into a semiconductor, and over various potential barriers in transistors and diodes.

2.10 Constant-energy surfaces

In 3-D structures, the parabolic-effective-mass concept leads naturally to

$$E - E_{C0} = \frac{\hbar^2}{2}\left[\frac{k_x^2}{m_x^*} + \frac{k_y^2}{m_y^*} + \frac{k_z^2}{m_z^*}\right], \qquad (2.32)$$

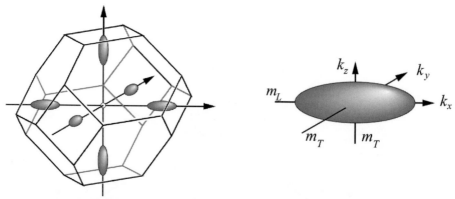

Figure 2.12 The six equivalent constant-energy surfaces for some energy $E > E_{C0}$ in the lowest conduction band of Si. On the right is a detail of one of the constant-energy surfaces, showing the longitudinal and transverse effective masses. From Davies [7, Fig. 2.19], © Cambridge University Press 1998, reproduced with permission.

where we have taken the example of the energy near the bottom of a conduction band for which the extremum is the energy $\hat{E} = E_{C0}$.

From Fig. 2.9 for Si, it can be seen that E_{C0} occurs about 80% of the way towards the X-points. In fact, because of the six-fold symmetry of the basically cubic lattice, there are six equivalent X-points. With reference to Fig. 2.12, consider the energy minimum in the direction to the right, and mark it as the k_x-direction. Moving away from this energy minimum in either of the two orthogonal directions, the same k-space environment is encountered, but this environment is different from that in the k_x-direction. Thus in Si, $m_y^* = m_z^* \neq m_x^*$. Usually, m_x^* is called the **longitudinal effective mass** (labelled m_L in Fig. 2.12), and the other two are called the **transverse effective mass** (labelled m_T in Fig. 2.12). Measurements of effective mass are obtained from cyclotron resonance experiments performed at very low temperatures. Commonly accepted values for 4 K are given in Table 2.1. The constant-energy surface around each of the six equivalent conduction band minima is a prolate spheroid.

From Fig. 2.9 for the band structure of GaAs, it can be seen that E_{C0} occurs at the Γ-point. This is the central point of the Brillouin zone and, with reference to the primitive unit cell in reciprocal space for the diamond/sphalerite structure (Fig. 2.7), it can be seen that moving away from this point in the k_x-direction, traverses exactly the same k-space environment as would be encountered on moving away from the Γ-point in the other two, orthogonal directions. Thus, for GaAs, at the bottom of the lowest conduction band, the parabolic effective mass is isotropic, i.e., $m_x^* = m_y^* = m_z^*$. The actual value is $0.067m_0$. Thus, in this case, the surface in k-space for some value of $E > E_{C0}$ is a sphere centred on the Γ-point.

Turning now to the valence band, a complication arises in the cases of both GaAs and Si inasmuch as there are two bands with the same minimum hole energy $\hat{E} = E_{V0}$. Due to this degeneracy, there are interactions between the electrons in each of the bands near the extremum, and approximations for the band structure beyond the order of the

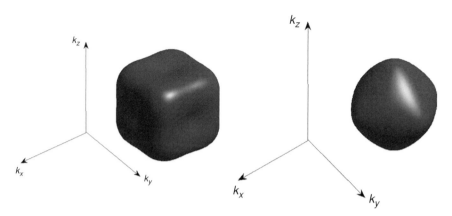

Figure 2.13 Constant energy surfaces for holes in Si. Heavy holes on the left, and light holes on the right. Courtesy of Parham Yaghoobi, UBC.

parabolic are necessary. It transpires that a good approximation is

$$E(k) - E_{V0} = \frac{\hbar^2}{2m_0} \left\{ Ak^2 \mp \sqrt{B^2 k^4 + C^2 (k_x^2 k_y^2 + k_y^2 k_z^2 + k_z^2 k_x^2)} \right\}, \qquad (2.33)$$

where the $+$ sign is for the **light-hole** band, i.e., the band with the greater convexity, and the $-$ sign is for the **heavy-hole** band. Values for A, B, and C for GaAs and Si are given in Table 2.1. Examples of the constant energy surfaces are given in Fig. 2.13. The shape depends strongly on the energy, but usually the surfaces are described as those of warped spheres. Approximating them as actual spheres allows effective masses for the heavy and light holes to be identified, and these values are often used in calculations. Typical values for m_{hh}^* and m_{lh}^* at 4 K are given in Table 2.1.

2.11 Effective-mass Schrödinger equation

As we have seen in Section 2.9, useful information about the band structure of a crystalline material is distilled into a single parameter, the effective mass $m^*(E)$. Here, we state how the parabolic-band effective mass m^* can be incorporated into the Schrödinger Wave Equation, thereby simplifying this formidable equation in situations where the potential energy is a superposition of that due to the periodic lattice U_L and some macroscopic, engineered potential energy U_M. Examples of the latter are: the potential energy due to an applied electric field; the potential energy due to a variation in ionized impurities in the crystal, such as occurs in a *p-n* junction.

In the presence of this additional potential energy $U_M(x)$, the time-independent Schrödinger Wave Equation from (2.3) becomes

$$\left[-\frac{\hbar^2}{2m_0} \frac{d^2}{dx^2} + U_L(x) \right] \psi(x) + U_M(x)\psi(x) = E\psi(x), \qquad (2.34)$$

where the square brackets denote the Hamiltonian for a single electron moving through a perfectly periodic crystal lattice, in which it experiences only the potential energy U_L.

The equation we now present assumes that the conduction-band energy can be described by a parabolic relationship

$$E_\nu(k) = E_{C0} + \frac{\hbar^2 k^2}{2m^*},$$ (2.35)

where E_{C0} is the energy at the bottom of the band, and ν is the band index.[10] The new equation takes the microscopic details of the semiconductor into account via m^* and E_{C0}, and it is called the **single-band, effective-mass Schrödinger Wave Equation**:

$$\left[-\frac{\hbar^2}{2m^*} \frac{d^2}{dx^2} + U_M(x) \right] F(x) = (E - E_{C0}) F(x),$$ (2.36)

where F is the **envelope function** of the actual wavefunction ψ; the two functions can be approximately related by

$$\psi(x) = u_{k0}(x) F(x),$$ (2.37)

where u_0 is the periodic part of the Bloch wavefunction, evaluated at the bottom of the conduction band, at which we have taken the Bloch wavevector to be $k0$.

The conditions under which (2.37) is a reasonable solution to (2.34) are [10]:

- Only one band is involved. This will obviously have to be relaxed for the valence band, at the top of which both heavy and light holes are present in separate bands.
- u_k is independent of k in the neighbourhood of $k0$. This condition stems from the need to attribute most of the variation in k of the Bloch wavefunction in the perfectly periodic case to the plane-wave part of the wavefunction.
- $F(x)$ varies slowly with x, i.e., when compared to the spatial variation of the potential energy $U_L(x)$ due to the periodicity of the crystal.
- The parabolic-band effective mass is applicable. This means that electron energies must be restricted to near the bottom of the conduction band.
- Information on the atomic-scale variation of the electron concentrations is not needed. This is because the sum of the probability densities of all the electrons $\sum F(x) F^*(x)$ involves the envelope functions, which produce a smoothed-out version of the true electron concentrations. The latter would be obtained from $\sum \psi(x) \psi^*(x)$, which would include atomic-level information, either by the use of the Bloch function in (2.37) after solving the effective-mass equation, or by direct use of the full-wave equation (2.34).

Despite this seemingly very restrictive set of conditions, the single-band, effective-mass equation is widely employed, and can give insightful results, even when not all of the above conditions are strictly satisfied.

---

[10] From hereon, we assume we are talking about the lowest band, so $\nu = 1$ and we drop the band index.

2.11.1   Boundary conditions for the effective-mass equation

The effective-mass Schrödinger equation can be recast as

$$-\frac{\hbar^2}{2}\frac{d}{dx}\left(\frac{1}{m^*(x)}\frac{dF}{dx}\right) + E_C(x)F(x) = EF(x),\qquad(2.38)$$

where $E_C(x) = U_M(x) + E_{C0}$ is defined following (2.41). No a-priori reason can be given for the greater correctness of this form of the equation than that of (2.36), but a posteriori there is good reason to choose (2.38): it suggests a boundary condition for the derivative of F that correctly conserves current (see Exercise 5.14). The boundary condition in question, taken to apply to the interface at $x = 0$ between two regions (1 and 2), is

$$\frac{1}{m_1^*(x)}\frac{dF_1}{dx}\bigg|_{x=0} = \frac{1}{m_2^*(x)}\frac{dF_2}{dx}\bigg|_{x=0}.\qquad(2.39)$$

The boundary condition for the envelope function itself is the same as for a true wave-function:

$$F_1(0) = F_2(0).\qquad(2.40)$$

These boundary conditions are used elsewhere in the book when tunnelling is examined.

2.12   Energy-band diagram

The band structure of a semiconductor gives information of the energy in k-space. Often, in diodes and transistors, we need information of the energy variation in real space. To convey this concisely, we return to the expression (2.35) for parabolic energy bands, and add to it the macroscopic potential energy

$$E = U_M(x) + E_{C0} + \frac{\hbar^2 k^2}{2m^*}$$

$$\equiv E_C(x) + \frac{\hbar^2 k^2}{2m^*},\qquad(2.41)$$

where E_C is the position-dependent **conduction band potential energy**, or, as it is usually called, the **conduction-band edge**.[11] With this interpretation, $\hbar^2 k^2/2m^*$ becomes the **kinetic energy of electrons in the conduction band**.

It is now possible to convey the spatial variation of $E_v(x)$ for the lowest conduction band by simply drawing $E_C(x)$, and imagining that energies above it at any position x represent the kinetic energy of electrons at that point. The resulting plot is called an **energy-band diagram**. The relationship between it and the parabolic dispersion relationship is illustrated in Fig. 2.14. The example is for the case of a uniform electric field, which would cause a linear change in the macroscopic potential energy $U_M(x)$.

---

[11] E_C differs from the electrostatic potential energy by a material constant called the electron affinity, as we show in Chapter 6.

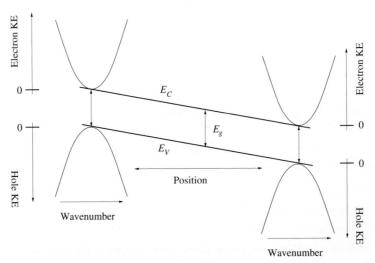

Figure 2.14 Representation of the relationship between the band structure and the energy-band diagram. The example is for the case of a homogeneous semiconductor in a uniform electric field. The change in potential is conveyed by a spatial change in the conduction- and valence-band edges, $E_C(x)$ and $E_V(x)$, respectively.

2.13   From microscopic to macroscopic

We have come a long way in this chapter. We started by treating the electron as a wave and by considering the microscopic nature of the semiconductor crystal in which the electron moves. This led to energy-wavevector plots and the concept of allowed states of crystal momentum in bands of energies separated by energy bandgaps. We ended by distilling the microscopic details into an effective mass and an energy band edge. This led to energy-position plots, and the prospect of being able to view the electron more classically as it moves through a device in response to external, macroscopic forces.

We take advantage of this macroscopic viewpoint whenever possible in the following chapters. However, its underlying microscopic basis should always be kept in mind, and there is no alternative to the microscopic viewpoint when considering events that are important in some devices, for example, photon absorption in solar cells, photon generation in LEDs, tunnelling and strain-engineering in MOSFETs, carrier confinement in HJFETs and carrier transport in 1-D nanotransistors.

Exercises

2.1   Consider Fig. 2.2. Remove the middle rows of atoms so that the resulting structure is no longer face-centred rectangular, but is simply rectangular. Construct the primitive unit cell for the new lattice. What is the basis?

2.2   Electrons propagating through periodic structures can be represented by Bloch functions.

Show that these functions properly account for the fact that an electron must have equal probability of being in any of the identical primitive unit cells in a perfectly crystalline material.

2.3 A new 1-D material, A, has N primitive cells/unit length, 1 atom/primitive cell and 1 valence electron/atom.

Another new 1-D material, B, has $N/2$ primitive cells/unit length, 2 atoms/primitive cell and 1 valence electron/atom.

Which of these two materials is a semiconductor?

2.4 Consider a simple 1-D semiconductor crystal comprising 6 primitive cells separated by a distance a. Each primitive cell contains a single atom; each atom has 4 valence electrons.

Make a rough sketch of the reduced-zone representation of the band structure. Identify the conduction band and the bandgap.

2.5 Compute the number of atoms per unit area on the (100) and (111) surface planes in the silicon lattice.

This information is relevant to devices where surface conditions are important, as in MOSFETs, for example.

2.6 Search beyond this book for a drawing of the primitive unit cell for a real-space, 3-D, body-centred, cubic lattice.

Compare this unit cell with that shown in Fig. 2.7 for the primitive unit cell in reciprocal space of a face-centred cubic lattice.

Sketch how these primitive unit cells nest together to fill up all of real space or reciprocal space.

2.7 In Section 2.8 we showed that $|k_{100}|_{max} = 2\pi/a$ for the Brillouin zone of Fig. 2.7 for a face-centred cubic real-space lattice of lattice constant a. This length is the distance between Γ and X in the direction Δ on the figure. Show that the distance Γ to L in the direction Λ in Fig. 2.7, which is $|k_{111}|_{max}$, is $\sqrt{3}\pi/a$.

The difference in the two vector lengths is evident from the x-axis of the band-structure plots of Fig. 2.9.

2.8 The E-k relationships for the conduction bands of two semiconductor materials, A and B, each with spherical constant-energy surfaces, can be expressed as

$$E_A = \alpha k^2 \quad \text{and} \quad E_B = 2\alpha(k - k')^2,$$

respectively, where α is a constant and $k' > 0$.

Which material has the higher electron effective mass?

2.9 Fig. 2.10b shows the valence band of a semiconductor with one unoccupied state below the top of the band.

If a positive electric field $\mathcal{E}_x$ is now applied, will the empty state move first to a position of higher hole energy or lower hole energy?

2.10 Write a short program (a .m MATLAB file, for example) to plot your own version of Fig. 2.4 from (2.17). For your numerical answer use $[m_0\beta a/\hbar^2] = 4$, rather than the value of 10 used in the text.

Plot the figure.

2.11 Imagine that your plot from the previous question applies to a material with a lattice constant of 0.5 nm, and a primitive unit cell that has a basis of 2 atoms, each of which has 2 valence electrons.

Estimate the bandgap of this material in eV.

2.12 The ga values for the first 5 allowed bands from Fig. 2.4 are listed in the table below. This data was used to generate Fig. 2.6.

(a) Plot your own version of this reduced-zone plot, and recall that $[m_0\beta a/\hbar^2] = 10$ in this case.

(b) Imagine, that this plot applies to a material with a primitive unit cell that has a basis of 2 atoms, each of which has 2 valence electrons.

Show a parabolic fit to the valence band, and give your estimate (in units of m_0) of the parabolic-band effective mass for holes.

ka/π	0	0.2	0.4	0.6	0.8	1.0
ga_1/π	0.8364	0.8500	0.8871	0.9367	0.9814	1.0000
ga_2/π	2.0000	1.9637	1.8804	1.7887	1.7171	1.6984
ga_3/π	2.5678	2.6099	2.7112	2.8343	2.9475	3.0000
ga_4/π	4.0000	3.9331	3.7981	3.6533	3.5285	3.4723
ga_5/π	4.3988	4.4681	4.6105	4.7701	4.9205	5.0000

2.13 In the previous question a value for the parabolic band effective mass was found by curve fitting. Here, an analytical approach is to be used via an expansion of (2.17) about the valence-band extremum.

Take the first two terms of the cosine expansion on the left-hand side of the equation, and perform a Taylor Series expansion to first-order of the right-hand side of the equation.

Making use of (2.6), show that

$$\frac{m_h^*}{m_0} = \frac{-1}{g^*a} \left[\sin g^*a - A \left(\frac{\cos g^*a}{g^*a} - \frac{\sin g^*a}{(g^*a)^2} \right) \right], \qquad (2.42)$$

where g^*a is ga at the extremum, and $A \equiv [m_0\beta a/\hbar^2] = 10$.

Compare your result with that from the previous question.

If the discrepancy bothers you, use your value for m_h^* from this question to make a new parabolic curve, and compare this with the 'true' band as given by the ga data in Exercise 2.12. The discrepancy should be seen to be merely due to how far away from the extremum you wish the parabolic fit to extend.

References

[1] S.M. Sze, *Semiconductor Devices, Physics and Technology*, 1st Edn., Fig. 1.3b, John Wiley & Sons Inc., 1985.

[2] C.M. Wolfe, N. Holonyak, Jr. and G.E. Stillman, *Physical Properties of Semiconductors*, Prentice-Hall, 1989.

[3] D.J. Griffiths, *Introduction to Quantum Mechanics*, pp. 52–55, 198–201, Prentice-Hall, 1995.

[4] C. Kittel, *Introduction to Solid State Physics*, 3rd Edn., Chap. 2, John Wiley & Sons Inc., 1968.

[5] D.L. Pulfrey and N.G. Tarr, *Introduction to Microelectronic Devices*, Fig. 2.7a, Prentice-Hall, 1989.

[6] S. Datta, *Quantum Transport: Atom to Transistor*, Sec. 5.3, Cambridge University Press, 2005.

[7] J.H. Davies, *The Physics of Low-dimensional Semiconductors*, Cambridge University Press, 1998.

[8] R.F. Pierret, *Advanced Semiconductor Fundamentals*, Chap. 3, Addison-Wesley, 1987.

[9] L. Reggiani, *Hot Electron Transport in Semiconductors*, Chap. 1: General Theory, Springer-Verlag, 1985.

[10] S. Datta, *Quantum Phenomena*, Sec. 1.1 and Chap. 6, Addison-Wesley, 1989.

3   Electron and hole concentrations

From the previous chapter we learned that the mobile charge carriers in a semiconductor can be categorized as electrons in the conduction band and holes in the valence band. The carriers occupy states, which define their energy and crystal momentum. One of the objectives of this chapter is to determine the density of these states in energy or momentum, as well as the densities of the carriers in real space. We begin, however, with a look at how the mobile carriers are created and destroyed.

3.1   Creation of electrons and holes

The four mechanisms by which electrons (in the conduction band) and holes can be created in the devices discussed in this book can be classed as thermal, optical, electrical, and chemical.

3.1.1   Thermal generation

For temperatures $T > 0\,\mathrm{K}$, the lattice atoms vibrate about their mean positions. Occasionally, the local amplitudes of vibration are sufficient to break a valence bond, i.e., to release an electron from the valence band. Obviously, in a perfectly periodic crystal, the energy gained by the electron must be sufficient to promote the electron into the conduction band, as there are no available states elsewhere. The situation is illustrated in Fig. 3.1, where an electron from state $\langle 1 \rangle$ in the valence band is promoted to state $\langle 1' \rangle$ by the absorption of **phonons**.

A phonon is a quantum of lattice vibrational energy. The lattice vibrations travel through the material as waves because the random vibrations of each atom are coupled to neighbouring atoms by the bonding forces. In Fig. 3.2a this interatomic coupling is represented by springs. However, the crystal lattice, being a mechanical structure, can only vibrate in specific modes. Examples of two vibrational modes are shown in Fig. 3.2. The displacements of the atoms are in the direction of the chain of atoms, so they are called **longitudinal modes**. The associated phonons are termed **acoustic** and **optic**, depending on whether the displacements of neighbouring atoms are in-phase (Fig. 3.2b) or out-of-phase (Fig. 3.2c), respectively. If the atoms are allowed to vibrate in the two directions perpendicular to the direction of the atomic chain, the dispacements give rise to **transverse modes** [1]. Again, acoustic and optic phonons are possible.

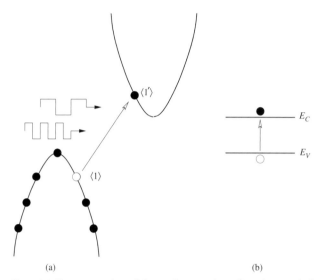

(a)   (b)

Figure 3.1 Representation of thermal generation of an electron-hole pair, (a) on an E-k plot, (b) on an energy-band diagram. The square waves are a symbolic representation of phonons of different frequency. The arrows pointing towards the band structure denote absorption.

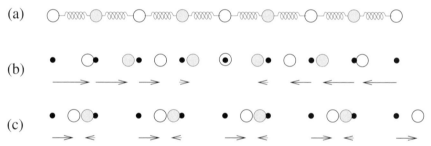

Figure 3.2 Examples of atomic displacements in a 1-D array in which the unit cell comprises a light atom (open circle) and a heavier atom (shaded circle). (a) Representation by springs of interatomic coupling. (b) The displacements of the light and heavy atoms are in-phase. This gives rise to longitudinal acoustic phonons. (c) The displacements of the two atoms are out-of-phase. This gives rise to longitudinal optic phonons.

In real 3-D crystals, the periodicity of the lattice imparts a band structure to phonons that has some similarities with that of electrons. The energy-momentum dispersion relationships for phonons in Si and GaAs are sketched in Fig. 3.3 for a particular direction. For these basically cubic materials the two transverse modes are the same, i.e., they are doubly degenerate, so we see only four branches in the band structure of Fig. 3.3, not six. Note that the frequencies are higher in Si than in GaAs because of the former's lighter atomic mass.

The elements of the phonon band structure (allowed bands (branches), bandgaps, reduced-zone representation) can be readily obtained by first viewing the displacements as constituting travelling waves of the form

$$u_{na} = A \exp[i(\beta na - \omega_\beta t)], \tag{3.1}$$

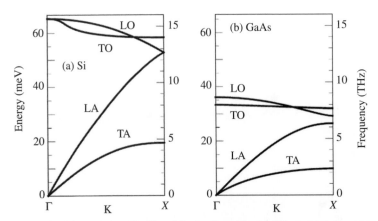

Figure 3.3 Phonon spectra for Si and GaAs. From Davies [2], © Cambridge University Press 1998, reproduced with permission.

where u_{na} is the displacement of the nth atom in a chain of atoms with spacing a, β is the phonon wavevector and ω_β is the radian frequency. Secondly, the spring representation of Fig. 3.2a can be used to formulate the equations of motion for this system of interconnected atoms. The allowed phonon energies (frequencies of vibration) predict acoustic and optic modes and, in the case of diatomic unit cells comprising atoms of different masses, bandgaps at the zone boundary [2, pp. 70–75].

Returning now to electron-hole-pair generation, let us take the thermal generation event in Fig. 3.1 to have occurred in silicon. Thus, the electron must have gained energy at least equal to that of the bandgap, i.e., 1.12 eV. A glance at Fig. 3.3 tells us that phonons have much lower energies than this; therefore, many phonons must be *simultaneously* absorbed for the electron to be excited to the conduction band. The likelihood of this happening is not great, so the number of electrons and holes produced in this way will be much smaller than the number of electrons in the valence band.[1] The latter is about 2×10^{23} cm^{-3} for silicon, while the former is around 1×10^{10} cm^{-3} at room temperature, and is assigned the label n_i, where the subscript i refers to intrinsic material, i.e., this thermal process is intrinsic to the material, and occurs irrespective of any other methods of excitation. Thus,

$$n_i = p_i,\tag{3.2}$$

where p_i is the concentration of holes created by this process.

The conservation of energy alluded to above can be expressed as

$$[E_e\langle 1'\rangle - E_e\langle 1\rangle] - \sum \hbar\omega_\beta = 0,\tag{3.3}$$

i.e., energy is gained by the electron at the expense of the overall phonon energy, or, the electron absorbs the energy of many phonons.

---

[1] The density of atoms in a material is given by the product of Avogadro's number and the material density, divided by the gram molecular weight. For Si, the value is 5×10^{22} atoms/cm^3. Each atom has 4 valence electrons.

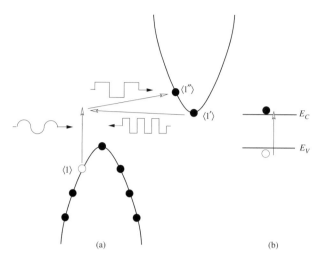

Figure 3.4 Representation of optical generation of an electron-hole pair, (a) on an *E-k* plot, (b) on an energy-band diagram. The sine wave indicates absorption of a photon. Instances of phonon absorption and emission are also shown.

Momentum must also be conserved during the process. This can be expressed as

$$[\hbar k_e \langle 1' \rangle - \hbar k_h \langle 1 \rangle] - \sum \hbar \beta = 0 \,. \tag{3.4}$$

The requirements of conservation of both energy and momentum determine the number and type of phonons that participate in the absorption process. Obviously, in the energy-band diagram representation of the event in Fig. 3.1b, information of the momentum exchange is not conveyed.

3.1.2 Optical generation

Generation of an **electron-hole pair** by absorption of optical energy is illustrated in Fig. 3.4. The energy- and momentum-balance equations are

$$E_e \langle 1' \rangle - E_h \langle 1 \rangle = \hbar \omega_{\text{photon}} + \sum \hbar(\omega_{\beta_a} - \omega_{\beta_e})$$

$$\hbar k_e \langle 1' \rangle - \hbar k_h \langle 1 \rangle = \hbar k_{\text{photon}} + \sum \hbar(\beta_a - \beta_e) \,, \tag{3.5}$$

where β_a and β_e are the wavevectors of phonons that are either absorbed or emitted, respectively. Phonons need to be involved if the electron transition involves a momentum change because photons carry very little momentum, as can be readily verified from

$$k_{\text{photon}} = \frac{E}{\hbar} \frac{n_r}{c} \,, \tag{3.6}$$

where c is the velocity of light in vacuum and n_r is the refractive index of the semiconductor. For a 1.12 eV photon in Si, for example, $k_{\text{photon}} \approx 3 \times 10^6 \,\text{m}^{-1}$. Contrast this with the wavevector difference between an electron at the top of the valence band at the Γ-point and one at the conduction band minimum, which is at $\approx 0.8 \times 2\pi/a$ in Si.

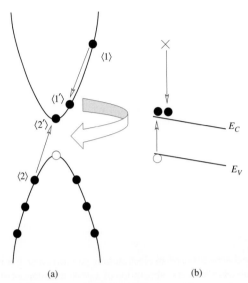

Figure 3.5 Representation of impact-ionization generation of an electron-hole pair, (a) on an E-k plot, (b) on an energy-band diagram. The energy from the electron transition $\langle 1 \rangle \rightarrow \langle 1' \rangle$ is absorbed by the lattice and promotes an electron from state $\langle 2 \rangle$ to $\langle 2' \rangle$. In (b), X marks the position of the impact-ionizing collision.

The difference in wavevector is $\approx 10^{10}$ m^{-1}. Clearly this cannot be met by the photon, so phonons must be involved in the transition.

In a direct bandgap material, interband electron transitions at energies near the bandgap need not involve any momentum change, so phonons need not be involved. It follows that photon absorption in direct bandgap materials occurs more readily than in indirect bandgap materials. However, the fact that silicon is widely used in solar cells and photodetectors indicates that absorption can be strong in indirect bandgap materials. This is because, for photon energies near to E_g, as shown in Fig. 3.4, the phonons involved need not have much energy (only momentum), and there are many acoustic phonons of this type available.[2]

3.1.3   Electrical generation

Generation of an electron-hole pair by the **impact ionization** of a lattice atom is illustrated in Fig. 3.5. A high-energy electron collides with a lattice atom, and there is a transfer of kinetic energy and momentum to the atom. If the energy involved is $\geq E_g$, then an electron can be 'freed' from the valence band, creating another electron in the conduction band, and leaving behind a hole in the valence band. The energy- and

---

[2] The average number of phonons in a vibrational mode of frequency ω_β follows from the statistical mechanics of a set of harmonic oscillators, and is given by $\langle n_\beta \rangle = 1/[\exp(\hbar\omega_\beta/k_B T) - 1]$, where k_B is Boltzmann's constant. Thus, low-energy phonons are very numerous. Another way of realizing this is to appreciate that phonons are bosons and, therefore, are not restricted by Pauli's Exclusion Principle to single occupancy of a state. Thus, many phonons can be expected to be found in the lower energy states.

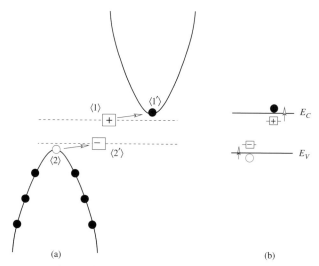

Figure 3.6 Generation of carriers by doping: electrons from donors, and holes from acceptors. The energy between the dopant ground state and the appropriate band edge (conduction band edge for donors and valence band edge for acceptors) is the ionization energy for the dopant.

momentum-balance equations are

$$[E_e\langle 1'\rangle - E_e\langle 1\rangle] + [E_e\langle 2'\rangle - E_h\langle 2\rangle] = 0$$

$$[\hbar k_e\langle 1'\rangle - \hbar k_e\langle 1\rangle] + [\hbar k_e\langle 2'\rangle - \hbar k_h\langle 2\rangle] = 0 \,. \tag{3.7}$$

Usually, electrons are confined to the lower energies of the conduction band by frequent collisions with atoms, defects, and any impurities in the lattice. However, if an electron is rapidly accelerated by an applied electric field, as is assumed to have happened in Fig. 3.5b, it could gain sufficient energy between collisions for impact ionization to occur. It is because of this electric-field origin of the kinetic energy that we give the label of "electrical generation" to this electron-hole-pair generation mechanism. It can lead to avalanche breakdown, as we discuss in Section 16.1.

3.1.4 Chemical generation

In chemical generation, *either* an electron *or* a hole is created by the substitution of a lattice atom with an appropriate impurity atom. In Si, for example, as it is a Group IV element, the addition of an element from Group V could create a 'free' electron, i.e., one that is not needed for bonding purposes. Contrarily, the incorporation of a Group III element would make the valence band deficient in electrons, i.e., a hole would be created. The controlled addition of impurities is known as **doping**. The doping is effective in creating extra electrons or holes if the ionization energy of the dopant is sufficiently small for charge exchange between the dopant atom and the energy bands to occur. Such is the case illustrated in Fig. 3.6.

Typical dopants used in Si devices are phosphorus and boron. These atoms have suitably low ionization energies (≈ 0.04 eV) for the dopant atom to have a very high probability of being ionized at room temperature. Additionally, these atoms have similar atomic masses to that of Si, so they can be accommodated in the host lattice without severely disrupting the local crystallinity.[3] Semiconductors doped with a few parts per million of certain impurities can have electron and hole concentrations that are markedly different from those in intrinsic material at the same temperature. In such cases the material is said to be **extrinsic**. It takes considerable energy to force a foreign atom into a lattice site, so doping usually involves a high-temperature process, such as diffusion, or a high-energy process such as ion-implantation [3]. In silicon, these procedures are performed on thin wafers, now as large as 300 mm in diameter, which are cut from single crystal boules that are pulled from molten silicon at about 1400°C, and can be several metres in length. An appropriate dopant is usually added to the melt so that the boule itself is extrinsic silicon.

In the energy band diagram on the right of Fig. 3.6 the dopant placements are shown as separate, short lines, to indicate their **localized** nature. For example, the dopant separation in silicon would be about 100 lattice points in any direction for a doping density of about 10^{17} cm^{-3}. At such large separations, there is no overlap between the wavefunctions of the electrons associated with the dopants, so there is no corresponding energy band with a range of permissible electron energies and momenta.[4] Thus, it is not strictly correct to represent dopants on an E-k diagram. Nevertheless, we have attempted to do so in Fig. 3.6a. The energy balance relation is straightforward

$$[E_e\langle 1'\rangle - E_e\langle 1\rangle] \geq E_a \,, \tag{3.8}$$

where E_a is the ionization, or activation, energy. However, the momentum balance relation is ill-defined because crystal momentum relates to Bloch waves, which are relevant to periodic structures, and this periodicity doesn't apply in the neighbourhood of a dopant. A dashed line has been drawn on Fig. 3.6a to emphasize this lack of knowledge of the precise momentum. Some measure of the uncertainty in momentum can be obtained from Heisenberg's Uncertainty Principle

$$\sigma_x \, \sigma_p \geq \hbar/2 \,, \tag{3.9}$$

where σ_x is the standard deviation in position, and σ_p is the standard deviation in momentum. Obviously, if the electron is highly localized, there must be a large uncertainty in its momentum.[5] Suffice to say that the dopant atom is localized to an extent that allows satisfaction of any momentum-change requirements for the transfer of an electron to the conduction band.

Thus, the phosphorus atom in the example that we have been considering acts as a **donor**: it donates one electron to the conduction band. In so doing it becomes a positive

---

[3] Crystalline defects are not wanted as they tend to encourage recombination.

[4] For very high doping densities, i.e., about 10^{20} cm^{-3}, the wavefunctions of the electrons in each dopant might overlap and cause a band of energies slightly below, or perhaps overlapping with, one of the band edges of the host material. The result can be a reduction in E_g; the phenomenon is known as **bandgap narrowing**.

[5] You can quantify this by doing Exercise 3.5, which shows how the uncertainty in momentum can explain how light emission is possible from an indirect bandgap material.

ion. So the material remains, overall, electrically neutral. Note, no hole is created, and the ion is stationary at normal temperatures, so the only new conducting entity is the electron. Thus, in this material $n > p$; the electron becomes the **majority carrier**, and the preponderance of negatively charged carriers leads to the label of **n-type**. Phosphorus and arsenic are the two donors most commonly used in silicon processing. In the III-V compound semiconductor GaAs, useful donors are silicon (Group IV), which substitutes for gallium, and selenium (Group VI), which substitutes for arsenic.

To make extra holes in silicon, a Group III element is introduced substitutionally. Boron is most commonly used for this purpose. In GaAs, zinc (Group II) is used in place of gallium, and silicon in place of arsenic. In the boron case, for example, the three valence electrons of this small dopant are insufficient to satisfy the full bonding requirements of the the four neighbouring silicon atoms that surround it. This deficiency in bonding can be repaired by the acceptance of a valence electron from elsewhere in the valence band. A small amount of energy is required to do this because the accepted electron is surplus to the needs of the boron atom, and, therefore, a small repulsive force has to be overcome. Once the electron has been accepted, the boron becomes a negative ion, and a new hole appears elsewhere in the valence band. Again, the ion is not mobile at the normal operating temperatures for a transistor or diode, so the boron-doped material has extra positively charged current carriers in the form of holes, and is labelled as **p-type material**. The boron atom is referred to as an **acceptor**. A representation of acceptor doping on energy diagrams is given in Fig. 3.6. The situation is analogous to that described for donors.

Because some thermal energy is required to 'free' the fifth electron from the P atom and to accept a fourth electron into the vicinity of the B atom, it can be anticipated that not all the dopant atoms are necessarily ionized at room temperature. However, for P and B in Si, the ionization energies are sufficiently low that complete ionization is a reasonable approximation to make for the normal operating temperatures of Si devices. We make this assumption in this book. Thus, the concentration of donor and acceptor ions, N_D^+ and N_A^-, respectively, can be taken as known, i.e., equal to the dopant concentrations (N_D and N_A) that are deliberately and precisely introduced during the wafer- or device-processing steps in the fabrication of a semiconductor device.

3.2 Recombination

Electrons that have been excited into the conduction band by any of the mechanisms just discussed will have a tendency to lose energy by 'falling into' holes in the valence band. This process is called **recombination**, and the four mechanisms of relevance to the devices treated in this book are: band-to-band (or radiative) recombination, recombination-generation-centre recombination, Auger recombination and surface recombination. We'll consider the first three, which are bulk-recombination mechanisms, in the following sections. Surface recombination is particularly relevant to solar cells, and is discussed in Section 7.4.1.

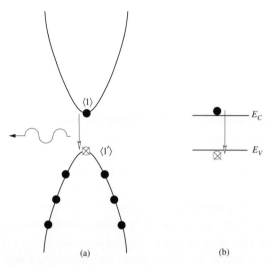

Figure 3.7 Band-to-band recombination.

3.2.1   Band-to-band recombination

Band-to-band recombination is the inverse of optical generation, and is illustrated in Fig. 3.7 for the case of a direct bandgap material. The energy lost by the electron usually creates a photon, resulting in light emission, rather than phonons, which would result in heating of the lattice. Thus, this type of recombination is often called **radiative recombination**. Phonons are not usually involved because their energies are small (see Fig. 3.3), and while the electron in the conduction band is waiting for 20 or more phonons to be simultaneously emitted, as would be required for a direct band-to-band transition in GaAs, a single photon is created instead.

The rate of recombination R_{rad} depends on there being both electrons and holes present, thus

$$R_{rad} = Bnp \,, \tag{3.10}$$

where B is the radiative recombination coefficient, and n and p are the electron and hole concentrations, respectively. For a direct bandgap material such as GaAs, $B \approx 10^{-10}\,\text{cm}^3\text{s}^{-1}$. For an indirect bandgap material such as Si, $B \approx 10^{-14}\,\text{cm}^3\text{s}^{-1}$. Radiative recombination is much less likely to occur in indirect bandgap materials because of the need to involve phonons in the process in order to take up the change in crystal momentum.

The energy- and momentum-balance equations for the transition in direct bandgap materials are

$$E_e\langle 1 \rangle - E_e\langle 1' \rangle = \hbar\omega_{\text{photon}}$$
$$\hbar k_e\langle 1' \rangle - \hbar k_e\langle 1 \rangle \approx 0 \,. \tag{3.11}$$

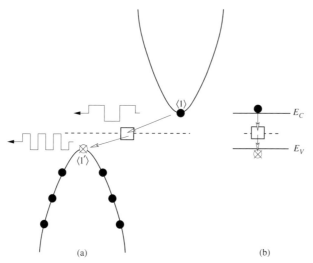

Figure 3.8 RG-centre recombination. The momentum is indeterminate, as in Fig. 3.6, so the state of the recombination-generation centre cannot be identified. The second step in the recombination process can be thought of as hole capture.

3.2.2   Recombination-generation-centre recombination

The recombination-generation (**RG**)-centre process takes advantage of the fact that crystals are rarely perfect. In practical crystalline materials, a few atoms will be missing from their regular lattice sites, and there will also be some distortion of the lattice in the vicinity of impurity atoms. Recall from Chapter 2 that perfect periodicity gives rise to energy bands and bandgaps. So, the presence of crystalline **defects** disturbs the periodicity of the structure and is manifest as localized energy levels within the bandgap. However, unlike the localized levels of deliberately introduced donors and acceptors, the energy levels of defects and gratuitous impurities are not confined to energies close to the band edges: they are distributed throughout the bandgap. The presence of these 'stepping stones' facilitates the recombination process by providing temporary (metastable) states for electrons. Each transition involves fewer phonons than would be required for band-to-band recombination in one step, and, consequently, is more likely to occur. Energy levels near to the middle of the bandgap (see Fig. 3.8) are most effective as recombination centres because the events of capture of electrons from the conduction band and of holes from the valence band are similarly probable.[6]

The analysis of this type of recombination event was first carried out by Hall [4] and by Shockley and Read [5], and is often referred to as **Shockley-Read-Hall**- or, simply, **SRH**-recombination. The intragap centres are usually treated as having a single (ground-state) energy level but, in reality, excited states will be present, thereby further reducing the number of phonons that have to be involved in any one transition, and, consequently,

---

[6] Shallow energy levels, i.e., ones close to the band edges, are not so efficient recombination centres because a captured carrier is more likely to be emitted back to its host band, than it is to be emitted into the other band.

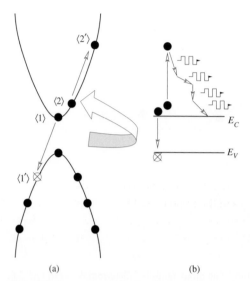

(a)   (b)

Figure 3.9 Auger recombination in an n-type semiconductor. The energy released by the transition $\langle 1 \rangle \rightarrow \langle 1' \rangle$ is absorbed by the electron in state $\langle 2 \rangle$. This energy could also increase the energy of a hole in the valence band, but this is less likely to occur in n-type semiconductors because of the fewer number of holes.

making RG-centre recombinaton even more likely to occur. For weak and moderate doping, it is the dominant recombination mechanism in Si, and the principal non-radiative recombination mechanism in direct bandgap semiconductors such as GaAs.[7]

The rate of recombination depends on the presence of electrons, holes and **traps**, as the sites of the favourable RG centres are often called. However, the rate-limiting step will be the capture of the minority carrier by the trap: the capture of the majority carrier is far more probable because of the greater number of carriers, e.g., holes in p-type material. Thus, for p-type material, for example, the recombination rate is

$$R_{RG} \approx r N_T n \equiv A n \qquad \text{p-type material} , \qquad (3.12)$$

where r is a temperature-dependent rate constant, N_T is the concentration of traps, and A is a trap-dependent recombination coefficient, and has units of s^{-1}.

The energy- and momentum-balance equations for this type of recombination are

$$E_e \langle 1 \rangle - E_h \langle 1' \rangle = \sum \hbar \omega_\beta$$

$$\hbar k_e \langle 1 \rangle - \hbar k_h \langle 1' \rangle = \sum \hbar \beta . \qquad (3.13)$$

3.2.3   Auger recombination

Auger recombination is illustrated in Fig. 3.9, and is the inverse of impact-ionization generation. The energy and momentum produced by the electron-hole recombination

---

[7] Note also that these excited states will also facilitate the thermal generation of electron-hole pairs.

Table 3.1 Recombination parameter values for Si and GaAs at 300 K. A, C and D as used in ATLAS, the semiconductor device simulator from Silvaco [6, p. B-8]. B from Schubert [7].

Semiconductor	A_e (s^{-1})	A_h (s^{-1})	B (cm^{3}s^{-1})	C (cm^{6}s^{-1})	D (cm^{6}s^{-1})
Si	1×10^7	1×10^7	3×10^{-14}	8.3×10^{-32}	1.8×10^{-31}
GaAs	1×10^9	1×10^8	2×10^{-10}	5×10^{-30}	1×10^{-31}

event are transferred to either a second electron or hole. These excited carriers then lose their energy and momentum by emitting phonons. The emission process occurs readily as there are many, many available states in the bands, to which the carriers can transfer with changes in momentum and energy that are compatible with the phonon dispersion relationship.

When the excited carrier is an electron, as shown in Fig. 3.9, the energy- and momentum-balance equations are

$$E_e\langle 1\rangle - E_h\langle 1'\rangle = E_e\langle 2'\rangle - E_e\langle 2\rangle = \sum \hbar\omega_\beta$$

$$\hbar k_e\langle 1\rangle - \hbar k_h\langle 1'\rangle = \hbar k_e\langle 2'\rangle - \hbar k_e\langle 2\rangle = \sum \hbar\beta \,. \tag{3.14}$$

In the above equations, $\hbar\omega_\beta$ and $\hbar\beta$ are the energy and momentum of each emitted phonon, respectively.

This type of recombination needs to be considered when one of the carrier concentrations is very high, yet both carriers are of importance, e.g., in the emitter of solar cells, in the space-charge region of LEDs, and in the base of HBTs. In each recombination event, two carriers of one type and one carrier of the other type are involved. Thus, the recombination rate for the sum of the process shown in Fig. 3.9, and the corresponding process involving the excitation of a hole, can be written as

$$R_{\text{Aug}} = Cn^2 p + Dp^2 n \,, \tag{3.15}$$

where C and D are Auger recombination coefficients with units of cm^{6}s^{-1}. Typical values for these coefficients, and for the other recombination coefficients are given in Table 3.1.

3.2.4   Recombination lifetime

Recombination is always occurring in a semiconductor, and, at equilibrium, its rate of occurrence must exactly equal the rate of generation. The net rate of recombination out of equilibrium is given, therefore, by the difference between the non-equilibrium- and the equilibrium-recombination rates. This net rate is often denoted by U, and has units of $m^{-3}s^{-1}$. For radiative recombination, for example, we have

$$U_{rad} = Bnp - Bn_0p_0 \,, \tag{3.16}$$

where the subscript '0' denotes thermal equilibrium (see next chapter).

Consider now the situation when Δn electron-hole pairs have been generated. The net recombination rate is proportional to

$$
\begin{aligned}
np - n_0p_0 &= (n_0 + \Delta n)(p_0 + \Delta n) - n_0p_0 \\
&= \Delta n^2 + \Delta n(p_0 + n_0) \\
&\approx \Delta n^2 + \Delta np_0 \qquad \text{p-type material} \\
&= \Delta np_0 \qquad\qquad \text{low-level injection} \\
&= \Delta n(p_0 + \Delta n) \qquad \text{high-level injection} \,.
\end{aligned}
\tag{3.17}
$$

In the above development, we have taken the case of p-type material ($n_0 \ll p_0$) and we have defined two cases of recombination, depending on the magnitude of the generated carrier concentration: when Δn ($=\Delta p$) is much less than the majority carrier concentration (p_0 in this p-type example), the situation is one of **low-level injection**; when this is not the case, we have **high-level injection**. Thus, the net rate of recombination for minority carrier electrons can be written as

$$U_{e,rad} \equiv \frac{\Delta n}{\tau_{e,rad}} \,, \tag{3.18}$$

where $\tau_{e,rad}$ is the **electron minority carrier radiative recombination lifetime**, and is given by

$$
\begin{aligned}
\tau_{e,rad} &= \frac{1}{Bp_0} \qquad\qquad\;\; \text{low-level injection} \\
&= \frac{1}{B(p_0 + \Delta n)} \qquad \text{high-level injection} \,.
\end{aligned}
\tag{3.19}
$$

For recombination-generation-centre recombination the majority carrier concentration is of little consequence, so we have, for both levels of recombination,

$$\tau_{e,RG} = \frac{1}{A} \,. \tag{3.20}$$

However, note from (3.12) that A depends on the density of traps, which will depend, in turn, on the dopant concentration because of the disruption to the lattice caused by the incorporation of foreign atoms. Thus, A is likely to have some dependence on dopant concentration. For Si, a simple, empirical relationship is often used for both electrons and holes [6, p. 3–81]:

$$\tau_{e,RG} = \frac{5 \times 10^{-7}}{(1 + 2N \times 10^{-17})} \,, \tag{3.21}$$

where N is the doping density in cm^{-3}. For GaAs, an empirical relationship for recombination-generation-centre recombination is [8, p. 68]

$$\tau_{e,\text{RG}} = \left(\frac{N_A}{1 \times 10^{10}} + \frac{N_A^2}{1.6 \times 10^{29}} \right)^{-1}$$

$$\tau_{h,\text{RG}} = \left(\frac{N_D^{0.693}}{5.4 \times 10^4} + \frac{N_D^{2.54}}{1 \times 10^{40}} \right)^{-1}, \tag{3.22}$$

where the doping densities are to be expressed in cm^{-3}.

For Auger recombination in p-type material under high-level injection conditions, the minority carrier lifetime reduces to

$$\tau_{e,\text{Aug}} = \frac{1}{C(\Delta n p_0 + 2n_0 p_0 + \Delta n^2)} + \frac{1}{D(2\Delta n p_0 + p_0^2 + \Delta n^2)}. \tag{3.23}$$

The overall, electron minority carrier lifetime is given by

$$\tau_e = \left[\frac{1}{\tau_{e,\text{rad}}} + \frac{1}{\tau_{e,\text{RG}}} + \frac{1}{\tau_{e,\text{Aug}}} \right]^{-1}. \tag{3.24}$$

This lifetime is crucial to the operation of bipolar devices, such as solar cells, LEDS, and HBTs, as discussed in later chapters.

3.3   Carrier concentrations

So far in this chapter we have talked about how electrons and holes are created and destroyed, and we also know that these carriers reside in states within their respective bands of allowed energies. We now derive expressions for the carrier concentrations, in preparation for their use in the next chapter.

The fundamental expression for the electron concentration at some position $\vec{r}$ at some time t is

$$n(\vec{r}, t) = \frac{1}{\Omega} \sum_{\text{filled states}}, \tag{3.25}$$

where Ω is the volume over which the concentration is defined: it depends on the dimensions of the system, e.g., it has units of m^3 in 3-D. To turn this sum into an integral over all momentum space $\vec{k}$, we first need to recall from Chapter 2 that the separation between states in one dimension is $2\pi/L$, where L is the length of the crystal in one dimension. Also, we must recognize that each momentum state in $\vec{k}$ space can contain 2 electrons, provided they are of different spin. Finally, if we are going to do an integral over all $\vec{k}$ states, some of which will be empty, we have to introduce a **distribution function** to account for the probability of a particular state being occupied. Doing all these things

$$n(\vec{r}, t) = \frac{1}{\Omega} \sum_{\text{filled states}} \Rightarrow \frac{1}{\Omega} \frac{2}{\frac{(2\pi)^i}{\Omega}} \int_{\vec{k}} f(\vec{r}, \vec{k}, t) \, d\vec{k}, \tag{3.26}$$

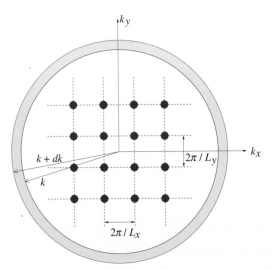

Figure 3.10 2-D representation of k-space spheres of radius k and $k + dk$. The circular shape implies equal lengths L_x, L_y and L_z (not shown) of the crystal in real space.

where $i = 1, 2$, or 3, is the dimensionality of the volume under consideration, and f is the distribution function.

To turn this general equation into something specific, we consider the electron concentration in the 3-D, time-independent case, and, to simplify the notation, we drop the position identifier. Thus, (3.26) becomes

$$n = \frac{1}{4\pi^3} \int_{\vec{k}} f(\vec{k}) \, d\vec{k} . \tag{3.27}$$

As we expect the electrons to fill-up states from the origin of $\vec{k}$ out to higher values of momentum, and, as the number of available states is huge, we can approximate the surface of the actual volume of $\vec{k}$-space being filled as that of a sphere of radius k (see Fig. 3.10), where $k = |\vec{k}| = \sqrt{k_x^2 + k_y^2 + k_z^2}$.

More formally, the simplification we are making is that the distribution function f depends only on the magnitude of k, i.e., in spherical coordinates

$$n = \frac{1}{4\pi^3} \int_0^\infty \int_0^\pi \int_0^{2\pi} f(k) k^2 \sin\theta \, d\phi \, d\theta \, dk . \tag{3.28}$$

In either case the result is

$$n = \frac{1}{4\pi^3} \int_k f(k) 4\pi k^2 \, dk$$

$$n(k) = g(k) f(k)$$

$$g(k) = \frac{k^2}{\pi^2} , \tag{3.29}$$

where $g(k)$ is the **density of states** in $\vec{k}$-space. In this particular case, $g(k)$ is quoted for 3-dimensional real space, and has units of $\mathrm{m}^{-3}[\mathrm{k}]^{-1} \equiv \mathrm{m}^{-2}$.

Often it is convenient to express the density of states with respect to energy, rather than with respect to wavenumber; in either case, the total number of states must be the same, e.g.,

$$\int_0^\infty g(E - E_C)\,d(E - E_C) = \int_0^\infty g(k)\,dk\,, \tag{3.30}$$

where $g(E - E_C)$, in the 3-D case, has units of $\mathrm{m}^{-3}\mathrm{J}^{-1}$.

To proceed further, we need a tractable expression for the dispersion relationship; this is where the parabolic-band approximation of Section 2.9.3 is helpful, namely

$$E - E_C = \frac{\hbar^2}{2}\left[\frac{k_x^2}{m_x^*} + \frac{k_y^2}{m_y^*} + \frac{k_z^2}{m_z^*}\right]$$

$$\equiv \frac{\hbar^2}{2}\frac{k^2}{m_{e,\mathrm{DOS}}^*}\,, \tag{3.31}$$

where $m_{e,\mathrm{DOS}}^*$ is the **density-of-states effective mass** for electrons. For materials such as GaAs, for which the constant-energy surfaces are spherical for electrons, $m_{e,\mathrm{DOS}}^*$ is simply the parabolic-band effective mass. For Si the situation is a little more complicated, as described in the next section.

Differentiating (3.31), $d(E - E_C)$ in (3.30) can be written in terms of dk. The integrands of each side of (3.30) can then be equated, which gives

$$g_C(E - E_C) = \frac{8\pi\sqrt{2}}{h^3}(m_{e,\mathrm{DOS}}^*)^{3/2}(E - E_C)^{1/2} \quad \text{for } E \geq E_C$$

$$g_V(E_V - E) = \frac{8\pi\sqrt{2}}{h^3}(m_{h,\mathrm{DOS}}^*)^{3/2}(E_V - E)^{1/2} \quad \text{for } E \leq E_V\,, \tag{3.32}$$

where we have added the analogous expression for the density of states in the valence band $g_V(E)$ in terms of the density-of-states effective mass for holes $m_{h,\mathrm{DOS}}^*$. More details on $m_{h,\mathrm{DOS}}^*$ and $m_{e,\mathrm{DOS}}^*$ are given in the next section. Meanwhile, please note that the densities of states in 3-D real-space increase parabolically with energy away from the band edges.

Using (3.32) in (3.30) and (3.29), the electron and hole concentrations (per unit volume) can be written as

$$n = \int g_C(E)f(E)\,dE$$

$$p = \int g_V(E)[1 - f(E)]\,dE\,, \tag{3.33}$$

where the integrals are over the conduction and valence bands, respectively.

To complete the calculation of the carrier concentrations we need to know the distribution function $f(E)$. Under the non-equilibrium operating conditions of a semiconductor device this is not an easy task (see Section 5.2). However, the distribution function under equilibrium conditions is well-known, and is presented in Section 4.2.

Table 3.2 Density-of-states effective masses for Si and GaAs. The values at 4 K come from using the parabolic-band effective masses from Table 2.1 in the equations given in this chapter. The estimated values at 300 K are from Pierret [9].

Semiconductor	$m^*_{e,\mathrm{DOS}}$ at 4 K (m_0)	$m^*_{h,\mathrm{DOS}}$ at 4 K (m_0)	$m^*_{e,\mathrm{DOS}}$ at 300 K (m_0)	$m^*_{h,\mathrm{DOS}}$ at 300 K (m_0)
Si	1.06	0.59	1.18	0.81
GaAs	0.067	0.53	0.066	0.52

3.4   Density-of-states effective masses in silicon

3.4.1   Electrons

In silicon, the constant energy surface for energies close to the bottom of the conduction band is a prolate spheroid (see Fig. 2.12), as described by

$$E - E_{C0} = \frac{\hbar^2}{2} \left[\frac{k_x^2}{m_l^*} + \frac{k_y^2 + k_z^2}{m_t^*} \right], \tag{3.34}$$

where m_l^* and m_t^* are the longitudinal and transverse parabolic-band effective masses, respectively. The density-of-states effective mass in this case is found by specifying the spherical volume, for which a single effective mass would apply, and that would contain the same volume of states as the prolate spheroid. In fact, the equality must be with 6 prolate spheroids, as there are this number of equivalent volumes in Si, as Fig. 2.12 shows. The procedure is well known [9], and you can discover it for yourself by doing Exercise 3.7. The result is

$$m^*_{e,\mathrm{DOS}} = 6^{2/3} \left(m_l^* m_t^{*2} \right)^{1/3}. \tag{3.35}$$

The value for Si is given in Table 3.2.

3.4.2   Holes

In both Si and GaAs, there is a band of light holes and a band of heavy holes, and a degeneracy at the extremum (see Fig. 2.9). The constant energy surfaces for each band are warped spheres (see Fig. 2.13), but are often approximated as true spheres in order to obtain values for the effective masses for light holes m_{lh}^* and heavy holes m_{hh}^* listed in Table 2.1. The density of states for a single, spherical band involves an effective mass raised to the power $3/2$ (see (3.32)). Thus, the density-of-states effective mass for holes in Si and GaAs is

$$m^*_{h,\mathrm{DOS}} = \left(m_{hh}^{*3/2} + m_{lh}^{*3/2} \right)^{2/3}. \tag{3.36}$$

Values for Si and GaAs are listed in Table 3.2.

Exercises

3.1 Suggest a combination of phonons, listing their number, type, energy and momentum, that, when simultaneously absorbed in silicon, would enable the generation of an electron-hole pair.

   Does your answer make you appreciate why the intrinsic carrier concentration in silicon is so low compared to the total number of electrons in the valence band?

3.2 Fig. 3.9 shows an Auger recombination event involving two electrons and one hole.

   Re-draw Fig. 3.9a and Fig. 3.9b for the case of an Auger recombination event involving one electron and two holes.

3.3 Obtain an appreciation of the rates of hole recombination in n-type silicon by plotting the rates for radiative-, recombination-generation-centre- and Auger-recombination against excess carrier concentration.

   Determine the excess-carrier concentration at which Auger recombination becomes dominant.

   Compare this value with the optical generation rate in Fig. 7.5 to understand why Auger recombination is important in the emitter of silicon solar cells.

3.4 Efficient LEDs and laser diodes are made from direct-bandgap semiconducting materials. On the other hand, photodetectors that are sensitive to radiation at the bandgap energy can be made from indirect-bandgap materials. Why is this so?

3.5 One exception to the general rule of LEDs requiring direct-bandgap materials is the green GaP LED that used to be widely employed in backlighting (on telephone keys, for example). GaP is an indirect-bandgap material with its conduction-band minimum occuring at the X-point. The lattice constant for GaP is 0.545 nm. Light emission comes from transitions involving optically active impurities, such as oxygen or nitrogen. The electron wavefunctions associated with these impurities are localized within a region Δx of real space.

   Make use of the Heisenberg Uncertainty Principle ((3.9)) to obtain an estimate of Δx.

3.6 The E-k relationships for the conduction bands of two semiconductor materials, A and B, each with spherical constant-energy surfaces, can be expressed as

$$E_A - 0.7 = \alpha k^2 \quad \text{and} \quad E_B - 1.4 = 2\alpha(k - k')^2,$$

respectively, where α is a constant, $k' > 0$ and the energies are in units of eV. Both materials have the same valence-band structure, with the top of the valence band at $E = 0$ and $k = 0$.

   Which material would have the higher intrinsic carrier concentration?

3.7 Derive (3.35) for the density-of-states effective mass for electrons in Si.

3.8 Plot the densities-of-states expressions in (3.32), with energy on the y-axis, for silicon.

   State the values for the densities of states in each band, in units of states $cm^{-3} eV^{-1}$, at an energy of $\frac{3}{2}k_B T$ away from each extremum. This energy is the mean kinetic energy of carriers in an equilibrium distribution at low and moderate doping densities.

References

[1] C.M. Wolfe, N. Holonyak, Jr. and G.E. Stillman, *Physical Properties of Semiconductors*, pp. 177–182, Prentice-Hall, 1989.

[2] J.H. Davies, *The Physics of Low-dimensional Semiconductors*, p. 75, Cambridge University Press, 1998.

[3] J.D. Plummer, M.D. Deal and P.B. Griffin, *Silicon VLSI Technology: Fundamentals, Practice and Modeling*, Chaps. 7 and 8, Prentice-Hall, 2000.

[4] R.N. Hall, Electron-hole Recombination in Germanium, *Phys. Rev.*, vol. 87, 387, 1952.

[5] W. Shockley and W.T. Read, Jr., Statistics of the Recombination of Holes and Electrons, *Phys. Rev.*, vol. 87, 835–842, 1952.

[6] ATLAS Users Manual, Silvaco Data Systems, Santa Clara, June 11, 2008.

[7] E.F. Schubert, *Light-emitting Diodes*, 2nd Edn., p. 53, Cambridge University Press, 2006.

[8] W. Liu, *Fundamentals of III-V Devices: HBTs, MESFETs, and HFETs/HEMTs*, p. 58, John Wiley & Sons Inc., 1999.

[9] R.F. Pierret, *Advanced Semiconductor Fundamentals*, Chap. 4, Addison-Wesley, 1987.

4 Thermal equilibrium

Thermal equilibrium in a semiconductor refers to the state when the temperature is uniform and has been steady for a long time, and when there are no sources of energy other than heat, e.g., no applied electric field nor optical irradiation. Obviously, semiconductor devices will not be in thermal equilibrium when they are in operation. However, it turns out that parts of a device often remain in a state very close to thermal equilibrium and, furthermore, knowledge of the carrier concentrations in thermal equilibrium is often a good starting point for understanding how a device works.

In this chapter, we briefly discuss the collision processes that tend to randomize the momenta of excited electrons and holes, then we introduce the thermal-equilibrium distribution function, develop some useful expressions for the carrier concentrations in equilibrium, and finish by considering the mean thermal velocity associated with an equilibrium distribution of electrons. The last property is a further step towards developing an understanding of current in diodes and transistors.

4.1 Collisions

In the previous chapter, we showed how the processes of recombination and generation alter the carrier concentrations in the conduction and valence bands. Under thermal-equilibrium conditions, the thermally activated band-to-band and chemical generation processes are operative, along with one or all of the following recombination mechanisms: radiative, RG centre, Auger. Obviously, for thermal equilibrium to be maintained, the net sum of these various generation- and recombination-rates must be zero, i.e., as heat is taken from the device to create an electron-hole pair, so it must be returned to the lattice on recombination.[1]

The other dynamical aspect of thermal equilibrium is the maintenance of the carriers in their lowest energy states within each band. As hinted at in Section 2.9, this is achieved by the carriers continually making collisions with atoms, ions, defects and other carriers. The situation is crudely illustrated in Fig. 4.1. The word 'collision' is familiar, and conjures up visions of physical contact, as befitting particle-particle interactions. If we

---

[1] You might like to ponder on how heat is eventually returned to the lattice after a radiative recombination event.

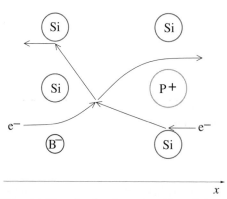

Figure 4.1 Very simple schematic representation of collisions: two electron-atom (Si) glancing interactions; one electron-electron near-'head-on' collision, and two electron-ion (P^+ and B^-) deflections.

were to take the wave viewpoint of an electron, a collision between an electron and a lattice atom, for example, would refer to the interaction of the electron wave with a phonon. Briefly, as the atoms vibrate about their mean positions, with an amplitude that depends on the temperature, the periodicity of the lattice is disturbed. Thus the potential in the system can no longer be considered as being exactly periodic. However, if the perturbations are small, the electron wavefunctions can still be considered to be Bloch wavefunctions, but with different values to those for a stationary, periodic lattice. Thus, if the eigenfunctions change, the eigenvalues will also change, i.e., the state of the electron will change on interacting with a vibrating atom. To get away from the particle-oriented 'collision', the word **scattering** is often used, i.e., the electron scatters to a new state on interacting with an atom, an ion or another electron.

At low temperatures, when the thermal vibrations of the atoms are small, **ionized-impurity scattering** tends to be the dominant scattering mechanism. If there are many carriers present, they can screen other carriers from the attractive or repulsive effect of the impurity ion's charge, leading to **screened ionized-impurity scattering**. If the concentration of carriers is extremely large, **carrier-carrier scattering** becomes important. In doped semiconductors at higher temperatures, phonon scattering dominates. Longitudinal acoustic phonons, for example, alternately compress and dilate the lattice (see Fig. 3.2b); the resulting strain deforms the band edges, leading to **deformation potential scattering**. In semiconductors for which there is no crystal inversion symmetry, such as GaAs, the strain may generate a potential via the piezoelectric effect, leading to **piezoelectric scattering**. Si and GaAs both have two atoms per unit cell. When these atoms move in opposite directions, optic phonons are generated (see Fig. 3.2c). When the two atoms become ions due to electron exchange, as is the case in the formation of compound semiconductors such as GaAs, electric fields are created and we have **polar optic phonon scattering**. This is the dominant scattering mechanism in GaAs at room temperature. More details on scattering are given in Wolfe *et al.* [1, Section 6.6] and in References [2, 3].

In this book, we don't treat scattering at the microscopic level. Instead, we allow all such details to be absorbed into a macroscopic quantity called **mobility**, as described in Chapter 5.

No matter how you view the interactions between the carriers and other entities in the crystal, they serve to randomize the carrier velocities. So, in equilibrium, the bands contain certain numbers of electrons and holes, and the velocities of these carriers are randomized: there is a lot of motion but no net current.

4.2   The Fermi level

The problem of finding the most probable distribution of electrons among the available states in the conduction and valence bands is a problem in statistical mechanics [1, Sections 4.1, 4.2]. Here, we follow Talley and Daugherty [4] and arrive at the desired distribution function in a relatively non-mathematical way by considering one of the types of collisions (elastic, electron-electron) that are instrumental in maintaining thermal equilibrium.

Let two electrons have energies E_1 and E_2 before a collision and energies E_1' and E_2' after the collision. No energy is lost to other processes in an elastic collision, so

$$[E_1' - E_1] + [E_2' - E_2] = 0. \tag{4.1}$$

For this collision to happen there must be filled states at energies E_1 and E_2, and empty states at E_1' and E_2'. The latter requirement is crucial to our argument, and is a recognition of Pauli's Exclusion Principle, i.e., no two electrons may exist in the same quantum state at the same time. Calling the probability of occupancy $f(E)$, we have for the rate of this collision

$$r_{1,1':2,2'} = C\, f(E_1) f(E_2)[1 - f(E_1')][1 - f(E_2')], \tag{4.2}$$

where C is the rate constant.

To maintain thermal equilibrium, elsewhere in the semiconductor a collision must be occurring, at the same rate as the above collision, involving electrons with energies E_1' and E_2', with the result that the new energies of these electrons are E_1 and E_2. For this collision to occur there must be filled electron states at energies E_1' and E_2', and empty states at E_1 and E_2. This leads to

$$r_{1',1:2',2} = C\, f(E_1') f(E_2')[1 - f(E_1)][1 - f(E_2)]. \tag{4.3}$$

Equating (4.2) and (4.3), dividing by $f(E_1)f(E_2)f(E_1')f(E_2')$, we end up with

$$\left[\frac{1}{f(E_1)} - 1\right] \left[\frac{1}{f(E_2)} - 1\right] = \left[\frac{1}{f(E_1')} - 1\right] \left[\frac{1}{f(E_2')} - 1\right]. \tag{4.4}$$

The solution to this equation, as can be verified by substitution, is

$$\left[\frac{1}{f(E)} - 1\right] = A e^{\beta E}, \tag{4.5}$$

where A and β are constants to be determined. Thus,

$$f(E) = [1 + Ae^{\beta E}]^{-1} , \qquad (4.6)$$

where we realize that both constants must be positive for $f(E)$ to be not greater than 1, and for the states with the lowest energy to be the ones most likely to be occupied.

To identify β, consider some large energy E such that there is very little chance of quantum states at this energy being filled by electrons, i.e.,

$$f(E) \approx \frac{1}{A} \exp(-\beta E) \ll 1 . \qquad (4.7)$$

In such a situation there's very little chance of two electrons vying to occupy the same state, i.e., the electrons are essentially non-interacting. A similar situation pertains to gas molecules in an ideal gas, for which it is well-known from Thermal Physics that the probability that a molecule has an energy E is $\propto \exp(-E/k_BT)$, where T is the absolute temperature and $k_B = R/N$ is Boltzmann's constant; R is the gas constant and N is Avogadro's number. Thus β can be seen to be a universal constant, given by $1/k_BT$. It remains to label the constant A in (4.6); it is defined as $\exp(-E_F/k_BT)$, where E_F is called, in the subject of semiconductor devices, the **Fermi level**.[2] Thus, from (4.7), the electron distribution function under conditions of thermal equilibrium is

$$f_{FD}(E) = \frac{1}{1 + \exp[(E - E_F)/k_BT]} . \qquad (4.8)$$

This is the **Fermi-Dirac distribution function**, the form of which is illustrated in Fig. 4.2. We can also refer to $f_{FD}(E)$ as the probability of an electron occupying a state of energy E at thermal equilibrium because it varies in value from 0 to 1, as we would expect from having invoked the Pauli Exclusion Principle in its derivation.

Note that the probability of occupancy of states with energy E falls off rapidly as E exceeds E_F. In n-type semiconductor material there are many electrons that must be accommodated in the conduction band by the filling-up of states. Clearly, in such a material, states will be filled-up to higher energies than in the case of p-type material, in which there are relatively few electrons. For higher energy states to have a higher probability of being filled, it follows that the Fermi level must be raised, i.e., the Fermi level is higher in energy for n-type material than it is for p-type material. Exactly how much higher depends on the relative doping levels, of course. This link between E_F and the **equilibrium electron concentration** n_0 is established in the next section.

---

[2] In Physics, E_F is called the Fermi level only at $T = 0\,\mathrm{K}$. At higher temperatures it is properly called the chemical potential (see Section 6.1).

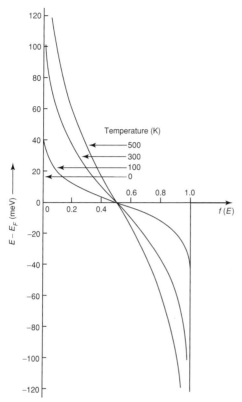

Figure 4.2 The Fermi-Dirac distribution function for various temperatures. Reproduced from Pulfrey and Tarr [5, Fig. 4.2].

4.3   Equilibrium carrier concentrations and the Fermi level

Adding bounds to the bands, (3.33) for the equilibrium concentrations of electrons and holes becomes

$$n_0 = \int_{E_C}^{\text{top of band}} g_C(E) f_{FD}(E) \, dE$$

$$p_0 = \int_{\text{bottom of band}}^{E_V} g_V(E)[1 - f_{FD}(E)] \, dE , \tag{4.9}$$

where the subscript '0' indicates equilibrium, and $f_{FD}(E)$ is the Fermi-Dirac distribution function.

To actually evaluate the carrier-concentration integrals it is convenient to assign the limits of $+\infty$ and $-\infty$ to the top of the conduction band and to the bottom of the valence band, respectively. This is not likely to cause significant error in the evaluation of the integrals because it turns out that E_F in semiconductors is usually either in the bandgap, or close to one of the band edges, so $f_{FD}(E)$ rapidly goes to zero for energies high in

Table 4.1 Effective densities-of-states and intrinsic carrier concentrations for Si and GaAs. The values are for $T = 300\,\text{K}$ and have been computed using the density-of-states effective masses for this temperature from Table 3.2.

Semiconductor	N_C (cm^{-3})	N_V (cm^{-3})	n_i (cm^{-3})
Si	3.2×10^{19}	1.8×10^{19}	9.5×10^{9}
GaAs	4.2×10^{17}	9.4×10^{18}	2.4×10^{6}

the conduction band, as does $[1 - f_{FD}(E)]$ for energies low in the valence band. The result from (4.9) for electrons is

$$n_0 = N_C \mathcal{F}_{1/2}(a_F),\qquad(4.10)$$

where all the material-constant terms have been collected together into N_C, which has the units of m^{-3} for a 3-D device and is termed the **effective density of states in the conduction band**, given by

$$N_C = 2 \left(\frac{2\pi m^*_{e,\text{DOS}} k_B T}{h^2} \right)^{3/2},\qquad(4.11)$$

and $\mathcal{F}_{1/2}(a_F)$ is called the **Fermi–Dirac integral of order one-half**, given by

$$\mathcal{F}_{1/2}(a_F) = \frac{2}{\sqrt{\pi}} \int_0^\infty \frac{a^{1/2}\,da}{1 + \exp(a - a_F)},\qquad(4.12)$$

where $a = (E - E_C)/k_B T$ and $a_F = (E_F - E_C)/k_B T$.

Fermi–Dirac integrals are listed as tabulated functions [1, Appendix B], and formulae exist for their approximate evaluation [6]. A very convenient approximation to $\mathcal{F}_{1/2}(a_F)$ arises if $a_F < -2$:

$$\mathcal{F}_{1/2}(a_F) \rightarrow \exp(a_F),\qquad(4.13)$$

which then enables (4.10) to be written concisely as

$$n_0 = N_C \exp\left(\frac{E_F - E_C}{k_B T} \right).\qquad(4.14)$$

Values for N_C and for N_V, the corresponding **effective density of states in the valence band**, are tabulated in Table 4.1.

Equation (4.14) is known as the **Maxwell–Boltzmann expression** for the equilibrium electron concentration. It is compared with the full Fermi–Dirac expression of (4.10) in Fig. 4.3. The limit of validity of (4.14) is seen to be ($E_C - E_F > 2k_B T$); this corresponds to $n_0/N_C < 0.4$, i.e., to doping densities of about 10^{19} cm^{-3} and 10^{17} cm^{-3} in Si and GaAs, respectively. The corresponding equation for holes, under conditions such that

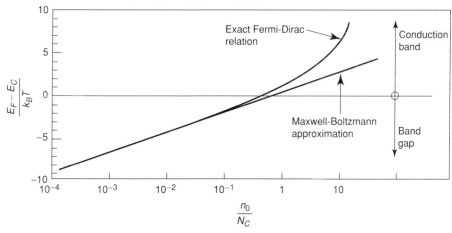

Figure 4.3 Comparison of the Fermi-Dirac ((4.10)) and Maxwell-Boltzmann ((4.14)) expressions for n_0. Reproduced from Pulfrey and Tarr [5, Fig. 4.3].

$(E_F - E_V) > 2k_B T$ is

$$p_0 = N_V \exp\left(\frac{E_V - E_F}{k_B T}\right). \tag{4.15}$$

The Maxwell-Boltzmann expressions are instructive: if n_0 is increased by adding more donors, for example, (4.14) informs that E_F will move closer to the conduction-band edge; if p_0 is increased by adding more acceptors, (4.15) informs that E_F will move closer to the valence-band edge. Thus, the inclusion of E_F on an energy-band diagram provides a convenient visual aid for ascertaining at a glance whether a region is doped n-type or p-type, and it also gives some indication of the extent of the majority carrier concentration.

The Maxwell-Boltzmann expressions (4.14) and (4.15) are so-named because their exponential dependence on energy and on the reciprocal of $k_B T$ is characteristic of the eponymous distribution function, namely

$$f_{MB}(E) = e^{-\frac{(E-E_F)}{k_B T}}. \tag{4.16}$$

Physically, a Maxwell-Boltzmann distribution function represents, in the case of semiconductors, the situation where the quantum states in the band can be filled without regard to Pauli's Exclusion Principle. This can only be a reasonable approximation when the number of electrons, for example, is far less than the number of available states in the conduction band, i.e., when the likelihood of more than one electron occupying the same state is very small. Evidently, as we have just seen, this is the case for $n_0/N_C < 0.4$. When the doping density is such that this condition does not hold, the material is said to be **degenerate**. In this circumstance, Fermi-Dirac statistics should be used; E_F will penetrate higher into the conduction band than predicted by Maxwell-Boltzmann statistics, as Fig. 4.3 shows.

4.4   Equations involving intrinsic properties

Often, it is useful to express the carrier concentrations in terms of their intrinsic, equilibrium values. For intrinsic material, (4.14) and (4.15) apply, and they give

$$n_i = N_C \exp\left(\frac{E_{Fi} - E_C}{k_B T}\right)$$

$$p_i = N_V \exp\left(\frac{E_V - E_{Fi}}{k_B T}\right),$$ (4.17)

where E_{Fi} is the Fermi energy for intrinsic material.

Equating these expressions, and substituting for the densities of states gives

$$E_{Fi} - E_V = \frac{E_g}{2} + \frac{3k_B T}{4} \ln\left[\frac{m^*_{h,\text{DOS}}}{m^*_{e,\text{DOS}}}\right],$$ (4.18)

where $E_g = (E_C - E_V)$ is the bandgap of the semiconductor. For Si, where $E_g \gg k_B T$ and the electron and hole density-of-states effective masses are not too dissimilar, it follows that E_{Fi} lies almost at the middle of the bandgap.

Multiplying n_i by p_i, and noting that $n_i = p_i$, gives

$$n_i = \sqrt{(N_C N_V)} \exp\left(\frac{-E_g}{2k_B T}\right).$$ (4.19)

This equation shows that n_i is a temperature-dependent material constant, and that the intrinsic carrier concentration drops drastically as the bandgap increases, as hinted at in Section 2.7 when discussing the difference between semiconductors and insulators. Note that the product of (4.14) and (4.15) is independent of the Fermi energy. It follows that

$$n_0 p_0 = n_i p_i = n_i^2.$$ (4.20)

However, remember that (4.14) and (4.15) are not correct when the doping is very high, i.e., *they only apply to non-degenerate conditions*. Check the text following (4.16) to appreciate the doping-density limitation to (4.20). The extent of the departure of the product $n_0 p_0$ from the value of n_i^2 at high doping levels can be appreciated by doing Exercise 4.2.

One example of the use of (4.20) is in the computation of the equilibrium carrier concentration in a uniform, field-free region of a semiconductor, e.g., in the bulk of the emitter of a diode or an HBT, or in the body of a MOSFET (see Exercise 4.3). In these circumstances it follows that local charge neutrality exists,[3] namely

$$q(p_0 - n_0 + N_D - N_A) = 0.$$ (4.21)

---

[3] More generally, out-of-equilibrium, but in steady-state operation, the semiconductor device will be *overall* neutral: $\int_{\text{volume}} q\,[p(x, y, z) - n(x, y, z) + N_D(x, y, z) - N_A(x, y, z)]\,dx\,dy\,dz = 0$. This can be appreciated from the facts that: the semiconductor starts out neutral; electrons and holes, or charge carriers and oppositely ionized dopants, are generated together; electrons and holes recombine in pairs; if there is a steady-state current, as one charge enters the device, another must leave it.

In conjunction with (4.20), (4.21) yields

$$n_0 = \frac{X + \sqrt{X^2 + 4n_i^2}}{2} , \qquad (4.22)$$

where $X = N_D - N_A$.

The carrier concentrations can be expressed in terms of the intrinsic carrier concentration by combining each of (4.14) and (4.15) with (4.17) to give alternative expressions for n_0 and p_0:

$$n_0 = n_i \exp \left(\frac{E_F - E_{Fi}}{k_B T} \right)$$

$$p_0 = n_i \exp \left(\frac{E_{Fi} - E_F}{k_B T} \right) . \qquad (4.23)$$

Again, please note that *these expressions only hold for non-degenerate conditions.*

4.5   Mean unidirectional velocity of an equilibrium distribution

The equilibrium spectral carrier concentration for electrons from (3.33) is

$$n_0(E) = g_C(E) f_0(E) . \qquad (4.24)$$

This expression can be evaluated easily when f_0 is expressed as a Maxwell-Boltzmann distribution:

$$
\begin{aligned}
f_{MB}(E) &= e^{-\frac{E - E_F}{k_B T}} \\
&\equiv e^{-\frac{E - E_C + E_C - E_F}{k_B T}} \\
&= \frac{n_0}{N_C} e^{-\frac{E - E_C}{k_B T}} .
\end{aligned}
\qquad (4.25)
$$

The result for the case of $n_0 = 10^{19}$ cm^{-3} is shown in Fig. 4.4, where the distribution has been split-up into two parts. The distribution on the right is that for the electrons with positive crystal momentum $+\hbar k_e$, or, equivalently, for example, with a velocity component in the positive x-direction, while the one on the left is its oppositely directed counterpart. Each half of the distribution contains exactly $n_0/2$ electrons, and is called a **hemi-Maxwellian** in this case.[4] At any plane in the material, the flow of electrons in the positive-going distribution will be opposed by the flow of negative-going electrons from a neighbouring distribution, so, as must be the case at equilibrium, there is no net current. However, if the conditions in a device were *near*-equilibrium, but the distributions were slightly different in neighbouring locations, due to a non-uniform

---

[4] If Fermi-Dirac statistics had been used, each half of the distribution would be a **hemi-Fermi-Diracian**.

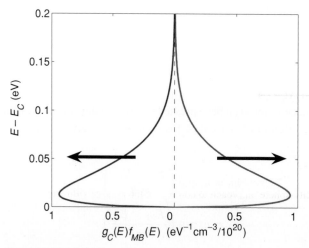

Figure 4.4 The full Maxwellian distribution, comprising two, counter-propagating hemi-Maxwellian distributions, for the equilibrium case of $n_0 = 10^{19}$ cm^{-3}.

doping density, for example, then the opposing charge flows at an intermediate plane would not cancel, and there would be a current. This is an example of diffusion. Other examples where the current due to a near-equilibrium hemi-Maxwellian or hemi-Fermi-Diracian might arise are in the injection of carriers into the base of an HBT or into the channel of a FET. Thus, it is useful to compute the current due to each of these hemi-distributions. To do so, we must first determine the mean speed of the electrons in the total distribution.

The **mean speed** of electrons in a distribution at thermal equilibrium is defined as

$$v_{th} = \frac{\int_0^\infty v n_0(v)\, dv}{\int_0^\infty n_0(v)\, dv}, \tag{4.26}$$

where $n_0(v)$ is the velocity-spectral concentration of electrons at equilibrium, i.e., the number of electrons per m^3 per unit velocity. The denominator is just the total electron concentration, as given by (4.10). To deal with the numerator we need an expression for $n_0(v)$. To obtain this, take $n(E) = g_C(E) f_0(E)$, substitute for the density of states from (3.32) and for the Fermi–Dirac function from (4.8), and then make the transformation from an energy distribution to a velocity distribution via the kinetic-energy expression (2.41):

$$(E - E_C) = \frac{1}{2} m_{th}^* v^2, \tag{4.27}$$

where m_{th}^* is an appropriate effective mass, which will be identified in the following subsection. The result is

$$n_0(v) = \frac{8\pi}{h^3} \frac{(m_{e,\mathrm{DOS}}^* m_{th}^*)^{3/2} v^2}{1 + \exp[m_{th}^* v^2 / 2k_B T] \exp[(E_C - E_F)/k_B T]}. \tag{4.28}$$

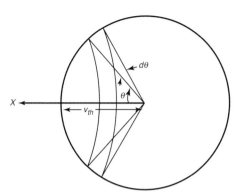

Figure 4.5 A 'velocity-sphere' approach to computing the average x-directed velocity. The radius v_{th} is the mean thermal speed. As the electron velocities are randomly distributed in direction, the number of electrons with velocities in the range θ to $\theta + d\theta$ is related to the area of a strip of length $2\pi v_{th} \sin\theta$ and of width $v_{th} d\theta$. To express this as a fraction of the total number of electrons, we divide by $4\pi v_{th}^2$, to get $\frac{\sin\theta d\theta}{2}$. Each stream of electrons has a velocity component $v_{th}\cos\theta$ in the x-direction. Integrating from 0 to $\pi/2$ completes the calculation of the mean, unidirectional velocity. Reproduced from Pulfrey and Tarr [5, Fig. 8.5].

Substituting this into the numerator of (4.26), integrating, and dividing by n_0, the result for the mean thermal speed from (4.26) is

$$
\begin{aligned}
v_{th} &= \frac{\int_0^\infty v^3 \left[1 + \exp(m_{th}^* v^2/2k_B T + (E_C - E_F)/k_B T)\right]^{-1} dv}{\int_0^\infty v^2 \left[1 + \exp(m_{th}^* v^2/2k_B T + (E_C - E_F)/k_B T)\right]^{-1} dv} \\[2mm]
&\equiv \frac{\int_0^\infty a \left[1 + \exp(a - a_F)\right]^{-1} da}{\int_0^\infty a^{1/2} \left[1 + \exp(a - a_F)\right]^{-1} \sqrt{m_{th}^*/2k_B T}\, da} \\[2mm]
&= \sqrt{\frac{8k_B T}{\pi m_{th}^*}} \frac{\mathcal{F}_1}{\mathcal{F}_{1/2}},
\end{aligned}
\tag{4.29}
$$

where, following the notation of (4.12), the substitutions $a = (E - E_C)/k_B T = m_{th}^* v^2/2k_B T$ and $a_F = (E_F - E_C)/k_B T$ have been made, and the final result has been cast in the form of a material-dependent constant and a ratio of Fermi–Dirac integrals. Note that for non-degenerate cases, the ratio of the two integrals is essentially unity.

The resulting v_{th} is to be understood as the speed (magnitude of the velocity) of an electron stream in any particular direction. As no one-particular direction is favoured in thermal equilibrium, the electron motion is random. However, if all the random velocities are resolved along a particular direction, as shown in Fig. 4.5, then the average velocity component, $\langle v_x \rangle$ for example, is found to be $v_{th}/4$, as explained in the figure caption.

Table 4.2 Mean, unidirectional thermal velocities for electrons at 300 K in Si and GaAs, for various doping densities. The conductivity effective-masses for electrons and holes are from the expressions given in Section 5.4.

Semiconductor	$m^*_{h,CON}$ (m_0)	$m^*_{e,CON}$ (m_0)	n_0, (cm^{-3})	$\mathcal{F}_1/\mathcal{F}_{1/2}$	$v_{R,e}$ (nm/ps)
Si	0.40	0.26	1×10^{18}	1.0	53
			1×10^{19}	1.0	53
			1×10^{20}	1.3	69
GaAs	0.39	0.066	1×10^{18}	1.2	125
			1×10^{19}	2.1	218
			1×10^{20}	4.5	468

This **mean, unidirectional velocity** is written as v_R.[5] Thus,

$$v_R = \sqrt{\frac{k_B T}{2\pi m^*_{th}} \frac{\mathcal{F}_1}{\mathcal{F}_{1/2}}}. \tag{4.30}$$

4.5.1   Effective mass and v_R

The effective mass in (4.30) must be related in some way to the unidirectional motion of electrons. For spherical constant-energy surfaces and a single band, m_{th} is simply the band effective mass. This is the case for electrons near the conduction-band minimum in GaAs. For the ellipsoidal constant-energy surfaces applicable to Si, the 6 equivalent conduction-band minima lie along the $\langle 100 \rangle$ directions (see Fig. 2.12). Thus, for a specific uni-direction within this set, m^*_l applies to two of the ellipsoids, and m^*_t applies to the other four. An effective mass that takes this anisotropy into account is the **conductivity effective mass**, m^*_{CON}. In Section 5.4 we derive this parameter, both for electrons and ellipsoidal constant-energy surfaces, and for holes and two, non-interacting spherical constant-energy surfaces. Suffice to say here, we take

$$m^*_{th} \equiv m^*_{CON}. \tag{4.31}$$

With this interpretation of m^*_{th}, some values of v_R for electrons in Si and GaAs are as given in Table 4.2. The Fermi-Dirac integrals were evaluated using the short-series approximations given in Ref. [6].

4.5.2   Current and v_R

Now that we know the concentration of carriers and their mean, unidirectional velocity, we can appreciate that the average flux of electrons in the x-direction, for example, is $n_0 v_R$. A more instructive way of writing this is $\frac{n_0}{2} 2 v_R$, as this makes it explicit that only

---

[5] The subscript R comes from Sir Owen W. Richardson, Nobel laureate for physics in 1928 for 'his work on the thermionic phenomenon and especially for the discovery of the law named after him'.

one-half of the electrons in the distribution actually have a component of velocity in the positive x-direction. Evidently, for each of the counter-directed hemi-distributions, comprising $n_0/2$ electrons and illustrated in Fig. 4.4, the mean unidirectional velocity is $2v_R$. Formally, this fact follows from the construction in Fig. 4.5 by simply dividing by the area of a hemisphere, rather than by that of a sphere. Thus, the current densities due to the right- and left-going hemi-distributions are

$$\vec{J}_{e,\rightarrow} = -q\frac{n_0}{2} 2v_R \hat{x}$$

$$\vec{J}_{e,\leftarrow} = -q\frac{n_0}{2} (2v_R)(-\hat{x}). \tag{4.32}$$

Each component can be of enormous magnitude ($\approx 8 \times 10^6$ A/cm^2 for $n_0 = 10^{19}$ cm^{-3} in Si, for example), but they cancel each other exactly, as required of an equilibrium distribution.

Exercises

4.1 Confirm the symmetry of the Fermi-Dirac distribution function about E_F by showing that the probability of a state of energy ΔE above E_F being occupied is the same as the probability of a state of energy ΔE below E_F being empty.

4.2 Consider p-type Si for 3 different cases of boron doping, i.e., 10^{16} cm^{-3}, 10^{18} cm^{-3}, and 10^{20} cm^{-3}.

   (a) Calculate the equilibrium carrier concentrations, p_0, n_0, and the Fermi-level position, $(E_F - E_V)$, for each case at 300K. For the majority carrier concentration use Fermi-Dirac statistics.

   (b) Determine the validity of $p_0 n_0 = n_i^2$, where the intrinsic carrier concentration is given by (4.19).

   (c) The reason why this equation is not satisfied at high doping densities is sometimes referred to as 'Pauli blocking'. Suggest what this term may mean.

4.3 A diffused-junction n-p diode is made by diffusing phosphorus into boron-doped silicon. The concentrations of both dopants can be taken to be uniform, and to have values of $N_A = 10^{16}$ cm^{-3} and $N_D = 10^{17}$ cm^{-3}.

   Evaluate the equilibrium carrier concentrations in the diffused region at 300 K.

4.4 Some semiconductor devices are rated to operate at temperatures of about 200°C. In GaAs at this temperature, is it reasonable to assume that $E_{Fi} - E_V \approx E_g/2$?

4.5 Taking into account the temperature dependence of the effective density of states N_C and N_V, but ignoring the slight reduction in bandgap with temperature, at what temperature does n_i in Si reach the value of a typical doping density, say 10^{17} cm^{-3}?

4.6 The E-k relationships for the conduction bands of two semiconductor materials, A and B, can be expressed as

$$E_A - E_C = \alpha k^2 \quad \text{and} \quad E_B - E_C = 2\alpha k^2,$$

respectively, where α is a constant, and E_C is the energy of the bottom of the conduction band.

Consider each material to be moderately doped n-type, with the same concentration of donors, all of which can be taken to be ionized.

Which material would have its Fermi energy closer to E_C?

4.7 Given that $m_{th}^* = 0.40m_0$ for holes in silicon, compute the mean, unidirectional velocity for holes at 300 K when the p-type doping density is: (a) 10^{18} cm^{-3}, and (b) 10^{20} cm^{-3}.

4.8 Use the value of $v_{R,h}$ calculated in the previous question for the case of $p_0 = 10^{18}$ cm^{-3} to evaluate the current density due to a hemi-Maxwellian distribution of these holes.

References

[1] C.M. Wolfe, N. Holonyak, Jr. and G.E. Stillman, *Physical Properties of Semiconductors*, Prentice-Hall, 1989.
[2] J.H. Davies, *The Physics of Low-dimensional Semiconductors*, Secs. 8.1–8.4, Cambridge University Press, 1998.
[3] M.S. Lundstrom, *Fundamentals of Carrier Transport*, Chap. 2, Cambridge University Press, 2000.
[4] H.E. Talley and D.G. Daugherty, *Physical Principles of Semiconductor Devices*, Chap. 5, Iowa State University Press, 1976.
[5] D.L. Pulfrey and N.G. Tarr, *Introduction to Microelectronic Devices*, Prentice-Hall, 1989.
[6] R. Kim and M.S. Lundstrom, *Notes on Fermi-Dirac Integrals, 3rd Edn.*, posted Sept. 23, 2008. Online [http://nanohub.org/resources/5475/].

5 Charge transport

The picture of charge carriers that should have emerged so far is one of electrons and holes moving through the semiconductor in momentum states. The carriers are continually being generated and annihilated, and they are also frequently scattered to new momentum states via collisions with vibrating atoms, ionized impurities and other carriers. In thermal equilibrium large fluxes are present, but there is no net current. To disturb this equilibrium, and obtain a non-zero net flow of charge, we need to establish some driving force within the semiconductor. Fundamentally, this driving force is a gradient in energy. It can be manifest as a gradient in the potential energy, which is related to the electrostatic potential ψ, and as a gradient in kinetic energy. If each of the n electrons per m^3 has a kinetic energy u, the total kinetic energy density W is nu. Gradients in both n and u can produce a current. The main objective of this chapter is to describe and characterize the currents due to each of these three gradients.

The above picture is one of **dissipative transport**, i.e., one in which the directed momentum of electrons injected into a region from a hemi-Maxwellian distribution, for example, is dissipated by scattering events. If the region is so short that the injected electrons can traverse it without being scattered, then we have **ballistic transport**. In most of the devices considered in this book, dissipative transport is more prominent, but ballistic transport does occur in tunnelling, for example, and is likely to occur in future nanoscale devices, as briefly discussed in Chapter 18.

5.1 Charge, current and energy

Charge, current and energy figure prominently in this chapter. Here, we give the fundamental definitions of the densities of these properties. To keep things simple, we assume parabolic bands with an isotropic effective mass, and write the expressions for the densities of current and kinetic energy in terms of only the x-directed component. The starting point in all three cases is the expression for carrier density n, as given for the 3-D case by (3.27). Multiplying this equation by $-q$ gives an expression for the electron charge density:

$$-qn = -q\frac{1}{4\pi^3}\int_{\vec{k}} f(\vec{k})\, d\vec{k}. \tag{5.1}$$

Multiplying (3.27) by $-qv_x$, where $v_x = (1/\hbar)dE/dk_x$ is the x-directed velocity, attributes a velocity to each occupied state, the sum of which leads to the x-directed current density:

$$J_{e,x} = -q \frac{1}{4\pi^3} \int_{\vec{k}} v_x f(\vec{k}) \, d\vec{k} .$$

(5.2)

Multiplying (3.27) by $m^* v_x^2/2$ gives the x-directed kinetic energy density:

$$W_{e,x} = \frac{1}{4\pi^3} \int_{\vec{k}} \frac{1}{2} m^* v_x^2 f(\vec{k}) \, d\vec{k} .$$

(5.3)

The distribution function, $f(\vec{k})$, or to give it its full set of dependencies, $f(\vec{r}, \vec{k}, t)$, must be found for the non-equilibrium conditions necessary for a net current to exist. This is a difficult task. To begin to solve the problem, we must first formulate an equation that describes how f might change in both real space and in k-space in response to relevant forces.

5.2   The Boltzmann Transport Equation

The full, 6-D $(\vec{r}, \vec{k})$-environment is called **phase space**. Here, for simplicity, we'll restrict our discussion to just two dimensions of this space, x and k_x. Recall from Section 2.9 that the crystal momentum $\hbar k$ is the momentum due to forces other than the periodic crystal forces. It is these other forces that we are interested in, so we can view $f(x, k_x, t)$ as the probability of an electron at position x having a momentum $\hbar k_x$ at time t. The fact that we are simultaneously specifying a position and a momentum means that we are treating the electron classically. This is consistent with the effective-mass formalism in which the group velocity of a wavepacket is involved. In fact, the treatment is semi-classical, because there is some quantum mechanics involved in getting the group velocity and in describing scattering processes microscopically.

With this understanding of the situation, under the action of some force F_x, an electron will move a distance $v_x \Delta t$ in time Δt, where $v_x = (1/\hbar) dE/dk_x$ is the x-directed velocity. The electron will also change its momentum according to $\hbar \Delta k_x / \Delta t = F_x$. As we are saying that the electron has moved from one state in phase space to another state, the probability of occupancy of these two states must be the same, thus

$$f(x, k_x, t) = f\left(x + v_x \Delta t, \; k_x + \frac{F_x}{\hbar} \Delta t, \; t + \Delta t\right).$$

(5.4)

Taking Δt as being very small permits a Taylor series expansion of the right-hand side to first order. Equation (5.4) then becomes

$$\frac{\partial f}{\partial t} = -v_x \frac{\partial f}{\partial x} - \frac{F_x}{\hbar} \frac{\partial f}{\partial k_x}.$$

(5.5)

Now, let us specify the force as being due to an applied electric field $\mathcal{E}_x$ and due to a scattering force $F_{x,\,coll}$. The former force is given by $-q\mathcal{E}_x$, and the latter is defined from

$$\left. \frac{\partial f}{\partial t} \right|_{coll} \equiv -\frac{1}{\hbar} \frac{\partial f}{\partial k_x} F_{x,\,coll} ,$$

(5.6)

where we have reverted to the label 'coll' for scattering events, in keeping with the more classical picture of collisions involving interactions of particles with atoms, ions and other charge carriers. Adding in the forces, we obtain the 1-D version of the **Boltzmann Transport Equation**

$$\frac{\partial f}{\partial t} = -v_x \frac{\partial f}{\partial x} + \frac{q\mathcal{E}_x}{\hbar} \frac{\partial f}{\partial k_x} + \frac{\partial f}{\partial t}\bigg|_{coll}. \tag{5.7}$$

This is a 'balance equation' for the distribution function: driving forces due to kinetic energy (related to velocity) and potential energy (related to field) are moderated by the restoring force of collisions. Much of the difficulty involved in solving this equation comes from having to specify in a tractable way the complicated, microscopic nature of scattering.

5.2.1   The Method of Moments

Rather than describe how to solve the Boltzmann Transport Equation (BTE) directly for f, we appeal to the Method of Moments to arrive at useful expressions for the properties of more interest to us: charge density, current density and kinetic-energy density. The method involves multiplying each term of the BTE by some factor Υ and integrating over $d\vec{k}/4\pi^3$. We include v_x in Υ, and label the moments according to the power to which v_x is raised. Thus, we seek to solve for: $-qn$ in (5.1) using the zeroth-order moment, J_e in (5.2) using the first-order moment and W_e in (5.3) using the second-order moment.

We follow the approach of Datta [1], and we make use of the general result of applying the Method of Moments to the BTE:

$$\frac{\partial}{\partial t}\Phi + \nabla \cdot J_\Phi = G_\Phi - R_\Phi, \tag{5.8}$$

where Φ is $-qn$, J_e, or W_e, according to the order of the moment being considered. In (5.8), $J_\Phi = \vec{v}\Phi$ is the flux of Φ, G_Φ is the rate at which Φ is generated by the electric field, and R_Φ is the rate at which Φ is lost due to scattering. It is permissible to write $\nabla \cdot J_\Phi$ in this manner, instead of as $\vec{v} \cdot \nabla\Phi$, because $\vec{v}$, the group velocity, is a function of k only. The order of the terms in (5.8) is different from that in the BTE, and the equation has been written in 3-D form to distinguish between divergence and gradient. With these changes, (5.8) has a very clear physical interpretation: the difference between the generation and loss of Φ in a volume serves to either increase Φ with time in the volume, or to result in a net flow of Φ out of the volume. The situation for steady-state conditions is illustrated in Fig. 5.1.

5.2.2   The continuity equations

The results for each term in (5.8) for the first three moments are tabulated in Table 5.1.

For the charge density, there is no generation term because the electric field does not generate any charge. Also, there is no loss due to scattering because charge is not lost in

Table 5.1 Terms in (5.8) for the first three moment-solutions to the BTE.

Υ	Φ	J_Φ	G_Φ	R_Φ	Continuity of:
$-qv_x^0$	$-qn$	$J_{e,x}$	0	0	charge density
$-qv_x$	$J_{e,x}$	$-2qW_{e,x}/m^*$	$q^2 n\mathcal{E}_x/m^*$	$J_{e,x}/\langle\langle\tau_M\rangle\rangle$	current density
$m^* v_x^2/2$	$W_{e,x}$	$S_{e,x}$	$J_{e,x}\mathcal{E}_x$	$(W_{e,x}-W_0)/\langle\langle\tau_E\rangle\rangle$	kinetic energy density

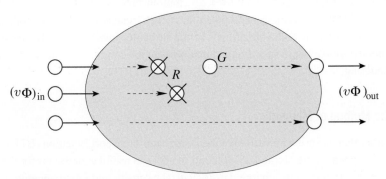

Figure 5.1 Schematic representation of the continuity equation (5.8), for the case of a 1-D flow of some property Φ. A flux of $v\Phi$ carries 3 units of Φ at a velocity v into the volume from the left. Two units are lost by scattering, and one unit is generated by the internal field. The situation shown here is for steady-state conditions, so 2 units of Φ exit to the right at velocity v.

the scattering process, it is merely redistributed among the allowed momentum states.[1] Thus, the **charge continuity equation** for electrons is

$$\frac{\partial n}{\partial t} = \frac{1}{q}\frac{\partial J_e}{\partial x}. \tag{5.9}$$

The **current-density continuity equation** describes the fact that current is generated by the field, and lost due to scattering. This loss is actually a loss in momentum, which is described by the momentum relaxation time τ_M. In general, one would expect the time dependence of the current, after the removal of the field, to be of the form $J(t) = J(0)e^{-t/\tau}$, which gives

$$\frac{dJ}{dt} = -\frac{J(t)}{\tau} \equiv -R_\Phi. \tag{5.10}$$

Substituting for R_Φ for the first-order moment term involving $\partial f/\partial t$ for collisions, reveals the precise form that τ must have:

$$\langle\langle\tau_M\rangle\rangle = \frac{-\int k_x f\,d\vec{k}}{\int k_x \left.\frac{\partial f}{\partial t}\right|_{\text{coll}} d\vec{k}}, \tag{5.11}$$

---

[1] Note that we are only considering conditions within one band, the conduction band in this case. If we were also to consider the valence band, then G_Φ would not be zero if impact ionization were occurring, and R_Φ would not be zero if recombination were happening.

where the double brackets signify the special averaging explicit in (5.11). The J_Φ term in the first-order-moment equation involves v_x^2, which is clearly related to the x-directed kinetic-energy density. Therefore, $\nabla \cdot J_\Phi$ is a flux of kinetic-energy density. Thus, the meaning of the first-order-moment equation is: if the loss of x-directed momentum due to scattering is greater than the gain of x-directed momentum due to the field, then, to maintain the steady state, there must be a net inflow of kinetic energy density.

In the **kinetic-energy-density continuity equation** the generation term is the product of the current density and the field, i.e., the electrons constituting $J_{e,x}$ are accelerated by $\mathcal{E}_x$, thereby gaining kinetic energy. This gain is countered by scattering events that randomize the kinetic energy. The latter are described by a properly averaged energy relaxation time $\langle\langle \tau_E \rangle\rangle$. The principal parameters in the flux term J_Φ are $v_x m v_x^2 n$. Check out the units of this and you'll find they are, as expected, J/m^{2}s. Equivalently, the units are W/m^2, which are more comprehensible because they inform that the flux of kinetic-energy density is just a power density. A practical example of this is the irradiance of the sun, which we'll encounter in Chapter 7. The symbol used here for the flux is $S_{e,x}$.

It is an elusive task to evaluate n, $J_{e,x}$ or $W_{e,x}$ from the moment equations that can be assembled by substituting the terms from Table 5.1 into (5.8). If we want to evaluate n, we need $J_{e,x}$. The next moment equation gives us $J_{e,x}$, providing we know $W_{e,x}$. The next moment equation would allow evaluation of $W_{e,x}$ if the flux of kinetic energy density $S_{e,x}$ were known. And so it goes on. To get closed-form solutions for the properties of interest, we must make some approximations. Two approximations that are widely used for the design and analysis of semiconductor devices are described in the next two subsections.

The Drift-Diffusion Equation

The electron kinetic-energy density arises from a concentration of n electrons, in which each electron possesses some kinetic energy. In 1-D

$$W_{e,x} = n\langle u_x \rangle, \tag{5.12}$$

where $\langle u_x \rangle$ is the mean, x-directed kinetic energy of an electron. The single brackets indicate that the average is performed over the distribution function, i.e., for the parabolic band case,

$$\langle u_x \rangle = \frac{1}{n}\frac{1}{4\pi^3}\int\left(\frac{1}{2}m^* v_x^2\right) f \, d\vec{k}. \tag{5.13}$$

Thus, the divergence term in (5.8) can be written as

$$-2\frac{q}{m^*}\frac{d}{dx}W_{e,x} = -2\frac{q}{m^*}\left[\langle u_x \rangle\frac{dn}{dx} + n\frac{d\langle u_x \rangle}{dx}\right]. \tag{5.14}$$

To obtain the **Drift-Diffusion Equation**, we make the assumption that the contribution to the kinetic-energy density gradient from the spatial change in the electron concentration is much greater than the contribution from any spatial change in the average kinetic energy of each electron. Thus, in steady-state, the current-density continuity equation yields

$$J_{e,x} = qn\mu_e\mathcal{E} + qD_e\frac{dn}{dx},\tag{5.15}$$

where the electron **mobility** and **diffusivity** are defined from:

$$\mu_e \equiv \frac{q}{m^*}\langle\langle\tau_M\rangle\rangle,\tag{5.16}$$

and

$$D_e \equiv \frac{2\mu_e}{q}\langle u_x\rangle,\tag{5.17}$$

respectively. The latter equation is often called the **Einstein Relation**, and reduces to the familiar $D_e/\mu_e = k_BT/q$ when the mean, x-directed kinetic energy is given by $\langle u_x\rangle = k_BT/2$. This occurs for low doping densities under near-equilibrium conditions, as can be verified by doing Exercise 5.3. Understand that the mobility is a macroscopic representation of the microscopic scattering processes that are specified by the average momentum relaxation time.

The Drift-Diffusion Equation (DDE) has been the 'workhorse' for device engineers for decades, and it continues to be useful for modern devices in many situations, as we show in later chapters of this book. Physically, the approximation that we made in order to get the DDE is justifiable in situations where the regions of a device are long enough for any gradients in $\langle u_x\rangle$, due to injection of energetic carriers from a neighbouring region, for example, to be essentially eliminated by randomizing scattering events. This situation may not apply to some modern transistors, for which the trend is towards smaller and smaller devices.

The Hydrodynamic Equations

When neither component of the gradient in kinetic-energy density can be ignored, we need to proceed to the next order of continuity equation before making any approximations. Looking at the continuity equation for W, we see that we need to know something about the flux of kinetic-energy density S, which, basically, is the product of W_e and some velocity that describes the electron ensemble. This velocity comprises a component due to the applied field, and a component due to the scattering processes [6]. Here, we simplify the situation by ignoring any component due to the field. Thus, we regard the electrons as a sort of ideal gas, with their kinetic energy being related solely to their random motion. From the Kinetic Theory of Gases, this kinetic energy can be described by a temperature. Here, we label the **electron temperature** as T_e, and note that it may differ from the **lattice temperature** T_L. At equilibrium, $T_e = T_L$, of course. However, if the electron gas is 'heated', by being accelerated in an electric field, for example, then it may gain energy faster than it can lose it to the lattice, so T_e will exceed T_L. It is in this context that one speaks of 'hot' electrons. Having approximated the kinetic energy as that of an ideal gas, we can write

$$W_{e,x} = n\langle u_x\rangle \approx n\frac{1}{2}k_BT_e.\tag{5.18}$$

Because we've neglected the drift component of W, the kinetic-energy-density flux (the power density) S is a diffusive flow, which depends on the gradient of T_e. Thus, we can write

$$S_{e,x} \approx -\kappa_e(T_e)\frac{\partial T_e}{\partial x}, \qquad (5.19)$$

where κ_e represents the thermal conductivity of the electron gas: it is a function of the electron temperature, and also of the electron mobility and concentration (see Exercise 5.4).

Applying (5.18) to (5.14), and then using the resulting expression for $\partial W_{e,x}/\partial x$ in the current continuity equation, leads to a revised expression for the steady-state current density:

$$J_{e,x} = qn\mu_e\mathcal{E}_x + \mu_e k_B T_e\frac{dn}{dx} + \mu_e k_B n\frac{dT_e}{dx}. \qquad (5.20)$$

The new term makes it explicit that a diffusive current can arise from a temperature gradient.

Applying (5.18) and (5.19) to the kinetic-energy-density continuity equation:

$$\frac{\partial W_{e,x}}{\partial t} - \kappa_e\frac{\partial^2 T_e}{\partial x^2} = J_{e,x}\mathcal{E}_x - \frac{k_B}{2}\frac{(T_e - T_L)}{\langle\langle\tau_E\rangle\rangle}. \qquad (5.21)$$

When this energy-density continuity equation is added to the continuity equations for charge density and for current density, the resulting set is a simplified form of the 1-D **Hydrodynamic Equations**:[2]

$$\frac{\partial n}{\partial t} = \frac{1}{q}\frac{\partial J_e}{\partial x}$$

$$\frac{\partial J_{e,x}}{\partial t} = \frac{2q}{m^*}\frac{\partial W_{e,x}}{\partial x} + \frac{q^2 n}{m^*}\mathcal{E}_x - \frac{J_{e,x}}{\langle\langle\tau_M\rangle\rangle}$$

$$\frac{\partial W_{e,x}}{\partial t} = \kappa_e\frac{\partial^2 T_e}{\partial x^2} + J_{e,x}\mathcal{E}_x - \frac{k_B}{2}\frac{(T_e - T_L)}{\langle\langle\tau_E\rangle\rangle}. \qquad (5.22)$$

5.3   The device equation set

To assemble a master set of equations that is useful for the design and analysis of 'semi-classical' devices, we start with the Hydrodynamic Equations (5.22). The first and third equations in this set describe the conservation of charge density and of kinetic-energy density, respectively. In their present form they apply to conservation of charge and kinetic energy of electrons only within the conduction band. To make these equations useful for actual devices, we need to add terms that allow for interband transfer of electrons. Specifically, we must include recombination and generation events.

The charge continuity equation is amended to account for the explicit generation of charge carriers due to the optical and impact-ionization processes discussed in Chapter 3.

---

[2] Equations of this form were used in Fluid Dynamics before their application to electron transport.

We label the sum of these as $G_{op,ii}$. Thermal generation and all the recombination mechanisms discussed in Chapter 3 are lumped together in a net recombination rate U. Thus, the right-hand side of the first equation in (5.22) is augmented by $(G_{op,ii} - U)$.

The continuity equation for kinetic-energy density must also take recombination and generation into account because these processes change the number of carriers in a band and, consequently, the kinetic-energy density. Recombination of one electron, for example, removes $3k_B T_e/2$ of kinetic energy from the electron ensemble in the conduction band. Impact-ionization generation involves a loss of kinetic energy by one electron, but the promotion of another electron to the conduction band, so now two electrons can gain energy from the field. It's not so obvious how to quantify this process, but it is not unreasonable to attribute to each generation event a kinetic energy gain equal to the bandgap energy [2]. Thus the change in kinetic-energy density due to recombination and generation could be expressed as

$$H = -\frac{3k_B T_c}{2} U + E_g G_{ii} ,\qquad (5.23)$$

where T_c is the carrier temperature.

Recombination/generation processes involve electrons and holes, so our master set of equations must also contain expressions for the conservation of hole charge, and of hole kinetic-energy density. The second equation in the hydrodynamic set describes the continuity of electron flow (current density). We need to add the corresponding equation for holes. Finally, we must allow for the fact that the concentrations of electron and hole charge densities, $-qn$ and qp, respectively, could lead to local space charge, which would then influence the electrostatic potential ψ. Thus, Poisson's Equation must be added to our set.

Expressing W and S in their approximate forms (see (5.18) and (5.19)), the master set of equations for steady-state and 3-D is:

$$-\frac{1}{q}\nabla \cdot J_e = G_{op,ii} - U_e$$

$$\frac{1}{q}\nabla \cdot J_h = G_{op,ii} - U_h$$

$$J_e = -qn\mu_e \nabla\psi + k_B T_e \mu_e \nabla n + k_B n \mu_e \nabla T_e$$

$$J_h = -qp\mu_h \nabla\psi - k_B T_h \mu_h \nabla p - k_B p \mu_h \nabla T_h$$

$$-\nabla \cdot (\kappa_e \nabla T_e) = -J_e \cdot \nabla\psi - \frac{3nk_B}{2}\frac{(T_e - T_L)}{\langle\langle\tau_E\rangle\rangle} - \frac{3k_B T_e}{2} U + E_g G_{ii,e}$$

$$-\nabla \cdot (\kappa_h \nabla T_h) = -J_h \cdot \nabla\psi - \frac{3pk_B}{2}\frac{(T_h - T_L)}{\langle\langle\tau_E\rangle\rangle} - \frac{3k_B T_h}{2} U + E_g G_{ii,h}$$

$$-\nabla^2\psi = \frac{q}{\epsilon}(p - n + N_D - N_A) .\qquad (5.24)$$

The generation and recombination terms G and U in the charge continuity equations depend on the carrier concentrations n and p, (see Section 3.2). The parameters μ and $\langle\langle\tau_E\rangle\rangle$ are material properties, so they can, in principle, be determined from experimental measurements. In practice, the energy relaxation time and the carrier thermal conductivities are unlikely to be independent of carrier temperature, and the mobility may be field-dependent because the momentum relaxation time is likely to depend on the applied force. These dependencies can be expressed empirically, or derived from more detailed theoretical treatments. Given this, it is emphasized that there are only two assumptions embedded in the set (5.24). The first is that carrier transport is semiclassical, as we implied when using Newton's Laws to formulate the BTE. The second is that, for the purpose of describing the kinetic-energy density, the contribution to W of the average velocity that the carriers might have due to the applied field is negligible. The latter assumption was made to allow us to simplify the expression for the flux of kinetic-energy density S, and, thereby, to truncate the moment equations into a closed-form set.

The set of equations (5.24) comprises 5 independent equations in 5 unknowns, n, p, T_e, T_h, and ψ. Generally, the set has to be solved numerically, and there are commercial solvers available to do this.[3]

Finally, it is noted that, even though the carrier temperatures T_e and T_h are functions of position, it is implicit in our master set of equations that the lattice temperature T_L is constant. Thus, our equations are an **isothermal** set. To allow for spatial variation in T_L, a lattice energy balance equation must be added to the set, and provision must be made for heat to leave the device, through a contact, for example. A simple expression for heat balance in the steady state is

$$\nabla \cdot (\kappa_L \nabla T_L) = -(\vec{J}_e + \vec{J}_h) \cdot \mathcal{E} \,, \tag{5.25}$$

where κ_L is the material thermal conductivity, and the right-hand side of the equation describes Joule heating. Addition of this equation to (5.24) provides a set of equations for **non-isothermal** conditions. In Section 16.3.1 we briefly examine the effect of such a non-isothermal condition on the DC and high-frequency characteristics of an HBT.

5.4   Mobility

Mobility is defined in (5.13) in terms of the average momentum relaxation time $\langle\langle\tau_M\rangle\rangle$, which is given by (5.11). Because an applied force changes momentum, and some scattering mechanisms are momentum dependent, we can expect the mobility to be field dependent. Another way of looking at this is to recognize that electrons are accelerated by an electric field, and move to new momentum states. In the parabolic-band approximation, this is the same as saying the field moves electrons to new velocities. The average of these field-related velocity changes is called the **drift velocity**, $\vec{v}_d$.

---

[3] For example, ATLAS from Silvaco Data Systems, and MEDICI and Sentaurus from Synopsys, Inc.

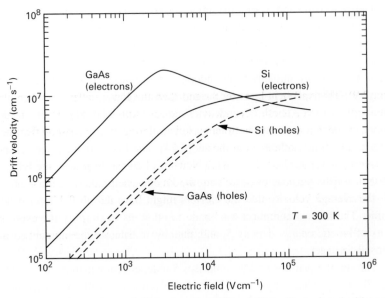

Figure 5.2 Drift velocity *vs.* field for low-doped Si and GaAs. Reproduced with permission from Sze [3, Fig. 2.23], © John Wiley & Sons, Inc. 1985.

From the current-density expressions in our master set of equations (5.24), it can be seen that the component of current that is directly related to the electric field is, using the electron current as an example, $qn\mu_e\mathcal{E}$. The electron drift velocity can be defined from this equation:

$$-qn\vec{v}_{de} \equiv -qn\mu_e\nabla\psi = qn\mu_e\vec{\mathcal{E}}, \qquad (5.26)$$

from which it should be noted that the drift velocity for electrons is in the opposite direction to the field, as dictated by the negative charge on the electron. This fact is hidden when magnitudes are used in the definition of drift velocity, as is often the case:

$$v_{de}(\mathcal{E}) = |\vec{v_{de}}(\mathcal{E})| \equiv \mu_e(\mathcal{E})\mathcal{E}. \qquad (5.27)$$

Note that the field-dependence of the mobility is included in this definition, so it is not restricted to low-field conditions, under which $\langle\langle\tau_M\rangle\rangle$ could be independent of field, and the relationship between v_d and $\mathcal{E}$ would be linear.

Experimentally, it is possible to measure v_d, and some results for Si and GaAs are shown in Fig. 5.2. This is a 'log-log' plot, but it can be inferred that the relationship is approximately linear at low fields, after which v_d tends to saturate, and even decrease for the case of electrons in GaAs.

In Si, an important scattering mechanism for electrons at room temperature is **inter-valley** scattering, in which the electrons scatter between the six equivalent conduction-band 'valleys', which result from taking a cross-section through the six equivalent prolate spheroids shown in Fig. 2.12. As $\mathcal{E}$ increases, the scattering rate for this process increases, i.e., $\langle\langle\tau_M\rangle\rangle$ decreases, causing a reduction in μ. Eventually, the scattering rate becomes so high that any further increase in field leads only to transfer of energy to the lattice; nothing is left to accelerate the electrons preferentially in the direction of the field, so v_{de} saturates at what is called the **saturation velocity**, v_{sat}.

In GaAs, when intervalley scattering occurs, it is not between equivalent valleys, as these don't exist in this material. Instead it is to a valley at slightly higher energy (see Fig. 2.9). This valley is less steep-sided than the lower valley, so when electrons are transferred to it their effective mass is increased (see (2.28)). Thus, at high fields in GaAs, μ_e decreases due to both increased scattering and increased effective mass: this leads to the decrease in v_{de} shown in Fig. 5.2.

The figure is compiled from steady-state measurements, and the transport under such conditions is sometimes called **stationary**. If measurements were made on a time-scale of the order of the momentum relaxation time, then, at high fields, we may anticipate higher drift velocities being recorded, due to there being insufficient time for collisions to establish a steady state. This phenomenon is called **velocity overshoot**, and can result in peak velocities that are several times higher than those shown in Fig. 5.2.

5.4.1   Empirical expressions for mobility

The mobility appears in each of the 3 terms for current in (5.24), so it is important to be able to characterize it in an easily usable form. For Si, this is not difficult because the saturating characteristic of Fig. 5.2 can be approximated simply by

$$\frac{1}{v_d(\mathcal{E})} = \frac{1}{\mu_0 \mathcal{E}} + \frac{1}{v_{\text{sat}}}, \tag{5.28}$$

where μ_0 is a field-independent mobility, which is taken to apply at low fields. Using (5.27) in this equation gives

$$\mu(\mathcal{E}) = \frac{\mu_0}{1 + \frac{\mu_0}{v_{\text{sat}}}\mathcal{E}} . \tag{5.29}$$

The measured low-field mobility μ_0 is shown for Si and GaAs in Fig. 5.3. The decrease at high doping density is due to increased ionized-impurity scattering. Useful empirical relationships for μ_0 exist, from which the mobility at 300 K can be readily computed. For Si [4],

$$\mu_{e0} = 88 + \frac{1252}{1 + 6.984 \times 10^{-18} N}$$

$$\mu_{h0} = 54.3 + \frac{407}{1 + 3.745 \times 10^{-18} N}, \tag{5.30}$$

and for GaAs [5],

$$\mu_{e0} = 8300 \left[1 + \frac{N}{3.98 \times 10^{15} + N/641} \right]^{-1/3}$$

$$\mu_{h0} = \frac{380}{\left[1 + 3.17 \times 10^{-17} N \right]^{0.266}}, \tag{5.31}$$

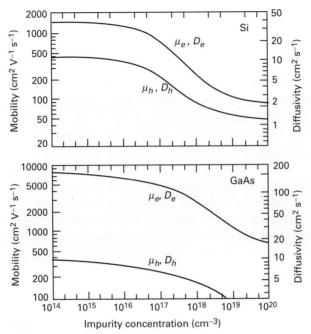

Figure 5.3 Dependence of low-field mobility μ_0 and diffusivity on total impurity concentration. Reproduced with permission from Sze [3, Fig. 2.3], © John Wiley & Sons, Inc. 1985.

where μ is in cm^2 V^{-1} s^{-1}, and the total impurity concentration $N = N_A + N_D$ is in cm^{-3}.

5.4.2   Conductivity effective mass

The definition of mobility in (5.16) involves the effective mass. For a single, spherical constant energy surface, as employed in the development of the drift-diffusion and hydrodynamic equations, the relevant effective mass is just the band effective mass m^*. For other situations, such as the multiple spherical surfaces in the valence band, and the multiple prolate spheroids in the conduction band of Si, some way of accounting for the multiple contributions to the mobility must be found. This is achieved by writing the expression for the **conductivity**, σ, in a suitable form. The units of σ are Sm^{-1}, where S is Siemens, and its value is given by the ratio of the drift current density to the electric field:

$$\sigma \equiv 1/\rho = q(n\mu_e + p\mu_h),   \qquad (5.32)$$

where ρ is the **resistivity**.

Conductivity effective mass for holes

In Si and GaAs, there are two hole bands, the heavy- and light-hole bands, that are degenerate at the tops of the bands (see Fig. 2.9). If the average momentum relaxation times are the same for holes in each of these bands, we can write the hole conductivity as

$$\sigma_h \equiv \frac{J_{h,\text{drift}}}{\mathcal{E}} = q^2 \langle\langle \tau_M \rangle\rangle \left[\frac{p_{hh}}{m_{hh}^*} + \frac{p_{lh}}{m_{lh}^*} \right] \equiv q^2 \langle\langle \tau_M \rangle\rangle \left[\frac{p_{hh} + p_{lh}}{m_{h,\text{CON}}^*} \right], \qquad (5.33)$$

where the band effective masses for the heavy holes and the light holes are identified, and a **conductivity effective mass**, $m_{h,\text{CON}}^*$ is defined. We know from the densities-of-states expressions (3.32) that the carrier concentrations are proportional to $m_{\text{DOS}}^{*\,3/2}$, and that, in the spherical case, m_{DOS}^* is simply the band effective mass. We can use these facts in the above equation to develop an expression for the hole conductivity effective mass:

$$\frac{1}{m_{h,\text{CON}}^*} = \left[\frac{p_{hh}}{m_{hh}^*} + \frac{p_{lh}}{m_{lh}^*} \right] \left[\frac{1}{p_{hh} + p_{lh}} \right]$$

$$= \frac{m_{hh}^{*\,1/2} + m_{lh}^{*\,1/2}}{m_{hh}^{*\,3/2} + m_{lh}^{*\,3/2}}. \qquad (5.34)$$

Values for $m_{h,\text{CON}}^*$ for Si and GaAs are listed in Table 4.2.

Conductivity effective mass for electrons in Si

Look at Fig. 2.12 for Si, and consider any direction from the set $\langle 100 \rangle$. In four of the six equivalent conduction bands the electrons would respond to an electric field in the chosen direction with the transverse effective mass m_t^*. In the other two bands the relevant effective mass would be the longitudinal effective mass m_l^*. If the n electrons in the material are distributed equally among the six bands, then we can write

$$\sigma_e \equiv \frac{J_{de}}{\mathcal{E}} = q^2 \langle\langle \tau_M \rangle\rangle \frac{n}{6} \left[\frac{2}{m_l^*} + \frac{4}{m_t^*} \right] \equiv q^2 \langle\langle \tau_M \rangle\rangle n \left[\frac{1}{m_{e,\text{CON}}^*} \right], \qquad (5.35)$$

from which it follows that

$$\frac{1}{m_{e,\text{CON}}^*} = \frac{1}{3} \left[\frac{1}{m_l^*} + \frac{2}{m_t^*} \right]. \qquad (5.36)$$

The value for $m_{e,\text{CON}}^*$ for Si is listed in Table 4.2.

5.5 Current

Current density J is defined by (5.2). Physically, it is the net flow of charge through a surface whose normal is parallel to the flow. Fundamentally, the flow is caused by some gradient in energy. One of the components of current in our master set of equations (5.24) arises from a difference in potential energy; this is the **drift current**. The other two are due to gradients in either of the two contributors to the kinetic energy: the carrier concentration and the carrier temperature. These components are the diffusion current and the thermal current, respectively. Some aspects of these currents are discussed in this section.

5.5.1   Drift current

From (5.24) and the discussion in the previous section on mobility, the drift-current densities for electrons and holes are

$$\vec{J}_{e,\text{drift}} = -qn\mu_e(\mathcal{E})\nabla\psi = qn\mu_e(\mathcal{E})\vec{\mathcal{E}} = -qn\vec{v}_{de}(\mathcal{E})$$
$$\vec{J}_{h,\text{drift}} = -qp\mu_h(\mathcal{E})\nabla\psi = qp\mu_h(\mathcal{E})\vec{\mathcal{E}} = qp\vec{v}_{dh}(\mathcal{E}). \qquad (5.37)$$

Note: for a given field, the electrons and holes drift in opposite directions, but the current densities have the same sign.

The actual distribution of the carriers in k-space is not explicit in these equations: it is implicitly taken into account in either $\mu(\mathcal{E})$ or $v_d(\mathcal{E})$. This is, perhaps, a reminder that we know nothing about the actual non-equilibrium distribution function for the electrons and holes. We do know, however, because we have expressed μ as a function of $\mathcal{E}$, and because we have allowed v_d to reach its saturation value (see Fig. 5.2), that the drift-current expressions can be used up to high values of field, i.e., in situations where we would expect the distribution function to be far from its equilibrium form. To appreciate this, note from Fig. 5.2 that v_{sat} for electrons in Si, and the peak v_d for electrons in GaAs, are around 10^7 cm s^{-1}, which is a very similar value to that of the mean, unidirectional thermal velocity for an equilibrium hemi-Maxwellian distribution (see Section 4.5). Thus, at such fields, v_d cannot be viewed as a minor perturbation to $2v_R$.

At low fields, perhaps around 10^2 V cm^{-1}, then it might be reasonable to view the distribution of carriers as being only slightly perturbed from its equilibrium value. Each electron in a velocity state (parabolic bands) of $\hbar k/m^*$ could be imagined to be shifted to a state $(\hbar k/m^* + v_d)$ due to acceleration by the field between collisions. Denoting the forward- and backward-going parts of the distribution by subscript arrows, we would have

$$\vec{J}_{e,\text{drift}} = \vec{J}_{e,\text{drift}\rightarrow} + \vec{J}_{e,\text{drift}\leftarrow}$$
$$= -q\frac{n}{2}(2\vec{v}_R + \vec{v}_{de}) + \left(-q\frac{n}{2}(-2\vec{v}_R + \vec{v}_{de})\right)$$
$$= -qn\vec{v}_{de}, \qquad (5.38)$$

where n, rather than n_0, is used to emphasize the non-equilibrium situation. However, bear in mind that near-equilibrium is implied by the employment of v_R in (5.38). This inconsistency is dealt with by characterizing the electron distribution via a **quasi-Fermi energy**, rather than by *the* Fermi energy E_F (see Chapter 6). The distribution in this low-field case is called a **displaced Maxwellian**, and is illustrated in Fig. 5.4. The velocity distribution function follows from (4.25) for $f_{MB}(E)$ and from the transformation of energy above E_C to velocity given by (4.27). Including the drift velocity v_{de}, the function is

$$f(v) = \frac{n_0}{N_C}\exp[-m^*_{\text{CON}}(v - v_{de})^2/2k_B T]. \qquad (5.39)$$

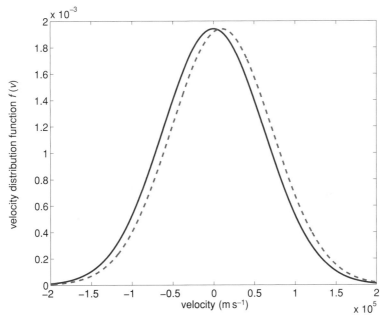

Figure 5.4 Velocity distribution functions for $n = 6.25 \times 10^{16}$ cm^{-3} in the equilibrium case (solid line) and for a current density of magnitude 10^4 A cm^{-2} (dashed line).

In the example shown, the displacement in velocity is $v_{de} = 10^4$ m s$^{-1} \approx 2v_R/10$, and this is sufficient to give a drift current density of 10^4 A cm^{-2}.

To emphasize that such a large current density can be borne by a near-equilibrium distribution, we consider the two hemi-Maxwellian velocity-spectral carrier concentrations for the conditions listed in Fig. 5.4. The concentrations in the Maxwell-Boltzmann case considered here follow from (4.28):

$$n_0(v) = \frac{8\pi}{h^3} \frac{(m^*_{e,\text{DOS}} m^*_{\text{CON}})^{3/2} v^2}{\exp[m^*_{th} v^2/2k_B T] \exp[(E_C - E_F)/k_B T]} . \qquad (5.40)$$

The results are plotted as a function of velocity in Fig. 5.5. Note how the symmetrical distribution in equilibrium is distorted by the transfer of carriers to the right-going distribution in the presence of a drift field.

5.5.2 Diffusion current

From (5.24), the equations for the **diffusion currents** are

$$\vec{J}_{e,\text{diff}} = k_B T_e \mu_e \nabla n$$

$$\vec{J}_{h,\text{diff}} = -k_B T_h \mu_h \nabla p . \qquad (5.41)$$

$k_B T_c$ is a measure of the mean kinetic energy of each carrier. Thus, even if the carrier temperature T_c is constant, a gradient in carrier concentration causes a flow of kinetic

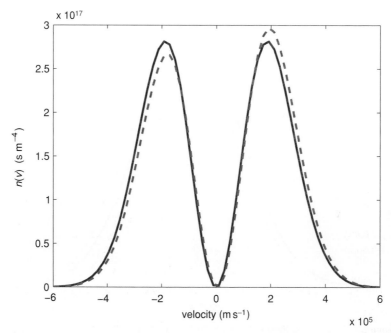

Figure 5.5 The hemi-Maxwellian portions of the carrier distribution in velocity for the two cases presented in Fig. 5.4: equilibrium case (solid line), non-equilibrium case (dashed line).

energy, and, consequently, a flow of charge. For a given concentration gradient, electrons and holes diffuse in the same direction, but the current densities have opposite signs.

As in the case of the expressions for drift current, the distribution function is not explicit, and nor are the diffusion currents limited to near-equilibrium conditions. However, it is instructive to appreciate that diffusion currents can issue from equilibrium distributions. To see this, consider Fig. 5.6, in which two Maxwellian distributions with different concentrations of electrons are separated by a distance $2\bar{l}$. The current density at some intermediate plane, e.g., at $x = 0$, is

$$\vec{J}_e = -q\frac{n_1}{2}\,2\vec{v}_R + (-q)\frac{n_2}{2}\,(-2\vec{v}_R) = -q\left(\frac{n_1}{2} - \frac{n_2}{2}\right)2\vec{v}_R\,. \qquad (5.42)$$

The two distributions have to be close enough together so that the electrons from each distribution can cross the plane at $x = 0$ without suffering collisions, as such events might change the distributions. The mean distance between collisions is called the **mean free path**, and it is denoted by $\bar{l}$ on Fig. 5.6. The concentration difference $\frac{1}{2}(n_1 - n_2)$ should be evaluated over a distance of this order. This can be done via a Taylor series expansion of each distribution about $x = 0$, and by keeping terms to first order. For example,

$$\frac{n_2}{2} = \frac{n(0 + \bar{l})}{2} \approx \frac{n(0)}{2} + \frac{1}{2}\frac{dn}{dx}\bar{l}\,.$$

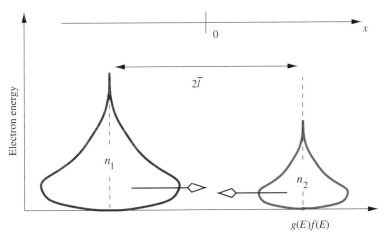

Figure 5.6 Schematic of diffusion due to the close juxtaposition of two hemi-Maxwellian distributions of different electron concentration.

Taking the corresponding expression for $n_1/2$, and substituting into (5.42) gives

$$\vec{J}_{e,\text{diff}} = q\bar{l}2v_R\frac{dn}{dx} \equiv qD_e\frac{dn}{dx}. \tag{5.43}$$

Expressing the diffusion current in terms of the diffusivity D is more usual than writing it in terms of either $\bar{l}$ or $k_B T \mu$.

You may be curious, or even be alarmed, that we've taken two Maxwellian distributions and produced a net current: this may appear to be inconsistent because Maxwellians are equilibrium distributions, and a net current is not allowed at equilibrium! What this actually means is that, if there were a diffusion current at equilibrium, then it would be exactly negated by a current due to some other mechanism, such as drift. This is precisely what happens in a np-junction at equilibrium (see Section 6.1).

5.5.3   Thermal current

From (5.24), the equations for the thermal currents are

$$\vec{J}_{e,\text{therm}} = k_B n \mu_e \nabla T_e$$

$$\vec{J}_{h,\text{therm}} = -k_B p \mu_h \nabla T_h. \tag{5.44}$$

These currents exist when neighbouring carrier ensembles have different average kinetic energies. One example where this might occur is the injection of high-energy electrons into a p-type region in which the existing carriers are at near equilibrium. As the hot electrons cool towards the near-equilibrium distribution by scattering and recombination, there is a gradient in T_e that drives the injected carriers forwards. This **thermal current** is, in this particular example, in the same direction as the diffusion current, and is often masked by it. One instance where the thermal current is evident is in HBTs operating at high currents (see Section 16.3.1).

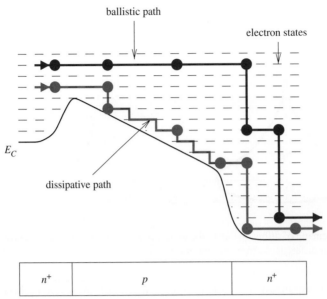

ballistic path

electron states

E_C

dissipative path

n^+	p	n^+

Figure 5.7 Illustration of electron transport through a region of electric field in a transistor. Bottom path: dissipative transport, creating near-equilibrium conditions. Top path: ballistic transport, creating non-equilibrium conditions in the uniform-field region. The energy of the ballistic electrons is eventually given to the lattice in the highly doped n-region, where ionized-impurity scattering and electron-electron scattering are large.

5.6   Ballistic transport

Drift, diffusion, and thermal currents all result from transport processes in which scattering tends to dissipate any momentum that is imparted by an applied force. Collisionless, or ballistic, transport is possible in regions of a device that are shorter than the mean free path $\bar{l}$. The difference between dissipative and ballistic transport is illustrated in Fig. 5.7.

We can estimate $\bar{l}$ in the near-equilibrium case from (5.43) and (5.17):

$$D_e = \bar{l}2v_R = \frac{k_B T_e}{q}\mu$$

$$\bar{l}_0 = \frac{k_B T_L}{q}\frac{\mu_0}{2v_R}, \tag{5.45}$$

where the subscript '0' indicates equilibrium or, in this case, near-equilibrium with a low applied field. Similarly, the electron gas is taken to be in equilibrium with the lattice. We'll now use this equation to estimate $\bar{l}_0$.

For very low doping densities, Fig. 5.3 informs that μ_0 is about 1300 cm^2 V^{-1} s^{-1} for Si, and about six times higher for GaAs. From Table 4.2, values of $2v_R$ at low doping and $T_L = 300$ K can be inferred to be about 10^7 cm s^{-1} for Si, and about twice this value for GaAs. Thus, under these conditions, $\bar{l}_0$ is about 30 nm for Si and about 100 nm for GaAs. These values will decrease as the doping density is raised, because of the associated decrease in μ_0 and increase in $2v_R$. Nevertheless, the estimated values of $\bar{l}_0$

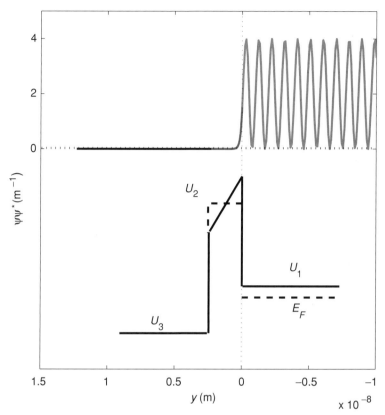

Figure 5.8 Bottom part: Potential energy profile to illustrate the concept of tunnelling from Region 1 to Region 3 of an electron with energy less than the potential energy U_2 of the barrier separating the two regions. The top part of the figure indicates the probability density for an electron entering the system from the right with an energy E such that $U_1 < E < U_2$. Specifically: $U_2 - U_1 = 2.7$ eV, $E - U_1 = 0.45$ eV, $U_3 - U_1 = -1$ eV, the barrier has thickness $d = 2.3$ nm, and is of a material with a relative permittivity of 3.9, and an effective mass for electrons of $0.3m_0$. The electron effective mass in Regions 1 and 3 is $0.9m_0$. All these parameter values are intended to represent tunnelling in a CMOS90 N-FET with $\psi_{ox} = 1$ V.

do give an indication of the device feature sizes at which we might need to consider the possibility of ballistic transport occurring. We do this specifically in Chapter 18. A special case of ballistic transport is tunnelling, which we now describe.

5.7   Tunnelling

5.7.1   Probability density current

Consider the potential profile shown in Fig. 5.8: two regions of constant potential energy, U_1 and U_3, are separated by a thin barrier. The top of this barrier is triangular, and could represent the potential profile in the oxide of a MOSFET, as illustrated in many of the

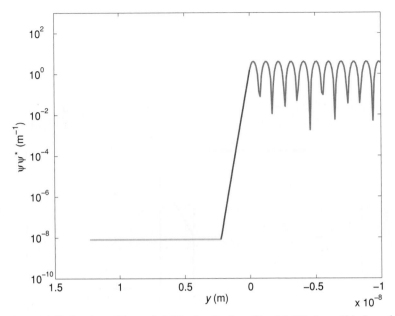

Figure 5.9 Redrawing of the probability density from Fig. 5.8. The logarithimic scale emphasizes the small ($\approx 10^{-8}$ m^{-1}) probability density of electrons tunnelling from Region 1.

band diagrams in Chapter 10. Here we simplify the barrier to that of a rectangular barrier with peak height U_2. Classically, an electron in Region 1 with energy $E_1 < U_2$ would be reflected from the interface with Region 2. Quantum mechanically, provided the barrier is not infinitely high, the electron has a finite probability of penetrating the barrier and appearing in Region 3. This phenomenon of penetrating an insurmountable barrier is known as **tunnelling**.

Recall from Chapter 2 that $\Psi(y, t)\Psi^*(y, t)$ is a probability density, where Ψ is the time-dependent electron wavefunction. We are interested in the flow of the probability density from one region to another

$$\frac{dP}{dt} = \Psi^* \frac{\partial \Psi}{\partial t} + \Psi \frac{\partial \Psi^*}{\partial t}$$

$$= \frac{1}{i\hbar} \left[\Psi^* \left(i\hbar \frac{\partial \Psi}{\partial t} \right) + \Psi \left(i\hbar \frac{\partial \Psi^*}{\partial t} \right) \right]. \tag{5.46}$$

The time-dependent Schrödinger Wave Equation is

$$i\hbar \frac{\partial \Psi}{\partial t} = -\frac{\hbar^2}{2m^*} \nabla^2 \Psi + U\Psi, \tag{5.47}$$

where U is potential energy, and the appearance of m^* means that we are using the effective-mass form of the Schrödinger equation, as discussed in Section 2.11. Furthermore, we have chosen to ignore the difference between wavefunctions and envelope functions (see (2.37)). Taking complex conjugates where necessary, and substituting into (5.46), some manipulation yields

$$\frac{dP}{dt} = -\frac{i\hbar}{2m^*}\nabla \cdot \left[\Psi\nabla\Psi^* - \Psi^*\nabla\Psi\right]$$

$$\equiv -\nabla \cdot \vec{J}_P \,, \tag{5.48}$$

where $\vec{J}_P$ is defined as the probability density current. The defining line of (5.48) follows from the basic idea of continuity, as discussed in Section 5.2.2: any spatial change in the flow of some property must result in a temporal change of that property.

5.7.2 Transmission probability

To begin to estimate $\vec{J}_P$ we need to evaluate Ψ. We'll assume a one-dimensional system and we'll limit U to having only a spatial dependence, so that we need only solve for $\psi(y)$, the spatial part of $\Psi(y, t)$.[4] For each of the three regions in Fig. 5.8 the Schrödinger Wave Equation can be written as

$$-\frac{\hbar^2}{2}\frac{d}{dy}\left(\frac{1}{m^*}\frac{d\psi}{dy}\right) + (U - E)\psi = 0 \,. \tag{5.49}$$

The boundary conditions for solving this equation follow from the discussion in Section 2.11.1, which can be summarized as:

- ψ is continuous across a boundary, and
- $\frac{1}{m^*}\frac{d\psi}{dy}$ is continuous across a boundary.

Within the three regions of Fig. 5.8, and for an electron entering from the right with an energy E such that $U_1 < E < U_2$, the solutions to (5.49) are

$$\psi_1 = Ae^{ik_1 y} + Be^{-ik_1 y}$$

$$\psi_2 = Ce^{k_2' y} + De^{-k_2' y}$$

$$\psi_3 = Fe^{ik_3 y} \,, \tag{5.50}$$

where $A - F$ are coefficients to be determined, and the wavenumbers are

$$k_1 = \frac{1}{\hbar}\sqrt{2m_1^*(E - U_1)}$$

$$k_2 = \frac{1}{\hbar}\sqrt{2m_2^*(E - U_2)} \equiv ik_2'$$

$$k_2' = \frac{1}{\hbar}\sqrt{2m_2^*(U_2 - E)}$$

$$k_3 = \frac{1}{\hbar}\sqrt{2m_3^*(E - U_3)} \,. \tag{5.51}$$

Note that k_1, k_2' and k_3 are all real. Therefore, $\psi_1\psi_1^*$ is oscillatory, $\psi_2\psi_2^*$ is exponentially damped if $C \ll D$, and $\psi_3\psi_3^*$ is a constant. These forms are illustrated in the top part of

---

[4] See Section 2.3 for how to separate $\Psi(y, t)$ into spatial and temporal components.

Fig. 5.8. Physically, a wave enters from the right, its magnitude is reduced exponentially as it traverses the insulator as an evanescent wave and, if the insulator is not too thick, some small fraction of the probability density wave exits on the left.[5] Applying the two boundary conditions to the two interfaces gives us four equations in the five unknowns $A - F$, so we can only evaluate relative amplitudes, such as FF^*/AA^*.

What we really want is the ratio of the output probability density current $J_{P,F}$ to the incident probability density current $J_{P,A}$. Using the definition of J_P from (5.48), we find that

$$J_{P,A} = \frac{\hbar k_1}{m_1^*} |A|^2$$

$$J_{P,F} = \frac{\hbar k_3}{m_3^*} |F|^2 .$$ (5.52)

So, the desired ratio, which is called the **transmission probability**, is

$$T = \frac{k_3}{k_1} \frac{m_1^*}{m_3^*} \frac{|F|^2}{|A|^2}$$

$$\equiv \frac{v_{k,3}}{v_{k,1}} \frac{|F|^2}{|A|^2} ,$$ (5.53)

where the second equation, written in terms of the state velocities, follows from our use of the parabolic-band approximation. It also follows from (5.52) that $|A|^2$ and $|F|^2$ must be charge densities. Note that T is a function of E.

The so-called asymmetric rectangular potential barrier of Fig. 5.8 allows an analytical solution for T. After considerable algebra, the solution is

$$T = \frac{4m_1 k_3/m_3 k_1}{(1 + k_3 m_1/k_1 m_3)^2 \cosh^2(k_2' d) + (k_2' m_1/k_1 m_2 - k_3 m_2/k_2' m_3)^2 \sinh^2(k_2' d)} .$$ (5.54)

The essence of this equation can be appreciated by making the assumption that $k_2' d \gg 1$. This assumption is satisfied when the wavelength of the electron in Region 2 ($\lambda_2 = 2\pi/k_2'$) is much less than the barrier thickness, and implies that the tunnelling probability is low. Under this approximation, both the hyperbolic functions in (5.54) reduce to $e^{2k_2' d}/4$, and T becomes

$$T = \frac{16}{4 + \left(\frac{k_2' m_1}{k_1 m_2} - \frac{k_3 m_2}{k_2' m_3}\right)^2} \exp\left[-\frac{2d}{\hbar}\sqrt{2m_2^*(U_2 - E)}\right] .$$ (5.55)

This equation brings out the importance in tunnelling of both the barrier thickness d and the barrier height U_2. The pre-exponential factor is often of order 1, in which case an

---

[5] The component $Ce^{ik_2'y}$ is that part of ψ_2 due to reflection of the evanescent wave at the boundary between materials #2 and #3.

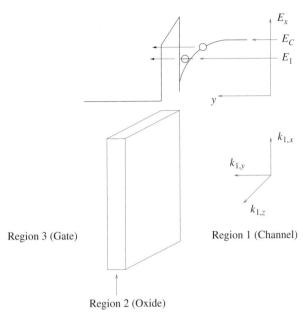

Region 3 (Gate)   Region 1 (Channel)

Region 2 (Oxide)

Figure 5.10 The top part illustrates tunnelling of electrons through an insulator from either the conduction band (top electron,) or from quasi-bound states (bottom electron), in a semiconductor. The bottom part of the figure shows the k-space axes for computation of the current due to tunnelling from Region 1 to Region 3. For application of this to tunnelling in a MOSFET, Region 1 would be the channel and Region 3 would be the gate.

even simpler expression results:

$$T \approx \exp\left[-\frac{2d}{\hbar}\sqrt{2m_2^*(U_2 - E)} \right].\tag{5.56}$$

Non-rectangular barrier

When the barrier is not rectangular, it can be divided up into a number of thin rectangular slices, each of width dy, and the approximate expression for the transmission probability (5.56) can be expressed in integral form

$$T \approx \exp\left[-\frac{2}{\hbar}\int_{y_1}^{y_2}\sqrt{2m_2^*(U(y) - E)}\,dy \right].\tag{5.57}$$

The same result can be obtained from the so-called **JWKB Approximation**, in which Schrödinger's equation is solved under the assumption that $U(y)$ varies slowly in comparison to the electron wavelength $\lambda = 2\pi/k$ [7].

5.7.3 Tunnel current

In this book the prime example of tunnelling is the so-called 'leakage current' through the insulator in a MOSFET. We derive an expression for this current in this subsection, but you may want to read about MOSFETs in Chapter 10 and Chapter 13 before trying to follow the details. Consider Fig. 5.10, which sets up the k-space axes in order to compute the current due to electrons tunnelling from Region 1 to Region 3. Here, we associate Region 1 with the channel of an N-MOSFET, Region 2 with the gate oxide, and Region 3 with the gate metal. We consider two cases: one where the electrons in Region 1 are not confined; and, secondly, where the electrons are confined in a potential well.

Tunnelling from a continuum of states

Here, we assume that electrons in Region 1 are not confined, and we treat E_C as being flat, as we did for U_1 in Fig. 5.8. The fundamental expression for the y-directed current density in Region 1 comes from (5.2). Multiplying this by the transmission probability gives the expression for the y-directed electron current density in Region 3:

$$J_{e,y} = -q \frac{1}{4\pi^3} \int_{\vec{k}} T(k_y) v_y f(\vec{k}) \, d\vec{k} . \tag{5.58}$$

Let us take $m^*_{1,x} = m^*_{1,z} \equiv m^*_\perp$: this enables the energy to be written as

$$E = U_1 + \frac{\hbar^2 k_y^2}{2m^*_{1,y}} + \frac{\hbar^2 k_\perp^2}{2m^*_\perp}$$

$$\equiv E_y + \frac{\hbar^2 k_\perp^2}{2m^*_\perp} , \tag{5.59}$$

where $k_\perp^2 = k_x^2 + k_z^2$.

Going to polar coordinates, we can write

$$\int_{-\infty}^{\infty} dk_x \int_{-\infty}^{\infty} dk_z \equiv \int_0^{2\pi} d\theta \int_0^{\infty} k_\perp dk_\perp . \tag{5.60}$$

Changing variables from $k_\perp$ to E using (5.59), substituting (5.60) in (5.58), and rearranging, we get

$$J_{e,y} = -q \frac{1}{2\pi} \int_0^{\infty} T(k_y) v_y \, dk_y \int_{E_y}^{\infty} \frac{2}{(2\pi)^2} 2\pi \frac{m^*_\perp}{\hbar^2} f(E) \, dE . \tag{5.61}$$

The reason for the rearrangement is to highlight the second integral. By examining its dimensions it can be appreciated that the integral represents an areal density. It is, in fact, the density of electrons n_{2D} in the 2-D sheet of electrons at the oxide/semiconductor interface having an energy $E = E_y$, i.e., each of the electrons spread across the surface that has energy E_y has a probability of tunnelling $T(k_y)$ with a velocity v_y. The integral is easily solved by variable substitution.[6] The result is

$$n_{2D}(E_F - E_y) = \frac{m^*_\perp k_B T}{\pi \hbar^2} \ln \left[1 + e^{(E_F - E_y)/k_B T} \right] . \tag{5.62}$$

Finally, making the variable change from k_y to E_y in (5.61) using (5.59), the electron tunnel current is

$$J = -\frac{q}{h} \int_{U_1}^{\infty} T(E_y) n_{2D}(E_F - E_y) \, dE_y . \tag{5.63}$$

---

[6] $b = (E_F - E)/k_B T$ is an appropriate substitution to make.

We will use this expression in Section 13.1.6 for estimating the gate leakage current in modern MOSFETs.

Tunnelling from quasi-bound states

In this case, which is also illustrated in Fig. 5.10, the volume in Region 1 is now $L_x L_z a$, where a is the width of the quantum well from which the electrons are tunnelling. Here we assume a rectangular well. The system can no longer be considered to be periodic in the y-direction, so we employ our usual conversion from a sum-of-states to an integral over all states only for the x- and z-directions. Thus, the equivalent to (5.61) is

$$J_{e,y} = -q\frac{1}{a}\sum_{k_y>0}T(k_y)v_y(k_y)\int_{E_y}^{\infty}\frac{2}{(2\pi)^2}2\pi\frac{m_{\perp}^*}{\hbar^2}f(E)\,dE\,, \qquad (5.64)$$

where E_y in this case is given by

$$E_y = U_1 + U_p\,, \qquad (5.65)$$

where U_p is the energy level of each of the p bound states. For an infinite well, for example, $U_p = p(\hbar^2\pi/2m_y^*a)$. Thus the expression for the current density corresponding to (5.63) is

$$J_{e,y} = -q\frac{2m_{\perp}^*}{ah\hbar}\sum_{p=1}^{\infty}T(k_{y,p})v_y(k_{y,p})\int_{U_1+U_p}^{\infty}f(E)\,dE\,. \qquad (5.66)$$

This expression, but perhaps without the simplifying assumption of an infinite well, could be used for the gate leakage current in a heterojunction FET (Section 11.3), or for an improvement upon (5.63) in the case of MOSFETs.

Exercises

5.1 Consider a 1-D crystal of length 1 cm. The effective mass for holes in this material is $0.5m_0$. What is the current density due to carriers in the valence band at the instant when the only unoccupied state has a wavenumber of $4.32 \times 10^8\ \mathrm{m}^{-1}$?

5.2 In our derivation of the Boltzmann Transport Equation, electrons were assumed to move 'classically' between scattering events. There is a limit to how small a device can be before this implicit assumption of precise knowledge of the position and momentum of an electron is rendered invalid by its violation of Heisenberg's Uncertainty Principle.

   (a) Consider that $k_B T/5$ is a reasonable value for the uncertainty in energy of an electron of energy $k_B T$, and derive an expression for the minimum allowable uncertainty in position.

   (b) Show that the transport of electrons in GaAs can be considered to be classical if the sample has dimensions that are greater than ≈ 25 nm.

5.3 In the balance equations deriving from the Boltzmann Transport Equation the mean x-directed kinetic energy of an electron $\langle u_x \rangle$ occurs frequently.

(a) Derive an expression for $\langle u_x \rangle$ in terms of $k_B T$ and Fermi–Dirac integrals.

(b) Evaluate $\langle u_x \rangle$ for carrier concentrations of 1×10^{16} and 1×10^{20} cm^{-3} in gallium arsenide.

5.4 Equation (5.19) is an expression for the x-directed flux of kinetic energy $S_{e,x}$ in terms of the electron-temperature gradient and an electron thermal conductivity κ_e.

Write $S_{e,x}$ as a product of kinetic energy density and electron velocity, and then perform a Taylor series expansion to get $S_{e,x}$ in terms of the mean-free-path length $\bar{l}$ and dT_e/dx. The procedure is the same as used for the diffusion current in Section 5.5.2. Use the relation between $\bar{l}$ and diffusivity to show that

$$\kappa_e = qn\mu_e \left(\frac{k_B}{q}\right)^2 T_e. \qquad (5.67)$$

5.5 Write down the equilibrium version of the 1-D BTE. This equation must be satisfied when f is the Fermi–Dirac distribution function.

Determine the implications of this as regards the spatial dependencies of E_F and lattice temperature T_L in the case of a semiconductor device in which there is a built-in electric field.

Recall that the BTE is an equation in phase space, so when dealing with just real space, k-dependent properties, such as kinetic energy, can be treated as constant.

5.6 Consider the electron current-density equation in (5.24), and take it to apply to the n-type region of a device in which there is negligible electric field. Electrons are injected into one end ($x = 0$) of the device and are extracted at the other end ($x = L$). The extraction process accelerates the electrons near $x = L$. The electron current density $J_{e,x}$ is constant throughout the device, and, for most of the length of the device, it is due entirely to the concentration-gradient term in (5.24).

Explain how the 'thermal-gradient' current alters the diffusion contribution to $J_{e,x}$ as the end of the device at $x = L$ is approached.

5.7 The E-k relationships for the conduction bands of two semiconductor materials, A and B, each with spherical constant-energy surfaces, can be expressed as

$$E_A - 0.7 = \alpha k^2 \quad \text{and} \quad E_B - 1.4 = 2\alpha(k - k')^2,$$

respectively, where α is a constant, $k' > 0$, and the energies are in units of eV. Both materials have the same valence-band structure, with the top of the valence band at $E = 0$ and $k = 0$.

Which material would have the higher intrinsic conductivity?

5.8 Look at the band structure of GaAs in Fig. 2.9 and consider n-type material. The electrons are occupying states in the conduction-band 'valley' centred at the Γ point.

An electric field $\mathcal{E}$ is now applied and steadily increased. Eventually, the electrons gain enough energy from the field to transfer into the conduction-band valley that is closer to the X point.

Sketch the $J - \mathcal{E}$ characteristic of the material, and give your reasons for its shape.

5.9 Consider Fig. 4.4 for the full Maxwellian distribution of 10^{19} cm^{-3} electrons in silicon. Now, let the equilibrium be very slightly disturbed so that the concentration of electrons in the left-going hemi-Maxwellian is reduced by a small amount. The concentration of electrons in the right-going hemi-Maxwellian distribution remains unchanged.

Calculate the change in concentration of electrons in the left-going distribution if a current density of magnitude 10^4 A cm^{-2} is to be supported.

Is this change small enough for our near-equilibrium assumption of hemi-Maxwellian distributions to be reasonable?

In support of your answer, it would be instructive to construct Fig. 4.4, and then to add to it the new left-going distribution. If there's no visible difference, then 'near-equilibrium' would be a reasonable description.

5.10 Consider a silicon sample doped with 10^{17} donors per cm^3. For a given electric field, what is the ratio of the drift-current densities of electrons and holes?

5.11 The donor doping density in a piece of silicon varies as $N_D(x) = N_0 \exp(-ax)$.
  (a) Find an expression for the electric field at equilibrium over the range for which $N_D \gg n_i$.
  (b) Sketch the energy-band diagram for this case, and indicate the direction of the electric field. Explain qualitatively why the electric field is in the direction you have shown.

5.12 Consider a sample of uniformly doped $(10^{17}$ cm$^{-3})$, n-type Si to which an electric field of 1000 V cm^{-1} is applied.
  (a) Use the BTE to show that the distribution function for this case can be represented by a 'displaced Maxwellian'. Assume that the collision term can be written as $(f_0 - f)/\tau_m$, where τ_m is a constant. This is known as the **Relaxation Time Approximation**.
  (b) Estimate the extent of the displacement.

5.13 The transmission probability for an asymmetrical rectangular barrier is given by (5.54).
  (a) Simplify this expression by: taking the barrier to be symmetrical and of height U_2; assuming the effective masses are the same in all three regions $(= m^*)$; assuming that $k_2' d \gg 1$. Show that the resulting expression is

$$ T = \frac{16}{4 + \left(\frac{k_2'}{k_1} - \frac{k_1}{k_2'} \right)^2} \exp \left[-\frac{2d}{\hbar} \sqrt{2m_2^*(U_2 - E)} \right]. \qquad (5.68) $$

  (b) Often, the pre-exponential factor in (5.68) is set to unity, as was done in arriving at (5.56). To examine the validity of this change, consider the particular example of $m^* = 0.3m_0$, $d_2 = 2.3$ nm, and $U_2 = 2.7$ eV, i.e., data from Fig. 5.8, and plot the energy dependence of the two terms in (5.68) (the pre-exponential factor and the exponent), and their product T.

5.14   In Section 2.11.1 one of the boundary conditions for solving the Effective-mass
Schrödinger Wave Equation made use of the 'fact' that $\frac{1}{m^*}\frac{d\psi}{dx}$ is continuous across
a boundary.

Actually, I don't know of any rigorous proof of this, but it can be justified
by examining the problem of a simple, step-like potential barrier. Write the
wavefunction on the left of the barrier as $\psi = e^{ik_1x} + re^{-ik_1x}$ and the wavefunction
on the right of the barrier as $\psi = te^{ik_2x}$.

Show that the probability density current is conserved at the boundary if the
above boundary condition is used, but not if the effective mass is not included in
the boundary condition.

This conservation of the probability density current is the quantum-mechanical
equivalent of Kirchoff's Current Law.

References

[1] S. Datta, *Quantum Phenomena*, Sec. 5.3.2, Addison-Wesley, 1989.

[2] ATLAS Users Manual, Sec. 3.4.3, Silvaco Data Systems, Santa Clara, June 11, 2008.

[3] S.M. Sze, *Semiconductor Devices, Physics and Technology*, 1st Edn., John Wiley & Sons,
Inc., 1985.

[4] N.D. Arora, J.R. Hauser and D.J. Roulston, Electron and Hole Mobilities in Si as a Function
of Concentration and Temperature, *IEEE Trans. Electron. Dev.*, vol. 29, 292–295, 1982.

[5] W. Liu, *Fundamentals of III-V Devices: HBTs, MESFETs, and HFETs/HEMTs*, p. 60, John
Wiley & Sons Inc., 1999.

[6] M.S. Lundstrom, *Fundamentals of Carrier Transport*, Chap. 5, Cambridge University Press,
2000.

[7] D.J. Griffiths, *Introduction to Quantum Mechanics*, Chap. 8, Prentice-Hall, 1995.

6   *np*- and *Np*-junction basics

When neighbouring regions of a homogeneous semiconductor are doped with different types of dopant, a *pn*- or *np*-junction is formed. When the junction is between different semiconductors, the junction is labelled either *Pn* or *Np*, where the capital letter denotes the doping type of the semiconductor with the higher bandgap. Semiconductor/semiconductor junctions play a crucial role in solar cells, LEDs, bipolar transistors, and HJFETs, and are prominent also in MOSFETs. As we demonstrate in this chapter, a potential energy barrier forms at this type of junction. In a solar cell, this barrier facilitates the separation of photogenerated electron-hole pairs into a current. In the other devices, the modulation of the junction barrier height by an applied voltage allows the current to be controlled by external circuitry.

In this chapter the focus is mainly on the *np*-junction; it is used to achieve an understanding of the properties of semiconductor junctions via the drawing of an **energy-band diagram**, the construction of which is explained here. We also introduce the concepts of **quasi-neutrality** and **quasi-Fermi levels**. The latter prove useful in describing carrier concentrations under non-equilibrium conditions. Finally, the *Np*-junction is described, and some consequences of the bandgap mismatch are noted.

6.1   *np*-junction at equilibrium

To construct the equilibrium energy-band diagram for an *np*-junction, consider first Fig. 6.1a, in which the separate *n*- and *p*- type regions are shown.[1] This band diagram follows from Fig. 2.14, and shows the band edges E_C and E_V: these were defined in (2.41), and they are related to the potential energies of electrons and holes, respectively. The diagram also introduces E_0, the **force-free vacuum level**, which is used as the energy reference level. It is assigned a value of zero, so all other energy levels on the band diagram are negative. Physically, E_0 is the energy that an electron would have if it were just removed from the semiconductor, under circumstances where there were no potential-energy gradients (electrostatic forces) in the semiconductor. The positive energy required to effect the removal ($E_0 - E_C$ in the force-free case) is called the **electron affinity** of the semiconductor, and takes the symbol χ.

---

[1] Unless otherwise stated, we will only consider **abrupt junctions**, i.e., junctions in which the doping density changes abruptly from N_D on the *n*-side, to N_A on the *p*-side.

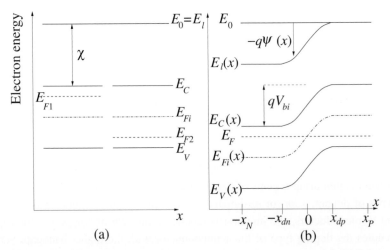

Figure 6.1 (a) Energy-band diagram for separate n- and p-type material. (b) Energy-band diagram for an np-junction at equilibrium.

Starting from (4.14), and referring to Fig. 6.1a, the Fermi energy can be written as

$$E_F = E_C + k_B T_L \ln n_0 - k_B T_L \ln N_C$$
$$= E_0 - \chi + k_B T_L \ln n_0 - k_B T_L \ln N_C$$
$$= -\chi + k_B T_L \ln n_0 - k_B T_L \ln N_C$$
$$\equiv \mu . \qquad (6.1)$$

The final equation in the set indicates that the Fermi energy is, in these field-free circumstances, equal to μ, which is the **chemical potential energy**[2]. As χ and N_C are material constants, E_F evidently increases as more electrons are added to the semiconductor.

The difference between the Fermi energies in the two separate materials of Fig. 6.1a can be written as a difference in chemical potential energy

$$E_{F1} - E_{F2} = k_B T_L \ln \frac{n_{01}}{n_{02}} = \Delta \mu . \qquad (6.2)$$

If the two materials are now joined (Fig. 6.1b), this difference in chemical potential energy acts as a driving force to equilibrate the two, differently doped regions. The minimum-energy state of thermal equilibrium is reached when the driving force vanishes, i.e., when sufficient electrons have been transferred from the n-side to the p-side so that $E_{F1} = E_{F2} \equiv E_F$.

The transferred electrons initially increase the np-product on the p-side near to the junction, leading to an enhanced recombination rate, and the annihilation of many electrons and a corresponding number of holes. This creates a region of negative space charge (acceptor ions) on the p-side of the junction. A region comprising an equal

---

[2] Invariably, the word 'energy' is dropped, leaving 'chemical potential'. The short term comes from thermodynamics, as does the choice of the symbol for it, which is regrettable because we've already used μ for mobility.

magnitude of positive space charge (donor ions) appears on the *n*-side from the departure of the transferred electrons. This **space-charge region** is a source of internal potential energy, characterized by an internal electrostatic potential difference called the **built-in voltage**, V_{bi}, which is always expressed as a positive quantity. The actual, position-dependent electrostatic potential $\psi(x)$ is defined in Fig. 6.1b. Note the direction of the arrow labelling the potential energy $(-q\psi(x))$: it indicates that a positive potential makes the **local vacuum level** E_l negative with respect to the reference energy E_0. For the reference potential $(\psi = 0)$, we arbitrarily set it to be zero at the end of the *p*-region, i.e., at $x = x_P$.

In the parts of the material where $\psi \neq 0$, it follows from Fig. 6.1b that the Fermi energy can be written as

$$E_F = -q\psi(x) + \mu(x).\tag{6.3}$$

Thus E_F is now more than just the chemical potential energy: it becomes the **electrochemical potential energy**, following the description from thermodynamics of a system possessing both chemical potential energy and electrostatic potential energy.[3] In a battery, for example, the difference in electrochemical potential energy between two electrodes provides the electro-motive force (voltage) to drive electrons (current) around an external circuit. It also manifests itself as a voltage that can be measured by a voltmeter. In an *np*-junction at equilibrium, the fact that there is no difference in electrochemical potential energy means that there can be no net current in any circuit to which the semiconductor is attached; nor is there a terminal voltage that can be recorded by a voltmeter. Thus, V_{bi} cannot be measured directly.

In an *np*-junction the built-in voltage develops to precisely the value that is required to ensure that the drift current of electrons down the potential gradient to the *n*-side, exactly cancels the diffusion current of electrons over the barrier from the side of higher electron concentration to the *p*-side (see Fig. 6.4c). The same cancellation process also occurs for the holes. Thus, as we noted in Section 5.5.2, if a diffusive flow of charge is present at equilibrium, then it must be exactly counterbalanced by a drift flow.

6.1.1   The built-in voltage

To illustrate the relationship between V_{bi} and the nullifying of the currents, use (5.37) and (5.41) for the drift and diffusion currents of electrons, respectively. At equilibrium, we have

$$\vec{J}_e = -qn\mu_e\nabla\psi + k_B T_L \mu_e \nabla n = 0.\tag{6.4}$$

Considering the edges of the space-charge region, and, for simplicity, one-dimension:

$$\int_{\psi(p-side)}^{\psi(n-side)} d\psi = \frac{k_B T_L}{q} \int_{n_{0p}}^{n_{0n}} \frac{1}{n} dn.\tag{6.5}$$

---

[3] Again, 'energy' is usually omitted, leaving 'electrochemical potential'.

Performing the integration yields the standard expression for the built-in voltage of a homojunction:

$$V_{bi} = \frac{k_B T_L}{q} \ln \frac{n_{0n}}{n_{0p}} \equiv V_{\text{th}} \ln \left[\frac{N_D N_A}{n_i^2} \right] , \qquad (6.6)$$

where, for convenience, we have defined the **thermal voltage** as

$$V_{\text{th}} \equiv \frac{k_B T}{q} , \qquad (6.7)$$

where $T = T_L$ is the lattice temperature. At 300 K, $V_{\text{th}} \approx 26\,\text{mV}$, and for a Si homojunction with doping densities of 10^{19} and $10^{17}\,\text{cm}^{-3}$, for example, $V_{bi} = 0.95\,\text{V}$.

6.1.2   Constructing an equilibrium energy-band diagram

The algorithm for constructing the equilibrium energy-band diagram is summarized here.

1. Draw a solid horizontal line and label it E_0. This is the reference energy level.
2. Measure down from this the electron affinity of one distinct region of the structure, e.g., the *n*-type region. Draw a solid horizontal line at this energy and label it E_C.
3. Add another solid horizontal line E_V to indicate the bandgap E_g.
4. Add a dashed horizontal line for the Fermi level at a position that gives some indication of the doping density, i.e., make it close to E_C if the doping density is high.
5. Repeat the above steps for any other distinct regions. For example, if you have an *np*-junction, your diagram should now look like Fig. 6.1a.
6. Now, to connect the two separate parts of the structure, start with a dashed horizontal line across the width of the entire structure: this is E_F.
7. Draw E_C and E_V for each part of the device, leaving a gap at the junction, where there would otherwise be a discontinuity in these energy levels.
8. Choose one side of the device to be the reference side for the electrostatic potential, and add-in $E_l = E_0$ for this side of the device. In Fig. 6.1b we chose the *p*-side for the reference.
9. On the non-reference side, add-in a solid line for E_l at an energy χ above E_C.
10. Across the interfacial region between the two parts of the structure, join-up E_l with a smooth curve, and E_0 with a horizontal line. The transition in $E_l(x)$ must be continuous because it represents the change in electrostatic potential.
11. For each region in turn, extend the band edges to the metallurgical junction by drawing lines that are parallel to the vacuum level. If the two regions of the device structure are made from the same material, each band edge should join-up smoothly across the interfacial region. This is the case for the homojunction shown in Fig. 6.1b.

For a junction between dissimilar materials, there may be discontinuities in E_C and E_V at the interface, but there cannot be discontinuities in the vacuum level.[4]

6.1.3   Potential profile

The potential ψ is defined from Fig. 6.1b:

$$-q\psi(x) = E_l(x) - E_0 . \tag{6.8}$$

In a homogeneous system, $E_C(x)$ tracks $E_l(x)$,[5] so

$$-q\psi(x) \equiv E_C(x) - E_C(x_P) . \tag{6.9}$$

It follows from (4.14) that, when Maxwell-Boltzmann statistics apply,

$$
\begin{aligned}
n_0(x) &= n_0(x_P)e^{\psi(x)/V_{th}} \\
p_0(x) &= p_0(x_P)e^{-\psi(x)/V_{th}} .
\end{aligned} \tag{6.10}
$$

Employing these in Poisson's Equation from our master set of equations (5.24), and considering one dimension for simplicity:

$$
\begin{aligned}
-\frac{d^2\psi}{dx^2} &= \frac{\rho}{\epsilon} \\
&= \frac{q}{\epsilon} \left[p_0(x_P)e^{-\psi(x)/V_{th}} - n_0(x_P)e^{\psi(x)/V_{th}} + N_D - N_A \right],
\end{aligned} \tag{6.11}
$$

where ρ is the volumetric charge density.

This non-linear equation can be solved numerically to find $\psi(x)$, provided n_0 and p_0 at the end of the p-side of the device are known. We'll discuss the nature of metal/semiconductor contacts in Section 11.1, but a common case is that of an **ohmic contact**. Such a contact allows both electrons and holes to flow easily into and out of the semiconductor, so the carrier concentrations at the semiconductor/metal interface are maintained at their equilibrium values. In this case, $n_0(x_P)$ and $p_0(x_P)$ are easily found (Section 4.4), and the boundary conditions are simply, $\psi(-x_N) = V_{bi}$ and $\psi(x_P) = 0$. Applying these conditions to a silicon np-junction with $N_D = 10^{18}$ cm^{-3} and $N_A = 10^{17}$ cm^{-3}, the solution for $\psi(x)$ from (6.11) is shown in Fig. 6.2. The built-in voltage is 0.9 V in this case, and it can be seen that it is dropped mostly across the lesser-doped portion of the space-charge region.

---

[4] A discontinuity in the vacuum level would imply a discontinuity in electrostatic potential, which would mean an infinite electric field. This would create an infinitely strong force on neighbouring charges, which would then be rapidly re-arranged to reduce the field.

[5] In a heterogeneous system, as mentioned in the previous subsection, $E_C(x)$ does not necessarily track $E_l(x)$ throughout the device, so (6.8) must be used for the potential, rather than (6.9).

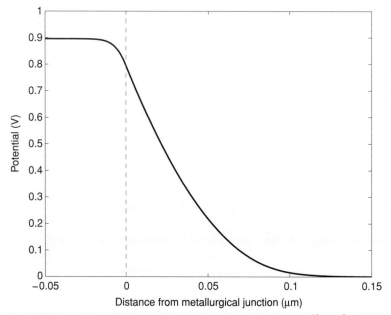

Figure 6.2 Potential profile for a silicon *np*-junction with $N_D = 10^{18}$ cm^{-3} and $N_A = 10^{17}$ cm^{-3}, under equilibrium conditions. The actual metallurgical junction between the *n*- and *p*-regions is shown by the dashed line. The solution is from the master set of equations (5.24), as implemented by the commercial solver ATLAS. At equilibrium, this set of equations reduces to (6.11) for ψ.

6.2   The Depletion Approximation

The space-charge region at the *np*-junction plays such an important role in the operation of many semiconductor devices that it is helpful to have a way of estimating its width without resorting to numerical methods of computation. This is achieved by making the **Depletion Approximation**, namely

$$n(x), \, p(x) \ll N_D \qquad -x_{dn} \leq x \leq 0$$

$$n(x), \, p(x) \ll N_A \qquad 0 \leq x \leq x_{dp}, \tag{6.12}$$

where the *x*-coordinates are from Fig. 6.1b. This approximation turns the smooth variations in $n(x)$ and $p(x)$ across the junction into abrupt changes at $-x_{dn}$ and x_{dp}, respectively, as shown in Fig. 6.3. The beauty of the approximation is that it turns (6.11) into two linear equations that are easily solved, and which enable the width of the so-called **depletion region** ($W = x_{dn} + x_{dp}$) to be determined. The equations are

$$\frac{d^2\psi}{dx^2} = \frac{-qN_D}{\epsilon} \qquad -x_{dn} \leq x \leq 0$$

$$\frac{d^2\psi}{dx^2} = \frac{qN_A}{\epsilon} \qquad 0 \leq x \leq x_{dp}. \tag{6.13}$$

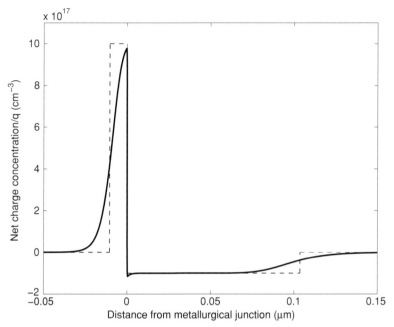

Figure 6.3 Profile of the charge density $\rho = (p + N_D - n - N_A)/q$ for a silicon np-junction with $N_D = 10^{18}$ cm^{-3} and $N_A = 10^{17}$ cm^{-3}, under equilibrium conditions. The full numerical solution from (5.24), as implemented by ATLAS, is shown by the solid line. The profile from the Depletion Approximation is shown by the dashed line.

A further aspect of the Depletion Approximation (DA) is that it assumes that there is no voltage drop in the regions outside of the depletion region. This assumption provides the boundary conditions required for the solution of (6.13):

$$-\frac{d\psi}{dx} = 0 \qquad x = -x_{dn}$$

$$-\frac{d\psi}{dx} = 0 \qquad x = x_{dp}$$

$$\psi = 0 \qquad x = x_{dp} . \qquad (6.14)$$

The solution is

$$\psi(-x_{dn}) - \psi(x_{dp}) \equiv V_J = \frac{q}{2\epsilon} \left[N_D x_{dn}^2 + N_A x_{dp}^2 \right], \qquad (6.15)$$

where V_J is the potential difference across the junction region; it equals V_{bi} in the equilibrium case. It should be clear that the magnitudes of the charges in the two parts of the depletion region are equal:[6]

$$q N_D x_{dn} A = q N_A x_{dp} A , \qquad (6.16)$$

---

[6] Formally, this follows from Gauss's Law and the continuity of the dispacement $\vec{D}$ in the absence of free charges at the actual interface.

where A is the cross-sectional area of the device. Putting all these facts together, the width of the depletion region can be written as

$$W = \sqrt{\frac{2\epsilon}{q} V_J \left(\frac{1}{N_D} + \frac{1}{N_A} \right)}. \tag{6.17}$$

For our Si example with $N_D = 10^{18}\,\mathrm{cm}^{-3}$ and $N_A = 10^{17}\,\mathrm{cm}^{-3}$, and considering the equilibrium case, we find that $V_{bi} = 0.9\,\mathrm{V}$, $W = 114\,\mathrm{nm}$, $x_{dn} = 10\,\mathrm{nm}$, and $x_{dp} = 104\,\mathrm{nm}$. The validity of the DA in the equilibrium case for this junction can be assessed from Fig. 6.3, where the charge densities predicted by the DA and the full numerical calculation are compared. Notice that the majority carrier concentrations do not change in the abrupt manner assumed by the DA. This is to be expected on physical grounds: if it were otherwise there would be infinite diffusion currents. Also note that, for the moderately high donor doping density used in this example, n is still significant at the **metallurgical junction** $(x = 0)$, so the charge density ρ on the n-side of the junction never attains the value of $q\,N_D$ assumed by the DA. Also, the electron injection into the p-type region is sufficient to push ρ just inside the p-region to a value slightly below the value of $-q\,N_A$ assumed by the DA. This feature would be more pronounced if N_D were higher.

Despite these discrepancies, the DA is widely employed, primarily to estimate the width of the space-charge region. We will use it for this purpose in all of the devices we study.

6.3   *np*-junction under bias

The equilibrium energy-band diagram of Fig. 6.1b is repeated in Fig. 6.4a for convenience. The Fermi level has been extended on either side of the device to represent the metallic contacts, to which leads would be attached and taken to the external circuit. Let us apply a voltage V_a between the two contacts. Taking the p-side contact as our zero reference for electrostatic potential, V_a is negative and the total potential difference across the device is

$$\psi(-x_N) - \psi(x_P) = V_{bi} + V_a. \tag{6.18}$$

In metallic elements with an odd number of valence electrons per primitive unit cell, as discussed in Section 2.7, the highest occupied band is half-filled with electrons. The number of electrons is huge,[7] and means that any electron exchange between a metallic contact and its adjoining semiconductor can occur without significantly disturbing the metal from its thermal-equilibrium state. Thus, the application of bias to an *np*-junction is manifest as a difference in electrochemical potential between the metallic end-contacts, and can be expressed as

$$E_F(-x_N) - E_F(x_P) = -q\,V_a, \tag{6.19}$$

[7] For Al, the number of electrons 'available for conduction' is $1.8 \times 10^{23}\,\mathrm{cm}^{-3}$.

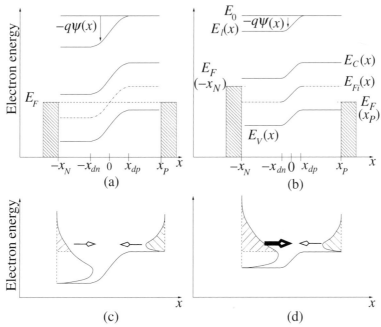

Figure 6.4 (a) Equilibrium band diagram of Fig. 6.1, showing the Fermi levels in the metallic contacts. Note that, by not connecting the band edges in the semiconductor to the band edges in the metallic contacts, we have tacitly assumed that none of the applied voltage is dropped across the contact/semiconductor interfaces. This is a good assumption for the ohmic contacts considered here. (b) Energy-band diagram under an applied forward bias $V_a = -[E_F(-x_N)$ $-E_F(x_P)]/q$. (c) Equal and opposite flows of electrons across the junction at equilibrium. (d) Net flow of electrons from left to right due to lowering of the potential barrier at the junction. By drawing hemi-Maxwellian distributions we are assuming that near-equilibrium conditions are applicable.

where the x-coordinates refer to Fig. 6.4b. $E_F(x_P)$ is taken as the reference energy for the electron potential energy $-q V_a$ due to an applied voltage V_a. It follows that Fig. 6.4b is drawn for the case of $V_a < 0$. Such a situation, when the applied voltage across an *np*-junction is such that the applied potential on the *n*-side contact is negative with respect to that on the *p*-side contact, is said to be one of **forward bias**. The converse situation is one of **reverse bias**.

We now have to consider how this applied voltage is distributed (dropped) across the device. For an initial guess, refer to Section 5.4.2, and make use of the fact that the conductivity is much lower in the space-charge region than in the other regions of the device, due to the relative lack of carriers in the former region. Thus, we would expect that most of V_a would be dropped across the more resistive space-charge region. For the forward-bias case illustrated in Fig. 6.4, the potential barrier at the junction would be lowered by $\approx |V_a|$. This lowering would enable more electrons to pass from the *n*-side to the *p*-side of the junction, resulting in a net current. Now, let's consider a case where this current density is rather large, say 10^4 Acm^{-2}. The flow of electrons constituting this current exists also in the *n*-type region of the device to the left of the depletion region. Let us take a doping density of 10^{19} cm^{-3} for this region and compute the field needed to support the chosen current density. If you do this, (see Exercise 6.1), then you should find that the field is only ≈ 6 mV μm^{-1}. Thus, in microelectronics, the actual voltage dropped across this region will be very small compared to the value for V_a of about 0.8–1.0 V, which would be typical for a Si *np*-junction delivering the stated current density. The guess of V_a being dropped entirely across the space-charge region is, therefore, a very good one.

6.3.1 Constructing a non-equilibrium energy-band diagram

As an example of how to turn a band diagram at equilibrium into one at non-equilibrium, consider the differences between Fig. 6.4a and Fig. 6.4b. The former is constructed according to the steps in Section 6.1.2. Proceed as follows:

1. Copy the portion of the band diagram on the reference side of the junction (the *p*-side in our case), but don't draw-in E_F.
2. On the *n*-side, draw a new horizontal line for E_l at a depth of $-q V_a$ below the previous position of E_l. In our forward-bias example this will raise E_l on the *n*-side above its equilibrium position.
3. Complete the *n*-side of the diagram by adding-in E_C at χ below E_l, and E_V at E_g below E_C.
4. You know that E_l must join-up smoothly across the junction, but before you actually do this, consider how the width of the space-charge region has changed because of the applied bias. Forward bias reduces the potential drop across the junction, so less charge is needed to support the potential difference, and the space-charge region shrinks. So, extend the horizontal portions of E_l a bit further towards the metallurgical junction before you join them up. Repeat this for the band edges. In reverse bias, the procedure is the same, but the larger voltage drop across the junction means that the depletion region is widened relative to its equilibrium value.

6.3.2 Quasi-neutrality

The fact that there is only a minuscule field in the regions of an *np*-diode outside of the space-charge region suggests that there is very little deviation from charge neutrality within them. To examine this suggestion, consider the response of the majority carrier holes to the injection of electrons from the *n*-side of the junction (see Fig. 6.5). The presence of new negative charge at some point beyond the edge of the depletion region on the *p*-side attracts holes, which flow in from the right-hand contact. Eventually, the holes, driven by the tiny electric field in this region, reach the site of the excess electrons. The charge of $+q \Delta p$ that they add to this site, balances the charge of $-q \Delta p$ at the contact from which they originated. Thus, the region of length L between the contact and the excess electrons is a bit like a parallel-plate capacitor of capacitance C. This region also has resistance R. The associated RC time constant defines the **dielectric relaxation**

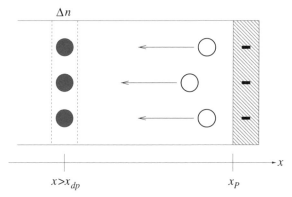

Figure 6.5 Showing the movement of holes towards an excess electron concentration in order to restore charge neutrality in a *p*-type region.

time τ_D:

$$RC|_{\text{QNR}} = \frac{L}{\sigma A} \frac{\epsilon A}{L}$$

$$= \frac{\epsilon}{\sigma} \equiv \tau_D , \qquad (6.20)$$

where A is the cross-sectional area, σ is the conductivity, ϵ is the permittivity, and QNR stands for **quasi-neutral region**. For a more rigorous derivation of this result see Fonstad [1].

The implication of this is that the majority carriers will respond to any changes in charge in a time of the order of τ_D. For example, in *p*-type silicon with a doping density of $N_A = 10^{18}$ cm^{-3}, the dielectric relaxation time is ≈ 50 fs. This is a very short time, and it means that unless we are interested in what happens during time periods of this scale, then the regions outside of the depletion region can be considered as being effectively neutral. We generally consider them to be so in this book. Thus, in a quasi-neutral region under low-level injection of minority carriers, the majority carrier concentration remains very close to its equilibrium form, and is subjected to a very small field. The distribution of the majority carriers, therefore, fits the prescription of a displaced Maxwellian (see Fig. 5.4).

To return to our question: where is the applied voltage dropped? We can now say that in an *np*-diode of the bulk type we have been considering, we have $V_{aj} \approx V_a$, where V_{aj} is the portion of the applied voltage that is actually dropped across the junction. This usually holds in bipolar devices, for which V_a is coupled resistively to the actual junction. An exception would be when the current were so high that the IR-voltage drop in the quasi-neutral regions could not be neglected. This can happen in high-power bipolar transistors (see Section 16.3.1). In MOSFETs, the coupling of V_a to the junction occurs capacitively, and we'll find in Chapter 10 that $V_{aj} \neq V_a$ in the ON-condition. To allow for all possibilities, we define V_{aj} as

$$V_{aj} + V_{bi} \equiv V_J = \psi(-x_{dn}) - \psi(x_{dp}) . \qquad (6.21)$$

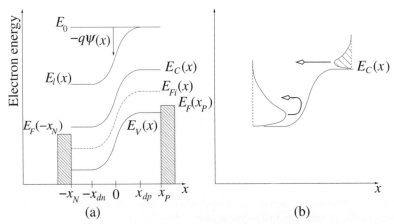

Figure 6.6 The *np*-junction in reverse bias. (a) Note the positive applied bias $(-qV_a = E_F(-x_N)$ $-E_F(x_P))$. (b) Illustrating the blocking action of the junction as regards the electron flux from the left.

6.3.3   Reverse bias

In the reverse-bias case, a positive potential is applied to the *n*-side contact with respect to the *p*-side contact, which we take to remain grounded. Thus, the barrier height of the junction $q(V_{bi} + V_{aj})$ is increased. The associated potential difference is supported by a widening of the space-charge layer, as illustrated in Fig. 6.6. Mathematically, the actual width follows directly from an extension of (6.17):

$$W = \sqrt{\frac{2\epsilon}{q}(V_{bi} + V_{aj})\left(\frac{1}{N_D} + \frac{1}{N_A}\right)}. \tag{6.22}$$

This expression also holds for forward bias ($V_{aj} < 0$).

If the barrier becomes so high that few electrons on the *n*-side have enough energy to diffuse over the barrier to the *p*-side, then the electron current is carried mainly by the equilibrium concentration of minority carriers on the *p*-side drifting down the potential barrier (see Fig. 6.6b). There are not many of these electrons, so the reverse-bias current is very small. To a first approximation, it is given by $q\frac{n_{0p}}{2}2v_R$, and is bias-independent. For a *p*-side doping density of 10^{17} ionized acceptors per cm^3, the current density is less than a nanoampere per cm^2. This should be contrasted with the forward-bias current, which is exponentially dependent on V_{aj}, and can be very large (see Section 6.6).

6.4   Quasi-Fermi levels

Our reasoning so far has led to the conclusions that, in the quasi-neutral regions, even in an out-of-equilibrium situation when large currents are present, the potential profile is essentially flat, and the majority carrier concentration is close to its equilibrium value. These 'facts' suggest that, in the quasi-neutral regions, the carrier concentrations out of equilibrium can be described by equations of an equilibrium form. Of course, *the Fermi level cannot be used* in this description of the semiconductor because equilibrium conditions do not apply. Instead, the Maxwell-Boltzmann equations for carrier concentrations, (4.23), are written in terms of two reference energy levels, called **quasi-Fermi levels**:

$$n = n_i \exp\left(\frac{E_{Fn} - E_{Fi}}{k_B T_L}\right)$$

$$p = n_i \exp\left(\frac{E_{Fi} - E_{Fp}}{k_B T_L}\right), \tag{6.23}$$

where E_{Fn} is the quasi-Fermi level for electrons and E_{Fp} is the quasi-Fermi level for holes. The concept of quasi-Fermi levels was first introduced by Shockley [2, p. 308].

It is emphasized that these quasi-Fermi levels are *defined* by (6.23), and do not have a basis in thermodynamics, as the Fermi level does. E_{Fn} and E_{Fp} are artefacts, the values for which must be consistent with the out-of-equilibrium carrier concentrations, whereas E_F helps define an equilibrium concentration by dictating the probability of states being filled. Thus, if n is known, then E_{Fn} is defined. However, it must be stated that (6.23) is approximate in the sense that it implies a distribution of carriers that can be described by a simple exponential relationship involving a reference energy and the 'Boltzmann energy' $k_B T_L$. Such a distribution is an equilibrium distribution, so it follows that quasi-Fermi levels should strictly only be used in situations where the carrier distributions do not stray too far from their equilibrium forms.

We have argued that this condition is met by the majority carrier concentration in the quasi-neutral region. The 'leap-of-faith' that is often made is that (6.23) applies not only to majority carriers in the QNR, but to *all* carriers *everywhere* in the device. In the devices considered in this book, scattering usually ensures that this assumption is not seriously violated. Equation (6.23) is widely used in commercial device simulators as its Maxwell-Boltzmann form is very appealing for computational purposes.[8]

One insightful result that follows immediately from (6.23) concerns the *pn*-product:

$$n(x)p(x) = n_i^2 \exp\left(\frac{E_{Fn}(x) - E_{Fp}(x)}{k_B T_L}\right). \tag{6.24}$$

This equation indicates that in regions where the $n(x)p(x)$ product differs from its equilibrium value of n_i^2, then there will be a separation of the quasi-Fermi levels. A large separation can be expected in the depletion region under forward-bias conditions because an excess of both electrons and holes will arise, due to increased injection from the *n*-side and *p*-side, respectively, on account of the lowering of the potential barrier at the junction (see Exercise 6.2).

Another utility of the quasi-Fermi level concept is that it allows a convenient extension of the $n_0(\psi)$ and $p_0(\psi)$ expressions to the non-equilibrium case. From (4.14), (6.10),

---

[8] For degenerate conditions, (6.23) is often modified in commercial device simulators to better represent Fermi-Dirac statistics.

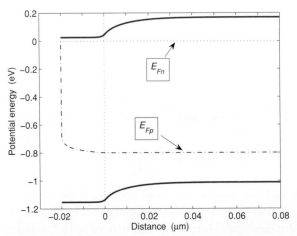

Figure 6.7 Energy-band diagram for the numerical simulation from (5.24) by ATLAS of an *np*-junction with *n*-side length of 0.02 μm and doping density 10^{19} cm^{-3}, and *p*-side doping density 10^{17} cm^{-3} and length > 0.08 μm. The applied bias is $V_a = -0.8$ V. The solid lines are the conduction- and valence-band edges. Note the near-constancy of E_{Fn} across the junction depletion region. E_{Fp} is separated from E_{Fn} in the junction region by $qV_{aj} \approx qV_a$.

and (6.23) it follows that

$$n = n_0(x_P)\exp\left(\frac{\psi}{V_{\text{th}}}\right)\exp\left(\frac{E_{Fn}-E_F}{k_B T_L}\right)$$

$$p = p_0(x_P)\exp\left(-\frac{\psi}{V_{\text{th}}}\right)\exp\left(\frac{E_F - E_{Fp}}{k_B T_L}\right). \tag{6.25}$$

Defining the **quasi-Fermi potentials** for electrons and holes as

$$-q\phi_n(x) = E_{Fn}(x) - E_F$$

$$-q\phi_p(x) = E_{Fp}(x) - E_F, \tag{6.26}$$

allows the out-of-equilibrium carrier concentrations to be written in a compact form:

$$n = n_0(x_P)\exp\left(\frac{\psi - \phi_n}{V_{\text{th}}}\right)$$

$$p = p_0(x_P)\exp\left(\frac{\phi_p - \psi}{V_{\text{th}}}\right). \tag{6.27}$$

An immediately useful application of these equations is to substitute them into the expression for the total current, as given by the sum of the electron and hole currents from the Drift-Diffusion Equation (5.15). Doing this leads to

$$\vec{J}_T = \vec{J}_e + \vec{J}_h = -q\mu_e n\nabla\phi_n - q\mu_h p\nabla\phi_p. \tag{6.28}$$

This emphasizes that if there is no gradient in either of the quasi-Fermi levels, there is no current.

To illustrate the profile of the quasi-Fermi levels in a forward-biased np-junction, consider Fig. 6.7. The left-hand side contact to the n-region was defined as being 'ohmic' (see Section 7.4.1). This means that n and p at the contact are maintained at their equilibrium values. Thus, the two quasi-Fermi levels come together at this point. Note how the majority-carrier quasi-Fermi levels stay essentially constant at their equilibrium levels through their respective quasi-neutral regions. Note also the constancy of the band edges in the QNRs. These facts confirm that the majority carrier concentrations remain essentially at their equilibrium values throughout the QNRs (at least for low-level injection). Note also that the constancy of the quasi-Fermi levels persists through the space-charge region.

6.5   Shockley's Law of the Junction

The statements about the near-constancy of the quasi-Fermi levels when there is an applied bias may bother you, as we stated via (6.28) that a gradient in quasi-Fermi levels is associated with a net current. But note, in that equation the gradient $\nabla\phi$ always appears in a product term with the carrier concentration. So, when the carrier concentration is very high, such as for electrons in the quasi-neutral n-region, only a very small gradient in ϕ_n is needed. This is another way of saying the field must be small in this region, as we noted in Section 6.3, wherein we used this fact to say $V_{aj} \approx V_a$. In the space-charge region, one or both of the carrier concentrations are also quite high, even though they may be small relative to the doping concentrations. Therefore, again, it is not unreasonable that the quasi-Fermi levels be nearly constant across the depletion region. Shockley recognized this [2, p. 312], and exploited the fact to derive an insightful expression for the minority carrier concentration injected across an np-junction:

$$n(x_{dp}) = n_0(x_P)\exp\left(\frac{\psi(x_{dp}) - \phi_n(x_{dp})}{V_{th}}\right)$$

$$= n_0(x_P)\exp\left(\frac{\psi(x_{dp}) - \phi_n(-x_{dn})}{V_{th}}\right)$$

$$= n_0(x_P)\exp\left(\frac{\psi(x_{dp}) - \psi(-x_{dn}) + V_{bi})}{V_{th}}\right)$$

$$= n_{0p}\exp\left(-\frac{V_{aj}}{V_{th}}\right), \tag{6.29}$$

where n_{0p} is the equilibrium electron concentration on the p-side. Shockley called this result 'the key equation of the rectification theory'. It is a fair description because the observed exponential dependence of current on voltage in an np-junction follows directly from it.

Recalling that $V_{aj} < 0$ for forward bias, this equation informs that injection across the lowered barrier leads to an exponential rise in the minority carrier concentration at the edge of the junction. Arriving at the minority carrier boundary condition in this way is said to be invoking Shockley's **Law of the Junction**.

Finally, one other way to accept the near-constancy of E_{Fn} and E_{Fp} across the space-charge region is to realize that the current in a forward-biased *pn*-junction diode, although potentially large absolutely, is generally small relative to the massive current of which a majority-carrier hemi-Maxwellian distribution is capable (see Section 4.5.2). So, the departure from equilibrium is usually slight, and it is reasonable to view the situation as one of quasi-equilibrium, for which the term 'quasi-Fermi level' is appropriate.

6.6   The ideal-diode equation

One useful application of Shockley's Law of the Junction is to set one of the boundary conditions for deriving an expression for the *J-V* characteristic of an **ideal *np*-junction diode**. Such a diode has the following features:

- no gradients in carrier temperature. This means that the Drift-Diffusion Equation can be used for the current.
- negligible fields in the quasi-neutral regions. This means that minority carrier transport in these regions is due only to diffusion. It also means that the applied voltage V_a is dropped entirely across the space-charge region.
- quasi-neutral regions that are so long that the injected minority carriers all recombine before reaching the end contacts.
- injection is sufficiently low that the quasi-Fermi levels are constant across the space-charge layer. This means that (6.29) can be used as a boundary condition.
- no recombination-generation in the space-charge region. This means that the electron current at $x = x_{dp}$ and the hole current at $x = -x_{dn}$ can be added to get the total current.
- no generation other than thermal generation.
- no doping density gradients on either side of the junction. This means that $n(x_{dp}) \equiv n_{0p}$, the equilibrium electron concentration in the *p*-side quasi-neutral region.

Applying these conditions to the electrons injected into the *p*-region, for example, we have, from the time-dependent form of the master set of equations (5.24)

$$J_e = q D_e \frac{\partial n}{\partial x} \quad \text{and} \quad \frac{\partial n}{\partial t} = \frac{1}{q} \frac{\partial J_e}{\partial x} - \frac{n - n_{0p}}{\tau_e}. \tag{6.30}$$

Thus, at steady-state, we have

$$0 = \frac{d^2 n}{dx^2} - \frac{n - n_{0p}}{L_e^2}, \tag{6.31}$$

where $L_e = \sqrt{D_e \tau_e}$ is the electron **minority-carrier diffusion length**. It is a measure of how far the injected electron travels in the *p*-region before it recombines. The general solution to (6.31) is

$$n(x) - n_{0p} = A e^{x/L_e} + B e^{-x/L_e}. \tag{6.32}$$

The relevant boundary conditions are:

$$n(x_P) \equiv n(\infty) = n_{0p}$$

$$n(x_{dp}) = n_{0p} e^{-V_a/V_{th}} . \qquad (6.33)$$

The resulting solution for the electron concentration is

$$n(x) - n_{0p} = n_{0p} \left(e^{-V_a/V_{th}} - 1 \right) e^{(x_{dp}-x)/L_e} . \qquad (6.34)$$

Thus, the electron current at the edge of the space-charge layer on the p-side is

$$J_e(x_{dp}) = -q n_{0p} \left(e^{-V_a/V_{th}} - 1 \right) . \frac{D_e}{L_e} , \qquad (6.35)$$

where the terms have been grouped to emphasize the general fact that the current density is a charge density multiplied by a velocity. The latter is D_e/L_e in this case, and is sometimes called the **diffusion velocity**.

The corresponding expression for the hole current density is

$$J_h(-x_{dn}) = -q p_{0n} \left(e^{-V_a/V_{th}} - 1 \right) . \frac{D_h}{L_h} , \qquad (6.36)$$

Thus the J-V relation for an **ideal diode** is

$$J = -q \left(n_{0p} \frac{D_e}{L_e} + p_{0n} \frac{D_h}{L_h} \right) \left(e^{-V_a/V_{th}} - 1 \right)$$

$$\equiv J_{00} \left(e^{-V_a/V_{th}} - 1 \right) , \qquad (6.37)$$

where J_{00} is the **saturation current density** for an ideal diode, and recall that $V_a < 0$ in forward bias. Despite the many assumptions that are invoked to obtain this expression, it is very informative because it captures the essence of a diode: an asymmetrical characteristic, with a reverse-bias current that is small and a forward-bias current that increases exponentially with voltage.

To get a 'feel' for the magnitudes of the current densities, consider an np-diode with $N_D = 10^{19}$ cm^{-3} and $N_A = 10^{17}$ cm^{-3}. To evaluate the diffusivities, use (5.30) for mobility and the near-equilibrium form of (5.17), the Einstein relation. To get the minority-carrier diffusion lengths, use $L = \sqrt{D\tau}$ after evaluating the minority carrier lifetimes from (3.21). To get the minority carrier concentrations use (4.20), after substituting for n_i from Table 4.1. Put the numbers into (6.37), and you should find that the reverse-bias current density ($\approx J_{00}$) is 1.7 pA cm^{-2}, and is dominated by the electron component. For this diode, (6.6) informs that $V_{bi} = 0.96$ V. For a forward bias of $-0.8V_{bi}$, J is 11.4 A cm^{-2}. That's quite some asymmetry!

The ideal-diode derivation is also useful in that it highlights the position-dependence of the electron and hole contributions to the total current density J_T. Fig. 6.8 illustrates the steady-state situation for a single-loop circuit in which the diode is so long that all carriers injected from the contacts recombine within the bulk semiconductor. J_T is

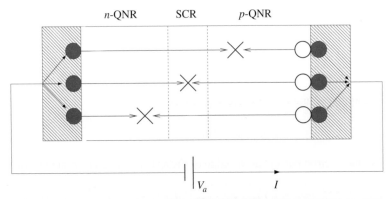

Figure 6.8 Illustration of the position-dependence of the electron and hole contributions to the total current in a forward biased *np*-diode. Recombination events in the two quasi-neutral regions, and in the space-charge region, are indicated by the crosses. The excess carriers consumed in the recombination events are replenished by carriers injected from the contacts.

constant, but is carried entirely by electrons near to the negative contact, and entirely by holes near the positive contact.

6.6.1   Deviations from ideality in diodes

Some features of practical diodes, and of the circumstances in which they operate, are not consistent with ideal-diode behaviour, as we now briefly indicate.

- Quasi-neutral regions are often short, so not all injected minority carriers recombine before reaching the contact. Appropriate boundary conditions must be used, and the saturation current density will be different from the ideal case. An example of this is in the solar cell (see Section 7.4.2).
- For very high currents, the quasi-Fermi levels in the space-charge region may deviate significantly from constant values. This will invalidate the use of Shockley's Law of the Junction to get the minority-carrier concentrations at $-x_{dn}$ and x_{dp}.
- High currents may also mean high-level injection of minority carriers, and a possible loss of charge neutrality in parts of the erstwhile quasi-neutral regions. This leads to local fields outside of the depletion region, in which case not all of the applied voltage V_a will appear across the junction. This can be represented in the diode equation by replacing V_a with V_a/γ, where $\gamma > 1$ is known as the **diode ideality factor**. Thus at high bias, the diode current still increases exponentially, but the dependence on bias is weaker. We encounter this situation in high-power bipolar transistors (see Section 16.3.1).
- At low bias, it may be appropriate to include the current due to the recombination of electrons and holes in the space-charge region. This recombination mechanism usually leads to values of saturation current density greater than J_{00}, and to a value of $\gamma \approx 2$ [3].

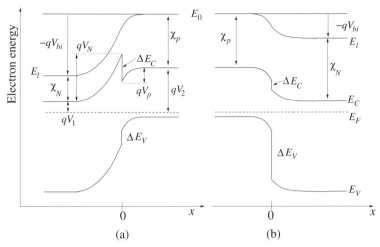

Figure 6.9 Equilibrium energy-band diagrams for two heterojunctions. (a) Type I *Np*-junction. (b) Type II *pN*-junction.

- Any electron-hole pairs generated in the space-charge region are likely to be separated by the junction field, leading to a current in the reverse-bias direction. At large reverse bias, the space-charge region may be so wide that this generation current is much larger than the ideal diode saturation current.
- At even higher reverse bias, carriers traversing the space-charge region may gain sufficient kinetic energy to impact ionize lattice atoms, leading to current multiplication and avalanche breakdown. We'll meet this phenomenon again in the chapter on high-power transistors.

6.7   *Np*-junction electrostatics

The two intrinsic properties of a semiconductor that are important for determining the properties of a semiconductor/semiconductor heterojunction are the electron affinity and the bandgap. The important extrinsic property is the doping density. When all of these properties are known, the energy-band diagram can be constructed following the procedures in Section 6.1.2 and Section 6.3.1. The results at equilibrium for two cases are shown in Fig. 6.9.

Considering isolated semiconductors, a **Type I** heterojunction is one in which the wider bandgap 'straddles' the smaller bandgap, and if the two bandgaps are 'staggered', the heterojunction is categorized as **Type II**. In the examples shown, $E_{gN} > E_{gp}$ and the difference between the two cases arises because in Fig. 6.9a $\chi_N < \chi_p$, whereas in Fig. 6.9b $\chi_N > \chi_p$. In both instances the built-in voltage is

$$q V_{bi} = q V_N + q V_p = q V_2 - q V_1 + (\chi_p - \chi_N), \tag{6.38}$$

where the various energies are noted on Fig. 6.9. Making use of (4.14), which implies Maxwell-Boltzmann statistics, and of (6.6) for the built-in voltage of a homojunction, it follows that

$$q V_{bi} = k_B T \ln \left[\frac{N_{Cp}}{N_{CN}} \frac{n_{0N}}{n_{0p}} \right] + (\chi_p - \chi_N)$$

$$\equiv q V_{bi}^{np,2} + k_B T \ln \left[\frac{N_{Cp}}{N_{CN}} \right] + (\chi_p - \chi_N), \qquad (6.39)$$

where $T \equiv T_L$, and $V_{bi}^{np,2}$ is the built-in voltage of a homojunction made from the lower bandgap material with the same doping densities as used for the heterojunctions under consideration. If the two semiconductors of the heterojunction have similar density-of-states electron-effective-masses, then the second term in (6.39) can be ignored, and we see that V_{bi} for the heterojunction will be greater than that for the corresponding homojunction in the Type I case, and lower in the Type II case.

6.7.1   Energy band offsets

Note from Fig. 6.9 that the local vacuum level $E_l(x)$ varies smoothly with position, as befits a parameter that indicates the electrostatic potential.[9] However, because of the difference in electron affinities, the band edges $E_C(x)$ and $E_V(x)$ show discontinuities at the interface between the two semiconductors of the heterojunction. Defining these discontinuities as the **band offsets**, and labelling them as ΔE_C and ΔE_V, respectively, it can be appreciated from Fig. 6.9, that

$$\Delta E_C + \Delta E_V = E_{gN} - E_{gp} \equiv \Delta E_g, \qquad (6.40)$$

where all the Δ's are taken to be positive quantities. Thus,

$$\Delta E_C = |\chi_2 - \chi_1|. \qquad (6.41)$$

6.7.2   Junction space-charge region

The potential V_J across the space-charge region at an *Np*-heterojunction is given by (6.15), but modified to allow for different permittivities of the two semiconductors. Equation (6.16), which equates the ionic charge on each side of the junction, also holds, provided there is no free charge density at the interface between the two semiconductors.[10] It then follows that the two components of the space-charge region are

$$x_{dN} = \sqrt{ \frac{2 V_J \epsilon_N \epsilon_p N_A}{q N_D (\epsilon_N N_D + \epsilon_p N_A)} } \quad \text{and} \quad x_{dp} = \sqrt{ \frac{2 V_J \epsilon_N \epsilon_p N_D}{q N_A (\epsilon_N N_D + \epsilon_p N_A)} }. \qquad (6.42)$$

---

[9] Recall that $E_l(x) - E_0 = -q\psi(x)$, where $\psi(x)$ is the potential.

[10] This is something that has to be considered in heterojunctions due to the possibility of charge being trapped at localized energy levels within the bandgap, which might arise from any interruption in periodicity of the crystal lattice.

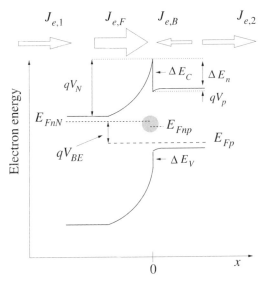

Figure 6.10 Energy-band diagram of a forward-biased Type I *Np*-junction, illustrating (within the shaded circle) the splitting of the electron quasi-Fermi level at the abrupt *Np* junction. $V_{BE} \equiv V_a$.

6.7.3   Quasi-Fermi-level splitting

In the case of heterojunctions with a conduction-band 'spike' at the interface, as in the Type-I *Np* junction shown in Fig. 6.9, the interesting phenomenon of **quasi-Fermi-level splitting** arises, as we now briefly explain.

In any junction, the application of a forward bias reduces both $|q V_N|$ and $|q V_p|$. In the junction under consideration, this has the effect of reducing the barrier to electron flow into the base, while increasing the barrier to electron flow from the base. This leads to a situation where the net electron flow may not be small compared to each of the counter-directed flows. In other words, unlike in a homojunction, the flow of electrons cannot be considered as a minor perturbation of the equilibrium condition. This significant departure from equilibrium is recognized by allowing the electron quasi-Fermi level (in this example) to be discontinuous at the interface, as illustrated in Fig. 6.10.

The forward and backward electron current densities injected over the potential barrier are written as hemi-Maxwellian fluxes

$$J_{eF} = -q\frac{n(0^-)}{2}2v_R \quad \text{and} \quad J_{eB} = q\frac{n(0^+)}{2}2v_R , \tag{6.43}$$

respectively, where the carrier concentrations are taken at either side of the top of the barrier at $x = 0$. Employing Maxwell-Boltzmann statistics, these concentrations are:

$$n(0^-) = n(-x_{dN})e^{-V_N/V_{th}} \quad \text{and} \quad n(0^+) = n(x_{dp})e^{(qV_p - \Delta E_C)/k_B T}, \tag{6.44}$$

where the edges of the depletion region are $-x_{dN}$ and x_{dp}.

At equilibrium, $n(0) = n(0^+)$, and $(V_N + V_p) = V_{bi}$. Further, assuming that low-level-injection conditions apply to the electron flux from region 2 to region 1, we can assert that $n(-x_{dN}) \approx n_0(-x_{dn})$. Putting these facts together we have for the current

$$J_{e,1} \equiv J_{eF} - J_{eB} = -q\gamma \left[n_0(x_{dp})e^{-V_a/V_{th}} - n(x_{dp}) \right] \tag{6.45}$$

where $\gamma = v_R e^{-\Delta E_n/k_B T}$.

Turning now to $J_{e,2}$ in Fig. 6.10, this diifusive current can be written as

$$J_{e,2} = q\frac{D_e}{W}[n(x_2) - n(x_{dp})]$$

$$\equiv -q\frac{D_e}{W}[n(x_{dp}) - n_0(x_2)], \tag{6.46}$$

where $x = x_2$ is the position of the end-contact to region 2, and $W = (x_2 - x_{dp})$ is the width of the quasi-neutral part of this region. The first version of the equation follows from assuming that there is no recombination in region 2, and the second version implies that the contact at the end of region 2 is ohmic.[11]

Assuming no recombination in the space-charge region, $J_{e,1}$ and $J_{e,2}$ can be equated, thereby yielding a boundary condition for $n(x_{dp})$:

$$n(x_{dp}) = n_{0p} \left[\frac{e^{-V_a/V_{th}} + \frac{D_e}{W\gamma}}{1 + \frac{D_e}{W\gamma}} \right], \tag{6.47}$$

where $n_{0p} = n_0(x_{dp}) = n_0(x_2)$, i.e., region 2 has been assumed to be uniform. This equation shows how (6.29), Shockley's Law of the Junction for a homojunction, has to be modified to describe a heterojunction. The new equation, via γ, considers both the finite velocity v_R of carriers crossing the junction, and the presence of an energy difference ΔE_n at the interface. In a homojunction, the latter is zero, and it is customary to imply that the former is infinite. Under these conditions, (6.47) reduces to (6.29). Thus, we now see that the ideal-diode treatment of the homojunction actually neglected any bottleneck to electron transport caused by the junction itself. Accounting for restricting features, such as a finite velocity and a band spike, reduces electron injection into the p-region.

Another way of characterizing the impeding effect of the junction itself is via quasi-Fermi-level splitting. To see this, note that

$$n(x_{dp}) = N_{Cp} e^{(E_{Fnp} - E_C(x_{dp}))/k_B T} \tag{6.48}$$

$$n_0(x_{dp}) = N_{Cp} e^{(E_{Fp} - E_C(x_{dp}))/k_B T},$$

where $E_{Fp} = E_F$ and we have assumed that there is no voltage drop in the quasi-neutral p-region. Noting from Fig. 6.10 that $(E_{FnN} - E_{Fp}) = -qV_a$, it follows from (6.47) and (6.48) that the splitting of the electron quasi-Fermi level at the interface is

$$\Delta E_{Fn} \equiv E_{FnN} - E_{Fnp} = -qV_a - k_B T \ln \left[\frac{n(x_{dp})}{n_{0p}} \right]. \tag{6.49}$$

---

[11] See Section 7.4.1 for the definition of an ohmic contact.

Using this in (6.46):

$$J_{e,2} = -q \frac{D_e}{W} n_{0p} \left[e^{-(qV_a + \Delta E_{Fn})/k_B T} - 1 \right].$$ (6.50)

Compare this current with the electron current component of the ideal homojunction diode in (6.37); recall that $V_a < 0$ in forward bias; and then it will be clear that quasi-Fermi-level splitting reduces the driving force for the diffusion current. Quasi-Fermi-level splitting can be reduced in a Type-I heterojunction by using materials with the same electron affinity, in which case $\Delta E_c = 0$. One such combination is N-In$_{0.49}$Ga$_{0.51}$P and p^+-GaAs; we encounter this in Chapter 9 as the emitter/base part of an HBT. Contrarily, ΔE_{Fn} can be very high in metal/semiconductor heterojunctions, leading to rectifying contacts called **Schottky barriers**, which we'll discuss in Section 11.1.

6.8   Emitter injection efficiency

In anticipation of using an Np-junction at the emitter/base interface of a heterojunction bipolar transistor (Chapter 9), we call the N-region the emitter and the p-region the base. Our purpose here is to estimate the **emitter injection efficiency** of a Type-I heterojunction diode: this is the ratio of the injected electron emitter current to the total current. The electron current is given by (6.50) and the hole current is taken from (6.37). The appropriate form of the latter is

$$J_h(-x_{dn}) = -q \frac{D_h}{L_h} p_{0N} \left[e^{-qV_a/k_B T} - 1 \right].$$ (6.51)

The bandgaps of the two materials enter via the equilibrium carrier concentrations, (4.19) and (4.20), and we will assume that the effective densities of states are the same in both materials.[12] In forward bias the '−1' terms in the expressions for the current components can be dropped, leading to the following expression for the emitter injection efficiency:

$$\eta_{emitter} = \left[1 + \frac{N_B D_h W_B}{N_E D_e L_h} e^{-(\Delta E_g - \Delta E_{Fn})/k_B T} \right]^{-1},$$ (6.52)

where the subscripts E and B refer to the emitter and the base, respectively, and the bandgap difference $\Delta E_g = (E_{gN} - E_{gp})$.

The dominant term in (6.52) is the exponential term; it can be appreciated that, even allowing for quasi-Fermi-level splitting, the presence of a large bandgap difference can enable attainment of extremely high emitter injection currents.[13] This holds true in modern HBTs, even though they generally have $N_B \gg N_E$ for reasons discussed in Section 14.6.

---

[12] This is a reasonable assumption for closely related materials such as AlGaAs/GaAs and InGaP/GaAs.
[13] Please do Exercise 6.10 to appreciate how high $\eta_{emitter}$ can be in modern HBTs.

Exercises

6.1   The current density in a forward biased n^+p diode is $10^4\,\mathrm{A\,cm^{-2}}$. In the n-type quasi-neutral region this current can be considered to be carried entirely by electrons. The length of this region is $1\,\mu\mathrm{m}$ and the donor doping density is $10^{19}\,\mathrm{cm^{-3}}$.

Estimate the potential difference across this region.

If the voltage drop across the p-type quasi-neutral region is similar, are you now convinced that the approximation of all the applied voltage being dropped across the space-charge region is a good one?

6.2   Making use of your conclusion from the previous question, and invoking Shockley's Law of the Junction, use (6.24) to show that the quasi-Fermi levels for electrons and holes are separated in the space-charge region by $-q\,V_a$, where V_a is the potential applied to the n-side contact, and the p-side contact is grounded.

6.3   Derive (6.28). This important equation indicates that a gradient in quasi-Fermi level must accompany a current.

6.4   If the quasi-Fermi levels are constant across the space-charge region, as Question 6.2 suggests, then how can this be consistent with (6.28)?

Substantiate your answer by computing $d\,E_{Fn}/dx$ at $x = x_{dp}$ for an ideal diode in forward bias, and then projecting this gradient across the depletion region to obtain an estimate of $(E_{Fn}(-x_{dn}) - E_{Fn}(x_{dp}))$.

Confirm that this change is small compared to $q\,V_J$.

6.5   A Si np diode has a very long quasi-neutral region (QNR) on the n-side, but a very short QNR on the p-side.

Is the current in this diode larger or smaller than that for the ideal diode discussed in Section 6.6?

6.6   A Si n^+p diode has very long QNRs on both sides of the junction.

Design A of this diode uses p-type material that is of much higher crystalline perfection than is used in Design B.

Which design would give the larger current at a given forward bias?

6.7   Consider the possibility of using an n^+p homojunction diode with a long p-region as the temperature sensor in an electronic thermometer. The current I is to be measured at a constant forward bias of 500 mV.

(a) Derive an expression for the fractional change in current $\Delta I/I$ resulting from a temperature change of ΔT.

(b) Evaluate this expression at $T = 300\,\mathrm{K}$. Assume that the diode current is dominated by recombination of electrons in the quasi-neutral p-region, and for simplicity, consider the temperature dependence to be due solely to those factors that have an exponential dependence on temperature.

6.8   Consider two diodes: A is an np GaAs homojunction diode, and B is an Np $\mathrm{Al_{0.3}Ga_{0.7}As/GaAs}$ diode. $N_E = 5 \times 10^{17}\,\mathrm{cm^{-3}}$ and $N_B = 10^{19}\,\mathrm{cm^{-3}}$ for both diodes.

Which diode has the wider depletion region at -1.25 V forward bias?

6.9 Consider an AlGaAs/GaAs Np^+ heterojunction with a p-type doping density of 10^{19} cm^{-3}. The Al mole fraction in the n-type material is 0.3, and $W_B = 50$ nm.

    Estimate the amount of electron quasi-Fermi-level splitting at a high value of forward bias.

6.10 Evaluate the emitter injection efficiency at a high value of forward bias for the heterojunction diodes of the two previous questions.

References

[1] C.G. Fonstad, *Microelectronic Devices and Circuits*, Appendix D, McGraw-Hill Inc., 1994. Online [http://hdl.handle.net/1721.1/34219].

[2] W. Shockley, *Electrons and Holes in Semiconductors*, D.Van Nostrand Co., Inc., 1950.

[3] D.L. Pulfrey and N.G. Tarr, *Introduction to Microelectronic Devices*, pp. 138–140, Prentice-Hall, 1989.

7 Solar cells

A solar cell is a large-area np-diode, designed to convert sunlight to electricity via the **photovoltaic effect**. Absorbed photons generate electron-hole pairs, which are separated within the diode, leading to a photocurrent. This current is directed to an external circuit where it develops a voltage across a load. Thus, there is conversion of optical power to electrical power. The elements of this conversion process are sketched in Fig. 7.1. Note that the top metal contact covers only a small fraction of the front surface so as to allow exposure of the semiconductor surface to the incident light.

In this chapter we look at the Sun as an electrical resource, and then consider the four features of the conversion process: absorption, generation, internal transport of photo-generated charges, development of photovoltaic power in an external circuit. The solar-cell example used in our treatment is a single-junction diode fashioned from a homogeneous, single-crystal semiconductor. Some other possibilities, which may be more practical, are then considered: multi-crystalline Si homojunction cells; thin-film compound-semiconductor heterojunction cells; multi-junction, multi-semiconductor tandem cells. We conclude with a brief discussion of the prospects for photovoltaics being widely used for terrestrial electric-power generation.

7.1   The Sun as an electrical resource

The Sun's radiant energy comes from the fusion reaction:

$$^2H_1 + {}^2H_1 \rightarrow {}^3He_2 + {}^1n_0 + E.   \tag{7.1}$$

The Earth 'circles' the Sun at a mean centre-centre distance of 149×10^6 km. We are fortunate that this distance is so large because it allows the power density to decrease from $\approx 6 \times 10^4\,\text{kW m}^{-2}$ at the Sun's surface to $\approx 1.35\,\text{kW m}^{-2}$ at the top of the Earth's atmosphere. After that, further reductions occur due to scattering of light by air molecules, aerosols and particulate matter, and by absorption by ozone, water, oxygen, and the 'greenhouse gas' carbon dioxide. The intensity of sunlight at a specific point on Earth depends on the times of day and year, and on the orientation of the irradiated surface. For standardization purposes, the solar spectrum at the Earth's surface is taken to be that when the Sun is at an angle of $\Theta = 48.19°$ from its zenith, and is incident on a south-facing surface (in the Northern Hemisphere) that is mounted at $37°$ to the

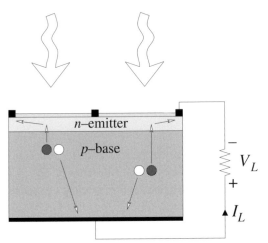

Figure 7.1 Schematic of the photovoltaic effect, illustrating how sunlight is converted to electricity via absorption, electron-hole pair generation and separation. The thin layer between the top contacts is an anti-reflection coating.

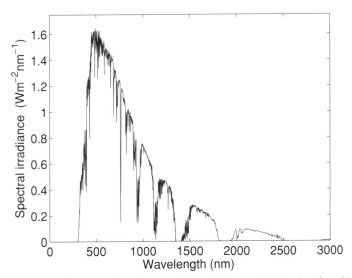

Figure 7.2 Solar spectral irradiance, global, AM1.5 on 37° tilted surface, Northern Hemisphere. Drawn from data in the ASTM G173-03e1 document [1], which can be purchased from ASTM.

horizontal.[1] The secant of the former angle is a measure of the path-length of the Sun's rays through the atmosphere. The actual number is known as the **Air Mass Number**, leading to the label AM1.5 for the angle cited. The solar spectral irradiance for these standard conditions is shown in Fig. 7.2. This spectrum is known as AM1.5G, where 'G' stands for global, i.e., the diffuse radiation from scattering by clouds and from reflections by the ground is added to the direct radiation from the Sun. Under these conditions,

---

[1] These angles have been chosen as reasonable yearly averages for the 48 contiguous states of the USA.

about 30% of the Sun's energy is lost in passing through the atmosphere, leading to a surface power density of $0.97 \, \text{kW m}^{-2}$. This number is rounded-up to $1 \, \text{kW m}^{-2}$ for the purposes of testing solar cells and listing their performance in product literature. The goal of solar cell design is to convert as much of this power as possible to electricity.

7.2   Absorption

Some of the light from the Sun that is incident on a solar cell is reflected, and some is transmitted into the device. In this section we consider the latter, and we write the spatial part of the electric-field component of sunlight's transverse electromagnetic wave in the form used in Chapter 2 for electron waves:

$$\mathcal{E}_y(x) = \mathcal{E}_0 e^{ikx} , \tag{7.2}$$

where the wavenumber $k = 2\pi/\lambda_{\text{semi}}$ in a semiconductor is given by

$$k = \frac{\omega}{c/n^*} = \frac{\omega(n_r + ik_r)}{c} \equiv \frac{2\pi(n_r + ik_r)}{\lambda} , \tag{7.3}$$

where c and λ are the velocity and wavelength of light in free space, respectively, and n_r and k_r are the real and imaginary parts, respectively, of n^*, the refractive index of the semiconductor. Substituting for k in (7.2) gives

$$\mathcal{E}_y(x) = \mathcal{E}_0 e^{i2\pi n_r x/\lambda} e^{-2\pi k_r x/\lambda} . \tag{7.4}$$

From this equation it is clear that the attenuation of the electromagnetic wave is related to the presence of an imaginary part in the refractive index of the semiconductor.

We are interested in the optical power density (W m^{-2}), and in electromagnetics this is given by the **Poynting vector**

$$\vec{S} = \vec{\mathcal{E}} \times \vec{H} , \tag{7.5}$$

where H is the magnetic field intensity, and the symbol S is consistent with that used for the energy-density flux in formulating the Hydrodynamic Equations in Section 5.2.2.[2] For a uniform plane wave, travelling in the x-direction, the magnitude of the power density is

$$S_x = \mathcal{E}_y H_z = \sqrt{\frac{\epsilon}{\mu}} |\mathcal{E}_y|^2 = \sqrt{\frac{\epsilon}{\mu}} \mathcal{E}_y(x)\mathcal{E}_y^*(x) = \sqrt{\frac{\epsilon}{\mu}} \mathcal{E}_0^2 e^{-\alpha x} , \tag{7.6}$$

where ϵ and μ are the permittivity and permeability of the semiconductor, respectively, and the **absorption coefficient** α is given by

$$\alpha = \frac{4\pi k_r}{\lambda} . \tag{7.7}$$

The absorption coefficients of some semiconductors used in the manufacture of solar cells are given in Fig. 7.3. Note how α for the indirect bandgap material Si increases

---

[2] Recall that the units of energy-density flux ($\text{J m}^{-2}\text{s}^{-1}$) are those of power density (W m^{-2}).

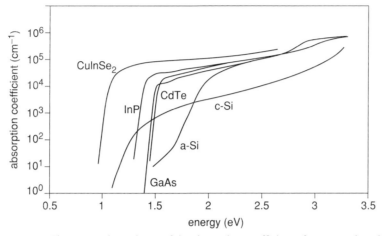

Figure 7.3 The energy dependence of the absorption coefficient of some semiconductors commonly used for solar cells. 'c-Si' is crystalline silicon, and 'a-Si' is amorphous silicon. From Markvart and Castener [2], © Elsevier 2005, Oxford, U.K., reproduced with permission.

less rapidly above its bandgap energy than does α for the other semiconductors shown, which are direct bandgap materials.

7.3   Generation

The power density S at some free-space wavelength λ can be translated into the flux of photons at that wavelength by dividing by the photon energy hc/λ. Let us call this photon flux $\Phi(\lambda)$; it has units of photons/m^2/s. From (5.8), the balance equation for monochromatic photons within the semiconductor can be written as

$$\frac{\partial P(\lambda)}{\partial t} + \nabla \cdot \Phi(\lambda) = R_{\mathrm{rad}}(\lambda) - \mathcal{A}(\lambda), \tag{7.8}$$

where P is the volumetric photon density, R_{rad} is the generation rate of photons within the semiconductor due to the radiative recombination of previously generated electron-hole pairs, and $\mathcal{A}$ is the rate of loss of photons due to absorption of photons within the volume. For a solar cell, we would like all of the absorbed photons to create separable electron-hole pairs, i.e., $\mathcal{A}(\lambda) = G_{\mathrm{op}}(\lambda)$, the generation rate discussed in Section 3.1.2. In practice, unwanted absorption mechanisms that could arise are: G_{fc}, excitation of already-free carriers (see Fig. 7.4a); and G_{ex}, generation of electron-hole pairs that remain bound together by Coulombic attraction (see Fig. 7.4b). The bound electron-hole pair is called an **exciton**. It is unwanted in a solar cell because it moves through the device as a neutral entity, so it does not contribute to the current. Thus

$$\mathcal{A} = G_{\mathrm{op}} + G_{fc} + G_{ex}, \tag{7.9}$$

from which an **internal quantum efficiency** can be defined:

$$G_{\mathrm{op}} = \eta_{\mathrm{int}}\mathcal{A}. \tag{7.10}$$

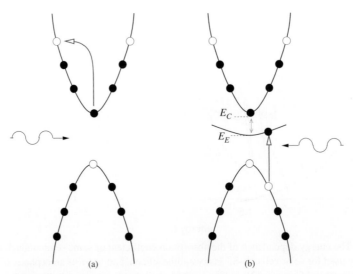

Figure 7.4 Unwanted results of photon absorption. (a) Free-carrier absorption. (b) Exciton formation. Coulombic attraction to the hole prevents the excited electron from gaining the conduction band. E_E can be viewed as the first excited state of a one-electron system. $(E_C - E_E)$ is the exciton binding energy; if it is much less than $k_B T_L$, then the exciton will be short-lived.

In steady-state, for a beam of sunlight travelling in 1-D, an expression for G_{op} can be obtained from (7.10), (7.8), and (7.6):

$$G_{op}(\lambda) = \eta_{int}(\lambda)\alpha(\lambda)\Phi_0(\lambda)e^{-\alpha x} , \qquad (7.11)$$

where $\Phi_0 = \sqrt{\epsilon/\mu}\mathcal{E}_0^2/hc/\lambda$ is the photon flux in $m^{-2}s^{-1}$ as it enters the solar cell.

Summing the monochromatic generation rates in silicon for the AM1.5G spectrum, Fig. 7.5 is obtained. Note the high generation rate near the surface of the cell. The roll-off is exponential for all wavelengths, but is particularly pronounced at shorter wavelengths where α is high. This has a big impact on the generation profile because the peak photon density occurs at the relatively short wavelength of about 500 nm (see Fig. 7.2).

7.4   Photocurrent

The current resulting from the photon absorption is called a photocurrent; to create it, the photo-generated electron-hole pairs must be separated, and made to flow in opposite directions. This is achieved by having a built-in field within the structure, and is most obviously implemented by making a *pn* diode. The major questions that need to be answered for the design of this diode are: how deep into the device should the metallurgical junction be placed; should the top part of the diode be *p*- or *n*-type; what should the doping densities be for the two regions?

The answers to these questions can be obtained by scrutinizing Fig. 7.5. Because the generation rate decays exponentially on moving into the cell, then the junction should be

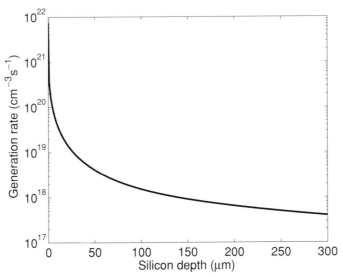

Figure 7.5 Generation-rate profile in Si for AM1.5G sunlight. Evaluated from (7.11) using spectral irradiance data from Fig. 7.2, and absorption coefficient data from Ref. [3].

placed very close to the front surface. Thus, the top part of the semiconductor, which is called the **emitter**, is thin. Recall that the current must take a lateral path to the metallic contacts on the front surface (see Fig. 7.1), so the emitter must be heavily doped to avoid unwanted, parasitic series resistance. The other, lower part of the diode is called the **base**; it needs to be relatively thick to collect the long-wavelength photons, and to provide mechanical strength for the solar cell. Minority carriers generated in this region may have to travel relatively long distances to the junction, so we should choose the doping type of the base to give us the more mobile minority carrier. For major solar cell materials such as Si and GaAs, this means that the base should be p-type (see Fig. 5.3). Hence, a solar cell is a shallow-junction, n^+p diode, where the superscript '$+$' means 'heavily doped'.

The energy band diagram for the solar cell is shown in Fig. 7.6. Because the generation rate decreases with distance into the cell, both electrons and holes in both of the quasi-neutral regions diffuse to the right. In the emitter, separation of the carriers occurs at the junction, where the electrons are reflected by the built-in potential barrier. To achieve the same effect for carriers generated in the base, a **back-surface field** is created by heavily doping the end of the p-type region. In fact, there is also a field at the front surface of the cell, often due to the donor density gradient, which occurs naturally during the diffusion doping process. In the remaining region of the cell, the space-charge region, carriers generated therein are automically separated by the built-in junction field.

7.4.1 Surface recombination velocity

To characterize the effect of the surface fields on minority carriers, the concept of a **surface recombination velocity** is used [4]. To paraphrase Shockley, the creator of this

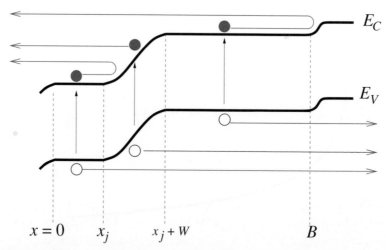

Figure 7.6 Energy-band diagram for a solar cell showing the separation of electron-hole pairs that are generated in each of the three major regions of the device: the emitter, the space-charge region, and the base. Minority-carrier-reflecting fields at the front and back surfaces, due to n^+n- and pp^+-junctions, respectively, are also shown.

concept: the rate of recombination of electrons at the surface is the same as if a flux of electrons of density $(n - n_0)$ were drifting with an average velocity S into the surface and being removed. Thus, in the quasi-neutral base at $x = B$ (see Fig. 7.6), the electron current is diffusive, and we have

$$-D_e \frac{d}{dx}(n - n_0) = -D_e \frac{dn}{dx} \equiv S(n - n_0), \qquad (7.12)$$

where S is the surface recombination velocity: it is always positive and has units of m/s.

Two extreme values of S neatly characterize **ohmic** and **blocking** contacts at some boundary $x = B$:

Ohmic: $S = \infty$, $n(B) = n_0$, but J_e is finite,
Blocking: $S = 0$, $n(B) > n_0$, but J_e is zero.

In solar cells we would like to have blocking contacts for the minority carriers so that electrons in the base and holes in the emitter are reflected towards the metallurgical junction.

7.4.2   Emitter photocurrent

To derive an expression for the component of the photocurrent due to generation in the emitter quasi-neutral region, we focus on the diffusion of minority carrier holes. Further, we assume that there are no large gradients in the kinetic energy per carrier. Thus, from our master set of equations (5.24), we have, at steady-state and for the region $0 < x < x_j$

identified on Fig. 7.6

$$0 = -\frac{1}{q}\frac{d J_h^E}{dx} + G_{op} - \frac{p - p_0}{\tau_h}$$

$$J_h^E(x) = -k_B T_L \mu_h \frac{dp}{dx} \equiv -q D_h \frac{dp}{dx}. \tag{7.13}$$

These equations are to be solved with the boundary conditions

$$D_h \frac{dp}{dx}\bigg|_{x=0} = S_F(p(0) - p_0)$$

$$p(x_j) = p_0, \tag{7.14}$$

where S_F is the front-surface recombination velocity, and the second boundary condition implies that all excess minority-carrier holes reaching the edge of the space-charge region are swept across the junction by the built-in field. Using (7.11) for G_{op} and assuming an internal quantum efficiency of 100%, the solution for monochromatic light of wavelength λ is

$$J_h^E(\lambda, x_j) = \frac{q \Phi_0 \alpha L_h}{\alpha^2 L_h^2 - 1}\left[- \alpha L_h e^{-\alpha x_j} \right.$$

$$\left. + \frac{H_h + \alpha L_h - e^{-\alpha x_j}(H_h \cosh Q_h + \sinh Q_h)}{H_h \sinh Q_h + \cosh Q_h} \right], \tag{7.15}$$

where $Q_h = x_j/L_h$, $H_h = S_F L_h/D_h$, with L_h being the hole, minority carrier diffusion length.

7.4.3   Base photocurrent

The photocurrent arising from the minority carrier electrons generated in the base can be derived in a similar manner to that described for the hole current in the emitter. The answer is

$$J_e^B(\lambda, x_j + W) = \frac{q \Phi_0 \alpha L_e e^{-\alpha(x_j + W)}}{\alpha^2 L_e^2 - 1}$$

$$\times \left[\alpha L_e - \frac{H_e(\cosh Q_e - e^{-\alpha B'}) + \sinh Q_e + \alpha L_e e^{-\alpha B'}}{H_e \sinh Q_e + \cosh Q_e} \right], \tag{7.16}$$

where $B' = B - (x_j + W)$, $Q_e = B'/L_e$, $H_e = S_B L_e/D_e$.

7.4.4   Space-charge-layer photocurrent

In the space-charge layer we can reasonably assume that all carriers generated therein are separated by the built-in field. Thus, in deriving the current due to photogeneration in this region we can consider either electrons or holes. The derivation is much simpler than for the photocurrents in the quasi-neutral regions because we can assume that the carriers are swept out of the space-charge region so quickly that there is no opportunity for recombination to occur. Thus, using electrons for our derivation, the relevant equation from our master set reduces to

$$0 = \frac{1}{q}\frac{dJ_e^D}{dx} + G_{op} = 0, \tag{7.17}$$

where the superscript 'D' indicates that we'll use the Depletion Approximation to estimate the width W of the space-charge region. Thus, integrating over $x_j < x < (x_j + W)$ and, once again, assuming perfect internal quantum efficiency, the result is

$$J_e^D(\lambda, x_j) = q\Phi_0 e^{-\alpha x_j}\left[1 - e^{-\alpha W}\right]. \tag{7.18}$$

7.4.5   Total photocurrent

In the above derivations, two photocurrents are specified at $x = x_j$, and the other (J_e^B) is specified at $x = (x_j + W)$. However, we can reasonably assume that J_e^B has the same value at $x = x_j$, because we don't expect any recombination to occur as the electrons are swept across the space-charge layer. Thus, we can add-up our three components of current to get the total photocurrent density J_{Ph} at any particular wavelength:

$$J_{Ph}(\lambda) = J_h^E(\lambda, x_j) + J_e^D(\lambda, x_j) + J_e^B(\lambda, x_j + W). \tag{7.19}$$

By summing over the wavelength range of terrestrial sunlight, we get the total photocurrent

$$J_{Ph} = \sum_{AM1.5G} J_{Ph}(\lambda). \tag{7.20}$$

An example of the three regional components of the spectral photocurrent for a Si solar cell is shown in Fig. 7.7. The specifications for the cell are listed in the caption to the figure. The total photocurrent density in this example is $39\,\text{mA}\,\text{cm}^{-2}$. It may surprise you to learn that the largest contribution comes from the region furthest from the front surface.

This calculated value of J_{Ph} is very close to the present (2009) world record of $42.2\,\text{mA}\,\text{cm}^{-2}$ for a Si solar cell [6]. We assumed the favourable conditions of no reflection of sunlight from the front surface of the cell, and 100% internal quantum efficiency for the conversion of light entering the cell. The fact that the experimental value of J_{Ph} is even higher than our calculated value indicates that the practical cell has outstanding minority-carrier- and photon-collection-properties. The latter is achieved by having a double-layer anti-reflection coating and by etching the front surface to give it an 'egg carton'-like topography (see Fig. 7.8). Sunlight hitting one sloping side is partially transmitted into the cell and partially reflected to the opposite sloping side, offering the light another opportunity for transmission into the cell. On the front and back semiconductor surfaces of the practical cell a thin insulating layer of silicon dioxide has been grown to provide a blocking contact, and to passivate the surfaces to reduce

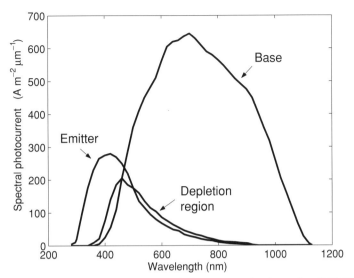

Figure 7.7 Components of the spectral photocurrent density under AM1.5G illumination for a Si solar cell with the following specifications: $x_j = 200\,\mathrm{nm}$, $B = 450\,\mu\mathrm{m}$, $N_D = 5 \times 10^{19}\,\mathrm{cm}^{-3}$, $D_h = 1.29\,\mathrm{cm}^2\,\mathrm{s}^{-1}$, $L_h = 7.2\,\mu\mathrm{m}$; $N_A = 1.5 \times 10^{16}\,\mathrm{cm}^{-3}$, $D_e = 27\,\mathrm{cm}^2\,\mathrm{s}^{-1}$, $L_e = 164\,\mu\mathrm{m}$, $\mathcal{S}_F = 1\,\mathrm{m\,s}^{-1}$, $\mathcal{S}_B = \infty$. Courtesy of Garry Tarr, ex-UBC.

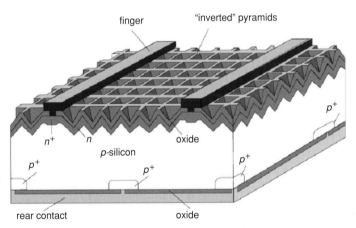

Figure 7.8 Schematic of the solar cell structure of the present holder of the world-record efficiency for a Si cell. From Jhao *et al.* [5], © 1997 IEEE, reproduced with permission.

recombination[3]. The minority carriers are collected through small 'windows' in the oxide via heavily doped regions, which form the surface-field regions discussed above. It is probable that the minority-carrier lifetimes in the practical cell are longer than the

---

[3] At the surface of a crystalline semiconductor there are bonds that are incomplete because of the interruption to the periodicity of the structure. This creates localized energy levels within the bandgap that serve as traps to facilitate electron-hole recombination. The completion of the bonds, often by hydrogen incorporated in the oxide, reduces the number of recombination sites, and the surface is said to be **passivated**.

values we have used in Fig. 7.7, giving further reason for the superior performance of the real cell.

7.5   Photovoltage

The electrons and holes constituting the photocurrent flow in the direction of a reverse bias current. When this current is led to an external circuit containing some load, R_L for example, a voltage is developed across the load such as to forward bias the diode, thereby generating a current that opposes J_{Ph}. Thus, the price paid for generating a voltage is the loss of some net current.

The question now arises: can the forward-bias current be estimated using a diode equation of the general form developed for the ideal 'dark' diode in Section 6.6? In other words, can the current due to an external bias applied to an unilluminated diode be simply added to the photocurrent to give the total current of a self-biased illuminated diode? The answer is 'yes', providing the illumination is not so intense as to (a) cause the quasi-Fermi levels in the space-charge region to deviate significantly from constant values, and (b) cause the minority carrier lifetime to change from its value 'in the dark'. It turns out that these provisos are met for unconcentrated sunlight at the Earth's surface [7]. So the diode dark current can be computed using the boundary conditions from (6.29) at the edges of the space-charge region, and a boundary condition like (7.12) at the contact-ends of the quasi-neutral regions. For example, for electrons injected into the base, we have

$$n(x_j + W) = n_{0p} e^{V_{Lj}/V_{th}}$$

$$-D_e \frac{dn}{dx}\bigg|_{x=B} = S_B(n(B) - n_0), \qquad (7.21)$$

where S_B is the back-surface recombination velocity, i.e., at $x = B$, and V_{Lj} is that part of the load voltage that is dropped across the junction: in the absence of series resistance we can take it to be the load voltage V_L. The result for the dark current due to electrons emitted into the base is of the general form

$$J_{e,D} = J_{0,e}[e^{V_L/V_{th}} - 1], \qquad (7.22)$$

where the subscript D identifies the dark current, and the saturation current density has been written with a subscript that is different from the ideal-diode case in order to recognize that different boundary conditions are appropriate. A similar expression can be derived for the holes injected across the forward-biased junction into the emitter. If we ignore any current change in the space-charge region due to recombination in that region, the total dark current is the sum of the two injected currents, and is of the form

$$J_D = J_0[e^{V_L/V_{th}} - 1]. \qquad (7.23)$$

Applying superposition, the load current J_L is simply

$$J_L = J_{Ph} - J_D, \qquad (7.24)$$

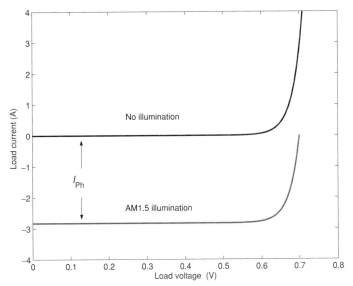

Figure 7.9 Solar cell *I-V* characteristic, illustrating the superposition of the photocurrent and the usual, exponential diode current. The polarity of the load current has been chosen to emphasize that the solar cell generates power ($IV < 0$). The load absorbs power, of course, ($IV > 0$).

and the output characteristic of the solar cell is just a displaced exponential, as illustrated in Fig. 7.9.

When the dark current exactly negates the photocurrent, we have effectively open-circuit conditions, and the load voltage is known as the **open-circuit voltage** V_{oc}. Thus,

$$V_{oc} = V_{\text{th}} \ln \frac{J_{\text{Ph}} + J_0}{J_0} . \qquad (7.25)$$

To examine the limits of V_{oc}, let us write

$$J_0 = C_E p_{n0} + C_B n_{p0} = \left(\frac{C_E}{N_D} + \frac{C_B}{N_A} \right) n_i^2 , \qquad (7.26)$$

where C_E and C_B include the minority-carrier properties of diffusion length and surface recombination velocity for the emitter and base, respectively, and N_D and N_A are the appropriate doping densities. The intrinsic carrier concentration n_i depends exponentially on the bandgap, so we can rewrite J_0 as

$$J_0 = C e^{-E_g / k_B T} , \qquad (7.27)$$

where the constant C includes the bracketed terms in (7.26) and the effective densities of states from (4.19). Thus, the open-circuit voltage becomes

$$V_{oc} \approx \frac{E_g}{q} - V_{\text{th}} \ln \left(\frac{C}{J_{\text{Ph}}} \right) , \qquad (7.28)$$

from which it is clear that the bandgap imposes a limit on V_{oc}, and that this limit is more closely realized for larger photocurrent densities.

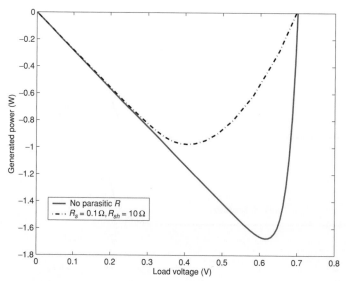

Figure 7.10 Power-voltage characteristic for a 10-cm diameter Si cell with $J_{Ph} = 40$ mA cm^{-2} and $V_{oc} = 700$ mV. The current and voltage at the point of maximum power generation are commonly called I_{mp} and V_{mp}, respectively. The presence of parasitic resistance in the cell affects the maximum power point, as illustrated here. R_s is series resistance and R_{sh} is shunt resistance.

7.5.1   Photovoltaic power

From Fig. 7.9, we see that for $0 < V_L < V_{oc}$, the solar cell's I-V curve is in the 4th quadrant, where the current is negative and the voltage is positive. Thus, the power is negative, indicating generation, rather than the more usual dissipation. The power characteristic is shown in Fig. 7.10. The power at the point of maximum power generation is P_{mp}; it is expressed as a density in W/m^2, and is given by

$$P_{mp} = J_{mp}V_{mp}$$
$$\equiv FF\,J_{sc}V_{oc}\,, \tag{7.29}$$

where J_{mp} and V_{mp} are the current density and voltage at the maximum power point and J_{sc} is the current at short circuit. The presence of parasitic resistances in the cell can cause $|J_{sc}|$ to be less than $|J_{Ph}|$, as Fig. 7.11 illustrates. Series resistance can come from the vertical path taken by the current through the thick base region, and from the horizontal path shown on Fig. 7.1 of the current through the thin emitter. The solar cell, being an inherently low-voltage device, is particularly sensitive to series resistance. Shunt resistance arises from internal imperfections at the junction, and from any conductivity along the vertical edges of the device. The former is taken to be more important in the equivalent-circuit representation of the solar cell in Fig. 7.12, as is evident from the placement of R_{sh}.

Equation (7.29) defines the **fill-factor** FF. This factor is somewhat less than unity because the exponential form of the solar cell's I-V characteristic is only an approximation to the perfect characteristic that would have $J_L = J_{sc} = J_{Ph}$ for all $V_L < V_{oc}$. In other words, FF recognizes that a real cell makes the transition from $J_L = J_{sc}$ to $J_L = 0$ in a

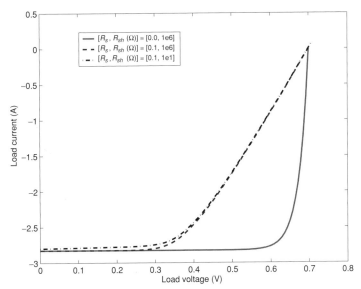

Figure 7.11 *I-V* characteristics when parasitic resistances are present. Same cell parameters as in Fig. 7.10. See the caption to Fig. 7.9 regarding the choice of sign for the current.

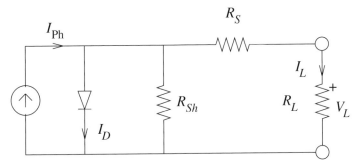

Figure 7.12 Solar cell equivalent circuit.

less abrupt manner. This 'softness' of the diode characteristic becomes less important as the bounding area $J_{Ph}V_{oc}$ is increased. Unfortunately, if one seeks to improve V_{oc} by choosing a material of high bandgap, then J_{Ph} will be reduced because of the increased transparency of the material to sunlight. The result of these competing requirements for high V_{oc} and high J_{Ph} is that the **photovoltaic conversion efficiency** η_{pv} peaks at some value of E_g. The results of detailed calculations are shown in Fig. 7.13, and the efficiency is given by

$$\eta_{pv} = \frac{FF\,J_{sc}V_{oc}}{S_{AM1.5G}}. \tag{7.30}$$

The peak in the maximum-efficiency curve occurs around a bandgap of $E_g \approx 1.4\,\text{eV}$, but the peak is quite broad, so silicon is not far from being the theoretically optimum material for a solar cell. This fact, coupled with the mature state of silicon-device processing has led to near-maximum efficiency Si solar cells being realized in practice

Table 7.1 World-record values (as of 2008) for photovoltaic conversion efficiencies, as measured in prototype solar cells under unconcentrated AM1.5G illumination. Si data from J. Zhao *et al.* [6]; CIGS data from I. Repins *et al.* [9]; tandem cell data from R.R. King *et al.* [10].

Semiconductor	E_g (eV)	J_{sc} (mA cm^{-2})	V_{oc} (V)	FF	η_{pv} %
Si	1.12	42.2	0.706	0.828	24.7
CIGS	1.04–1.7	35.4	0.690	0.812	19.9
GaInP/GaInAs/Ge	1.8/1.3/0.7	16.0	2.392	0.819	31.3

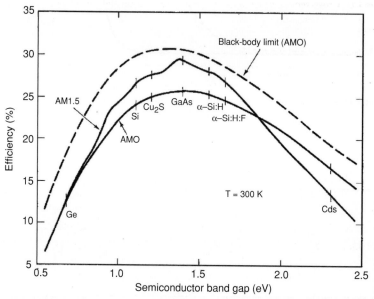

Figure 7.13 Estimates of maximum efficiency vs. bandgap under various illumination conditions. α-Si refers to amorphous silicon, from which thin-film solar cells can be made. Presently (2009), CIGS is a more promising thin-film material; its bandgap is in the range 1.4–1.7 eV (see Section 7.6.1). From Green [8], © Martin A. Green, reproduced with permission.

(see Table 7.1). In addition to its favourable theoretical properties and highly developed processing technology, silicon has the added advantage for a large-area device of being an abundant material.[4] Despite these attributes, solar-cell systems utilizing single-crystal solar cells are still quite expensive. The cost is largely due to the expense of purification and of growth of large-area single crystals. Many commercial silicon solar cells presently use multicrystalline material, which comprises crystallites of size $\approx 0.1 - 10$ cm in wafers that are sawn from large blocks of cast silicon, rather than being cut from Czochralski-pulled-ingots. The efficiency of experimental **multicrystalline Si** solar cells has reached $\approx 20\%$.

---

[4] Silicon is the second most abundant material in the earth's crust; it ranks below oxygen.

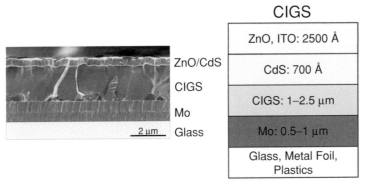

Figure 7.14 Cross-section of a CIGS solar cell. From Noufi and Zweibel [11], © 2006 IEEE, reproduced with permission.

7.6   Non-silicon solar cells

Two approaches to possibly economically viable solar cells involve **thin-film** cells or **tandem-junction** cells.

7.6.1   Thin-film solar cells

For many years the dream of thin-film solar cells that could be deposited *in situ* onto large-area surfaces has been pursued. Presently, an interesting contender is copper indium gallium diselenide, CIGS. Depending on the ratio of In to Ga in the material, the bandgap lies somewhere between 1.04 eV (no gallium) to 1.7 eV (no indium), i.e., it spans the optimum range shown in Fig. 7.13. Substitutional doping in this ternary compound is not easy to achieve, so excess carriers are created by deliberately encouraging the incorporation of vacancies in the deposited film. For example, if the growth conditions are adjusted so that there is a deficiency of Cu in the CIGS layer, then the Se atoms surrounding the Cu vacancies will be lacking in electrons. Thus, Se accepts electrons from elsewhere, and the material becomes *p*-type.

The reaction is described by

$$V_{Cu}^0 = V_{Cu}^- + h^+ \quad : \quad E_a = 0.03 \, \text{eV} \,, \tag{7.31}$$

where V_{Cu} denotes a copper vacancy, and the activation energy E_a is small. Unfortunately, formation of an *np*-junction is not easy because the different types of defect that would be needed tend to compensate each other. Thus, solar cells made from CIGS have a *p*-type base and use another semiconductor for the *n*-type emitter (see Fig. 7.14). The diode is, therefore, a heterojunction. The current in this type of junction is discussed in Chapter 9. In CIGS solar cells the *n*-type semiconductor has a large bandgap and, therefore, does not absorb much sunlight. Instead, it acts as a 'window' to allow the solar energy to penetrate directly into the absorbing CIGS film.

As Table 7.1 indicates, impressive efficiencies have already been obtained in laboratory specimens. There is presently much speculation about whether this performance

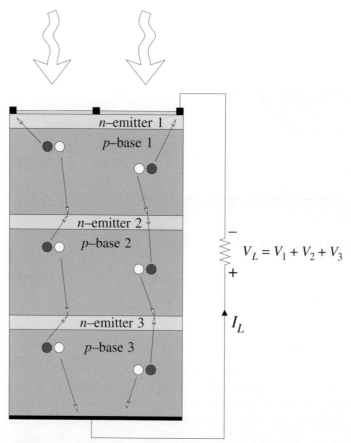

Figure 7.15 Tandem-junction solar cell concept. In this example, three solar cells of different materials are connected in series. The bandgaps are $E_{g1} > E_{g2} > E_{g3}$.

can be transferred to large-area cells and make photovoltaic power cost-competitive with conventional power.

7.6.2   Tandem-junction cells

In a tandem-junction solar cell, two or more solar cells of different materials are stacked together and connected serially (see Fig. 7.15). The bandgap of the materials decreases from top to bottom, so that photons that would normally pass through a high-bandgap cell and be lost are captured in the lower cells. In this way, more of the solar spectrum is absorbed, and it is absorbed more efficiently. For example, a 2 eV photon will be absorbed in both a semiconductor with $E_g = 1$ eV and a semiconductor with $E_g = 2$ eV. However, in the former case, the extra 1 eV of photon energy will be turned into heat because phonons will be emitted as the excited electron scatters and loses energy (much like the situation shown in Fig. 3.9b for Auger recombination). The rise in temperature will increase the dark current, thereby reducing V_{oc} and, consequently, η_{pv}.

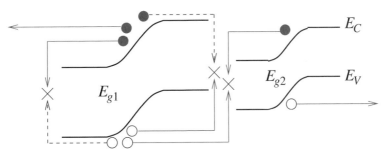

Figure 7.16 Two cells in tandem. Solid lines indicate the paths of the photogenerated carriers. In this example, more photogeneration occurs in the first cell, which has a bandgap E_{g1}. A dark current (dashed lines) is generated in this cell to reduce the net current to that of the photocurrent of the second cell. Recombination occurs at the sites marked with crosses.

Because of the series electrical connection of tandem cells, V_{oc} is additive, but J_{Ph} is limited to the value for that of the cell within the stack that generates the least photocurrent. Clever design would arrange for the thicknesses of the various parts of each cell to be such that J_{Ph} is the same in each cell. If this is not the case, then the cell producing the higher photocurrent becomes forward biased, so that the net current in each cell is forced to be the same (see Fig. 7.16).[5]

The cost of producing a stack of matched, single-crystal solar cells is likely to be relatively high, so one possible practical implementation would be to use only small-area cells, and then to collect the sunlight from a wider area using a concentrating lens. Such an arrangement would boost J_{Ph}, which, in turn, could lead to associated improvements in V_{oc} and FF. For example, the tandem cell listed in Table 7.1, when operated under a solar intensity equivalent to 240 Suns, yet kept cool at 25°C, shows improvements in V_{oc} and FF to 2.911 V and 0.875, respectively. The efficiency is increased to 40.7%.

7.7 Prospects for terrestrial photovoltaic power generation

The world's consumption of electricity is expected to double over the period 2004–2030, reaching an annual consumption close to 30,000 TWh. Where is all this electricity going to come from?

In 2006, the world's generating mix was as shown in Table 7.2. The fossil fuels (coal, oil, and gas) form the major generating source. These fuels do not naturally regenerate on a time-scale that would permit them to be called renewable sources of energy. Their resources are finite and diminishing. It is debatable whether oil and gas should even be merely burned to raise steam to generate electricity, rather than be used for some other purposes for which they are more uniquely suited, e.g., transportation fuel. Coal-fired

---

[5] This auto self-biasing of cells also occurs in series-connected strings of ordinary solar cells in a module if a shadow falls across one cell. The non-shaded cells develop a forward bias to reduce the net current. This undesirable feature can be protected against at the extra cost of incorporating by-pass diodes in the photovoltaic module.

Table 7.2 The installed capacity of various electricity-generating sources, expressed as a percentage of the world total. From the International Energy Agency [12].

Coal	Oil	Gas	Nuclear	Hydro	Renewables
40	7	20	16	15	2

electricity generation is growing enormously because of the abundant deposits in the rapidly developing nations of China and India. Coal plants are relatively cheap, due in part to the undesirable fact that their waste product (which contains a lot of CO_2) is usually dumped into the atmosphere without treatment. Regulations governing the disposal of waste products from nuclear generators are, thankfully, more stringent. However, nuclear-power generation is not without some well-known hazards. Hydro power is the most benign of the present generating methods, but many parts of the planet are not blessed with the rainfall and the terrain to facilitate the proliferation of this power source.

In view of the above shortcomings of conventional power sources, photovoltaic power generation would seem to be an attractive proposition. However, in 2006, it was lumped-in with 'renewables other than hydro', such as biomass, wind, geothermal, wave and tidal, which provided only 2% of the world's electricity. Obstacles to the widespread utilization of photovoltaics include:

- vested commercial interests in existing generating methods;
- large real-estate requirements. The daily yield is unlikely to exceed $2 \, \mathrm{kWh \, m^{-2}}$ (full sun for 10 hours per day at 20% conversion efficiency).
- high cost of manufacturing solar cells of reasonable efficiency ($\approx 20\%$). Silicon is the most developed solar-cell semiconductor, and a major component of this cell's cost derives from the need to produce material of sufficient quality to obtain long minority-carrier lifetimes.
- the diurnal nature of terrestrial insolation. This means that solar power plants must be backed-up with some storage capability (probably batteries), or be operated in conjunction with a more consistent and controllable source of power.

One solar option that may be cost-effective to implement is **building-integrated photovoltaics**. In this scheme, houses and commercial buildings are designed with roofs and walls that have solar cells incorporated into their structures, rather than being superficial additions. Each building would then become a distributed generating site, and would be under the control of the local electricity authority. The solar-generated electricity would not have to be used immediately by the owner of the host building, nor would it have to be stored in bulky batteries in the basement. The photovoltaic power would be transmitted around the grid, and used wherever it was needed. The solar power wouldn't necessarily be expected to meet the demand all of the time. Instead, it would add to power from a baseline source, such as hydro or nuclear.

To add significant amounts of photovoltaic power to the generation mix requires a big commitment. It is encouraging that the governments of at least two countries in the world, Germany and Japan, have begun to provide the incentives for citizens and industries to invest in photovoltaic power generation. Maybe you can start to lobby your government? Or perhaps you are too busy utilizing the material in this chapter to design a cheaper and more efficient solar cell?

Exercises

7.1 A solar cell, which operates in the dark like an ideal diode, is irradiated such that the optical generation rate G_{op} is uniform throughout the volume of the diode. Show that the photocurrent density is given by

$$J_{Ph} = q G_{op}(W + L_e + L_h),$$ (7.32)

where W is the depletion-layer width, and the L's are minority carrier diffusion lengths.

7.2 The E-k relationships for the conduction bands of two semiconductor materials, A and B, each with spherical constant-energy surfaces, can be expressed as

$$E_A - 0.7 = \alpha k^2 \quad \text{and} \quad E_B - 1.4 = 2\alpha(k - k')^2,$$

respectively, where α is a constant, $k' > 0$, and the energies are in units of eV. Both materials have the same valence-band structure, with the top of the valence band at $E = 0$ and $k = 0$.

Which material would make the better solar cell?

7.3 Fig. 7.7 shows the spectral photocurrent density for a Si np-junction solar cell with a base doping density of $N_A = 1.5 \times 10^{16}$ cm^{-3}. The particular values used for the basewidth and the electron back surface recombination velocity are $W_B = 450$ μm and $S_B = \infty$, respectively.

Would there be any significant effect on the photocurrent if the base properties were:

(a) $W_B = 450$ μm and $S_B = 0$?

(b) $W_B = 50$ μm and $S_B = \infty$?

7.4 Consider a silicon solar cell made from a wafer of diameter 10 cm. The top-contact metal covers 10% of the front-surface area. Under 1 Sun illumination, the photocurrent density is 40 mA cm^{-2}, and the open-circuit voltage is 0.7 V.

(a) Compute I_0, the diode saturation current.

(b) Plot the I-V characteristic under 1 Sun illumination.

(c) On a separate plot show the power-voltage characteristic.

(d) Evaluate the fill-factor FF and the conversion efficiency η_{pv}.

7.5 The emitter of the cell in the previous question is 200 nm thick and has a doping density of 5×10^{19} cm^{-3}. The top-contact grid pattern is such that the series resistance of the cell can be represented by a slice of the emitter material that is 14.66 μm long and 1 cm wide.

(a) Evaluate the series resistance R_s of the cell.

(b) Plot the new *I-V* characteristic for the cell on the same graph as for the cell with $R_s = 0$ from the previous question.

(c) Plot the new *P-V* characteristic for the cell on the same graph as for the cell with $R_s = 0$.

(d) Evaluate the new η_{pv}.

7.6 A photovoltaic module is made by connecting in series two of the cells from the previous question. Ignore series resistance.

A shadow falls across one of the cells so that 50% of its top surface is obscured. Plot the *P-V* characteristic of the module, and evaluate η_{pv}.

7.7 The situation described in the previous question leads to a loss of power at the load, but it also leads to the serious possibility of the shadowed cell burning out.

Explain why the temperature is likely to rise in the shadowed solar cell.

7.8 An ingenious approach to possibly reducing the cost of photovoltaic power generation using crystalline solar cells results in a Sliver© cell [13].

Imagine a conventional Si solar cell of diameter 10 cm and thickness 450 μm, with lines scribed on the front surface, 100 μm apart, to define 1000 strips. Dicing the cell into strips results in some kerf loss, so the final strips are 60 μm wide. These strips are then laid flat so that they can be exposed to solar radiation over the cross-section of the cell, e.g., at 90° to the direction shown in Fig. 7.1.

Evaluate the improvement in exposed front-surface area that this approach brings about.

References

[1] ASTM G173-03e1, *Standard Tables for Reference Solar Spectral Irradiance at Air Mass 1.5: Direct Normal and Hemispherical for a 37 Degree Tilted Surface.* Available from ASTM International, West Conshohocken, PA, [http://www.astm.org/Standards/G173.htm].

[2] T. Markvart and L. Castener, *Solar Cells: Materials, Manufacture and Operation*, Fig. 7, Elsevier, 2005.

[3] M.A. Green and M.J. Keevers, Optical Properties of Intrinsic Silicon at 300 K, *Progress in Photovoltaics: Research and Applications*, vol. 3, 189–192, 1995.

[4] W. Shockley, *Electrons and Holes in Semiconductors*, p. 321, D.Van Nostrand Co., Inc., 1950.

[5] Z. Jhao, A. Wang, P. Campbell and M.A. Green, 22.7% Efficient PERL Silicon Solar Cell Module with Textured Front Surface, *Proc. IEEE 26th Photovoltaic Specialists Conf.*, pp. 1133–1136, 1997.

[6] J. Zhao, A. Wang and M.A. Green, 24.5% Efficiency Silicon PERT Cells on MCZ Substrates and 24.7% Efficiency PERL Cells on FZ Substrates, *Progress in Photovoltaics: Research and Applications*, vol. 7, 471–474, 1999.

[7] N.G. Tarr and D.L. Pulfrey, An Investigation of Dark Current and Photocurrent Superposition in Solar Cells, *Solid-State Electronics*, vol. 22, 265–270, 1979.

[8] M.A. Green, *Solar Cells: Operating Principles, Technology and System Applications*, p. 89, Prentice-Hall, 1982.

[9] I. Repins, M.A. Contreras, B. Egaas, C. DeHart, J. Scharf, C.L. Perkins, B. To and R. Noufi, 19.9% Efficient ZnO/CdS/CuInGaSe$_2$ Solar Cell with 81.2% Fill Factor, *Progress in Photovoltaics: Research and Applications*, vol. 16, 235–239, 2008.

[10] R. R. King, D. C. Law, K. M. Edmondson, C. M. Fetzer, G. S. Kinsey, H. Yoon, R. A. Sherif and N. H. Karam, 40% Efficient Metamorphic GaInP/GaInAs/Ge Multijunction Solar Cells, *Appl. Phys. Lett.*, vol. 90, 183516, 2007.

[11] R. Noufi and K. Zweibel, High Efficiency CdTe and CIGS Thin-film Solar Cells: Highlights and Challenges, *IEEE 4th World Conference on Photovoltaic Energy Conversion*, pp. 317–320, 2006.

[12] International Energy Agency, *World Energy Outlook, 2006*, ISBN: 92 64 10989 7. (The IEA is the recognized authority on global energy growth; it publishes an outlook annually.)

[13] K.J. Weber, A.W. Blakers, P.N.K. Deenapanray, V. Everett and E. Franklin, Sliver © Solar Cells, *Proc. IEEE 31st Photovoltaic Specialists Conf.*, pp. 991–994, 2005.

8 Light-emitting diodes

The light-emitting diode (LED) is a *pn*-junction diode in which radiative recombination is encouraged to occur under forward-bias operating conditions. Thus, there is conversion of electrical energy to optical energy. In essence, the LED is the complement of the solar cell. In the example shown in Fig. 8.1, the internally generated photons escape through the top surface, which cannot, therefore, be covered entirely by the top metallic contact.

In the first part of this chapter, we develop an understanding of the LED by considering a number of efficiencies that relate to the various stages of the conversion of electrical energy to optical energy:

- voltage efficiency. This relates the applied voltage to the bandgap of the semiconductor. The latter would be chosen to obtain the desired colour of emitted light;
- current efficiency. This relates the current due to recombination in the desired part of the device to the current due to recombination elsewhere. Consideration of this efficiency leads to the heterostructural design that is a feature of modern, high-brightness LEDs;
- radiative efficiency. This relates to the relative amounts of radiative recombination and unwanted, non-radiative radiation. It leads to material selection (direct bandgap), and specifications on doping and purity;
- extraction efficiency. This relates to getting the photons out of the semiconductor in which they are generated. Consideration of this efficiency largely determines the substrate, contacts, and, in some cases, the shape of the device.
- wall-plug efficiency. This describes the overall efficency of the electrical-to-optical conversion process.

We then go on to discuss the features of white-light LEDs, and conclude with a short appraisal of the prospects of LEDs making an impact in the area of general lighting.

8.1 Voltage efficiency

At equilibrium, we know that electrons tend to reside in states near the bottom of the conduction band, and that holes preferentially reside near the top of the valence band. These dispositions will be altered somewhat by the application of a forward bias to a *pn*-junction, but we can expect that the energy of any photons resulting from radiative

Table 8.1 Bandgap and emitted-light properties for two material systems used for high-brightness LEDs. The bandgap is increased by increasing the mole fraction x of Ga in the GaInN system and of Al in the AlGaInP system.

Semiconductor	E_g (eV)	λ (nm)	Colour
$Ga_x In_{1-x} N$	2.64	470	blue
	2.36	525	green
	2.10	590	yellow
$(Al_x Ga_{1-x})_y In_{1-y} P$	2.03	610	orange
	1.98	625	red

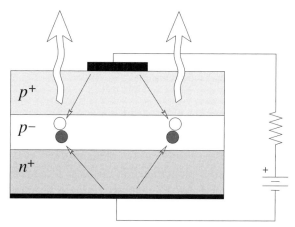

Figure 8.1 The basic LED, illustrating how injection of electrons and holes into the space-charge region leads to recombination and the generation of light.

recombination will be close to that of the bandgap E_g (see Exercise 8.6). We define a **voltage efficiency** as

$$\eta_V \equiv \frac{\hbar\omega}{qV_a} \approx \frac{E_g}{qV_a}, \qquad (8.1)$$

where V_a is the magnitude of the applied forward bias, and ω is the radian frequency of the generated light. Some properties of the two material systems from which today's high-brightness LEDs are made are listed in Table 8.1.

Equation (8.1) defines the applied voltage required to get a voltage efficiency of unity when using the semiconductor that has been chosen for a particular frequency of light output. Evidently, employment of $V_a < \hbar\omega/q$ would lead to higher voltage efficiencies, but, because of the exponential dependence of the carrier concentrations on potential (see (6.27)), the use of such low biases would not prove helpful in obtaining a high overall efficiency.

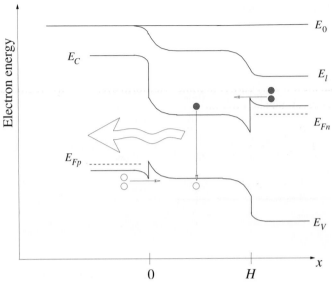

Figure 8.2 Energy-band diagram of a forward-biased $P^+ p N^+$ heterojunction diode. Radiative recombination is encouraged in the low-bandgap material.

8.2   Current efficiency

The rate of radiative recombination is proportional to the local $p(x)n(x)$-product (see Section 3.1.2). In a forward-biased diode, the minority-carrier concentrations will be elevated above their equilibrium values over distances that extend from the space-charge layer by several minority-carrier diffusion lengths (see Fig. 6.8). Thus, photon generation will occur over a length of the order of microns.

To obtain more intense photon generation, the minority carriers must be concentrated into a smaller region. This is achieved in modern high-brightness LEDs by using dissimilar materials for the p- and n-regions of the diode. An example of such a **heterostructure diode** is shown in Fig. 8.2. This structure uses a low bandgap material sandwiched between two higher bandgap materials to form a **potential well** in which the minority carriers are trapped, thereby increasing their concentrations and, consequently, the intensity of the generated light. The region of carrier confinement is called the **active layer**. We will discuss this heterojunction energy-band diagram in more detail in the next subsection, after we have defined the **current efficiency**.

Consider the current due to electrons that are injected from the N-region and recombine in the active region with holes that are injected from the P-region. From our master set of equations (5.24) the electron current density, in steady-state and with no non-thermal generation of electron-hole pairs, can be found from

$$\frac{1}{q}\frac{dJ_e}{dx} - \frac{\Delta n}{\tau_e} = 0 . \qquad (8.2)$$

Integrating over the active region, which we take to be bounded by $x = 0$ and $x = H$,

$$\int_0^H \frac{\Delta n}{\tau_e} \, dx = \int_0^H \frac{1}{q} \frac{dJ_e}{dx} \, dx \,. \tag{8.3}$$

Assume that the minority carrier lifetime is uniform over the active region, and introduce the radiative recombination lifetime

$$\int_0^H \frac{\Delta n}{\tau_{\mathrm{rad}}} \, dx = \frac{\tau_e}{\tau_{\mathrm{rad}}} \frac{1}{q} \int_0^H \frac{dJ_e}{dx} \, dx = \frac{\tau_e}{\tau_{\mathrm{rad}}} \frac{1}{q} [J_e(H) - J_e(0)] \,. \tag{8.4}$$

Ideally, we would like $J_e(0) \to 0$ as this would mean that there was no electron recombination in the P-region. Thus, the current efficiency of an LED is defined as

$$\eta_C = \frac{[J_e(H) - J_e(0)]}{J_D} \,, \tag{8.5}$$

where J_D is the total current density of the diode. To reduce $J_e(0)$ it is necessary to prevent electrons from escaping into the P-region. This is best accomplished by making a large energy barrier for electrons at the interface between the active region and the P-region. Similarly, it is desirable to have a barrier at the active/N-region junction to prevent the escape of holes from the active region. These attributes can be realized by implementing heterojunctions of the Type-I variety discussed in Section 6.7.1.

LEDs are usually p-on-n diodes, unlike solar cells, which are usually n-on-p diodes. In an LED it is not necessary to have the junction region extremely close to the surface because the photons are all of energy very close to the bandgap, and so the relevant absorption coefficient is low. For mechanical reasons the bottom layer of the LED is likely to be thick, and n-type material is preferred for this region in order to minimize series resistance. 'Vertical' resistance dominates in this device because there is little lateral current, and the LED has a much smaller cross-sectional area than a typical solar cell. Also, recall that the LED operates in forward bias, so the current is large and series resistance is to be avoided in order to obtain the desired junction voltage.

8.2.1   Heterojunction diodes

A prerequisite for a practical semiconductor/semiconductor heterojunction is that there should be a good match between the lattice constants of the two materials. Otherwise, there will be defects at the interface, causing recombination of electrons and holes via intra-bandgap states. Such recombination is particularly undesirable in an LED because it is likely to be non-radiative (see Section 3.2). For illustrative purposes we'll focus for the moment on the $Al_x Ga_{1-x} As$ system, which, as can be seen from Fig. 8.3, gives good lattice matching to GaAs over almost the entire compositional ratio. In fact, for LEDs we're only interested in mole fractions up to about $x = 0.4$, because after that the smallest bandgap in AlGaAs becomes indirect. For drawing the energy-band diagram, the

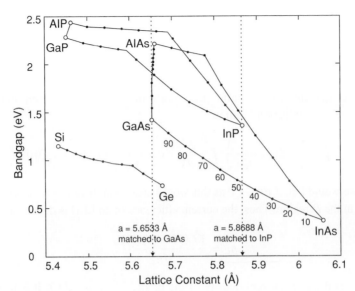

Figure 8.3 Bandgap dependence on composition of some binary, ternary and quaternary compound semiconductors. From Schwierz [1], © 2000 IEEE, reproduced with permission.

relevant properties are the bandgap and the electron affinity, as discussed in Section 6.7. The mole fraction dependencies of these two parameters for $x < 0.4$ are

$$E_g(x) = 1.424 + 1.247x \quad \text{eV}$$

$$\chi(x) = 4.07 - 0.79x \quad \text{eV} . \tag{8.6}$$

From these it follows that the band offsets between AlGaAs and GaAs are $\Delta E_C = 0.79x$ eV and $\Delta E_V = 0.46x$ eV. These offsets are discontinuities at the band edges, which give the energy-band diagram for a heterostructure a different appearance from that for a homojunction. The example shown in Fig. 8.2 is for a P^+-AlGaAs/p-GaAs/N^+-AlGaAs LED operating under forward bias. The procedures for constructing the band diagram are the same as listed in Section 6.1.2 and Section 6.3.1. Because the electron affinity for GaAs is greater than that of AlGaAs, the active region becomes a 'potential well'. The carriers injected from the neighbouring AlGaAs regions are confined by the interfacial potential barriers, thereby encouraging recombination of electrons and holes in the active region.

8.3   Radiative recombination efficiency

Now that we have engineered a situation in which recombination is confined to a well-defined active region, we need to ensure that the recombination is predominantly radiative. This means, first of all, that the active material should be a direct bandgap semiconductor (see Section 3.2.1). Secondly, to obtain a high rate of radiative recombination, high concentrations of both types of carrier are needed (see (3.19)). This means a high level of injection into the active layer of both electrons and holes must be established, which necessitates the forward-biasing of the device (see Fig. 8.2). The background doping density of the active layer is usually chosen to make the region p-type: in this way electrons are the minority carriers, and their superior minority-carrier mobility ensures a more uniform carrier distribution in the active layer and, consequently, a more spatially uniform generation of photons.

From (3.18), the net rate of recombination for electrons is given by $\Delta n / \tau_e$. The **radiative recombination efficiency** is given by

$$\eta_{\mathrm{rad}} = \frac{1/\tau_{\mathrm{rad}}}{1/\tau_e} = \frac{\tau_e}{\tau_{\mathrm{rad}}}, \tag{8.7}$$

where τ_e, the total minority-carrier lifetime for electrons, is given by (3.24).

The principal non-radiative contributors to τ_e are RG-centre recombination and Auger recombination, as discussed in Section 3.2. To reduce the former, the active layer must be of high crystalline perfection. This means that the active-layer material must be **lattice-matched** to the substrate material on which it is epitaxially grown. For example, from Fig. 8.3, the quaternary material $(\mathrm{Al}_x\mathrm{Ga}_{1-x})_y\mathrm{In}_{1-y}\mathrm{P}$ is lattice-matched to GaAs at $y \approx 0.5$. The matching is maintained for varying values of the Al mole fraction x. In practice, $x < 0.5$ as beyond that AlGaInP has an indirect bandgap. Incorporation of dopants inevitably introduces defects into a crystal, so the p-type doping density in the active layer is usually kept low ($\approx 10^{16}$ cm^{-3}). By comparison, the doping densities in the adjoining **confinement layers** are one or two orders of magnitude higher. This ensures that the space-charge region of the diode is contained largely within the active layer, and that the parasitic resistances of the quasi-neutral regions are low.

Some idea of attainable values for η_{rad} can be obtained from Fig. 8.4, where the radiative efficiency for GaAs is plotted as a function of the excess electron concentration. The various lifetimes were evaluated using the equations in Section 3.2.4 and the recombination parameters from Table 3.1. Notice that high-level-injection conditions have to be attained before η_{rad} becomes appreciable. The situation would be helped, of course, if the rate of RG-centre recombination could be considerably reduced, as the top curve indicates.

8.4 Extraction efficiency

The final stage in the electrical-optical conversion process is to get the internally generated photons out of the device. The **extraction efficiency** is defined as the ratio of the optical power that actually escapes from the structure to the optical power that is generated within the diode

$$\eta_{\mathrm{ext}} = \frac{S_{\mathrm{out}}}{S_{\mathrm{gen}}}. \tag{8.8}$$

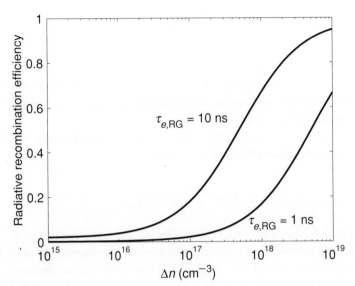

Figure 8.4 Radiative recombination efficiency for p-type GaAs of doping density 10^{16} cm^{-3}. The recombination parameters of Table 3.1 were used for the bottom curve. For the top curve, $\tau_{e,RG}$ was increased by a factor of 10.

Extracting the light is inherently difficult because the relatively large value of refractive index for most semiconductors means that the critical angle θ_c, beyond which total internal reflection occurs, is small. For example, GaAs has a refractive index of about 3.5, so, at the semiconductor/air interface, from Snell's Law

$$\theta_c = \arcsin\left(\frac{1}{3.5}\right) \approx 17°. \tag{8.9}$$

The spontaneous light emission from an LED is omnidirectional, and if the light is taken to emanate from a point source, it is easy to estimate the optical power that exits through a segment of a spherical surface defined by a polar angle of θ_c. The situation is illustrated in Fig. 8.5, from which the following expression can be deduced:

$$S_{out} = S_{gen} \frac{2\pi r^2 (1 - \cos\theta_c)}{4\pi r^2}. \tag{8.10}$$

For a point source in GaAs, this means that only about 2% of the optical power would escape into the surrounding air!

The semiconductor layers in an LED are usually deposited epitaxially onto a substrate, and the resulting structure is planar. Thus, the light source is more like a plane than a point, so the extraction situation is not as bad as just described. Shaping of the semiconductor die by sawing is possible, and wedge-shaped diodes, in which total-internal reflection is reduced by effectively increasing θ_c have been reported. Details are given by Schubert [2, Chapters 9,10], who also discusses designs with reflecting bottom surfaces to improve the light output from the top surface. Here, we briefly describe the innovative structure shown in Fig. 8.6.

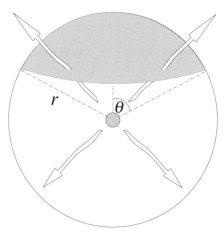

Figure 8.5 Light emitted from a point source and emerging from a spherical surface only through the shaded area. The fraction of emitted light to generated light is given by (8.10) with $\theta = \theta_c$.

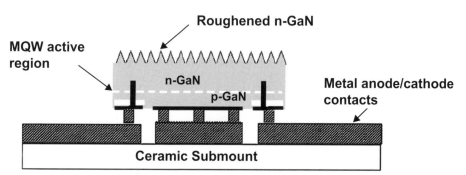

Figure 8.6 LED with pyramidal front surface and flip-chip mounting. Reused with permission from O.B. Shchekin, J.E. Epler, T.A. Trottier, T. Margalith, D.A. Steigerwald, M.O. Holcomb, P.S. Martin and M.R. Krames, Applied Physics Letters, **89**, 071109 (2006) [3]. Copyright 2006, American Institute of Physics.

This is an InGaN/GaN LED, the semiconducting layers of which are grown on a substrate that is subsequently removed by etching. The bottom layer (*n*-GaN) is then subjected to an anisotropic etch that results in a texturing of the surface that is similar to that employed in the high-efficiency solar cell of Fig. 7.8. The entire chip is then flipped over so that the roughened surface becomes the top of the LED. The internally generated light is guided by each cone via several internal reflections, after each one of which the angle of incidence becomes more nearly normal, until the light emerges near the tip of the cone. There are no contacts on the textured front surface, so it is desirably transparent. The contacts to the front and back regions of the device are made from the bottom, and are bonded to a ceramic substrate of high thermal conductivity. In the reference cited, very high brightness was reported for blue-green versions of this diode. We'll meet this impressive LED again in Section 8.7 in the context of white light emission.

8.5   Wall-plug efficiency

Combining all the efficiencies of the previous sections, the output optical power density can be written as

$$S_{out} = (V_D \eta_V)(J_D \eta_C) \eta_{rad} \eta_{ext}, \qquad (8.11)$$

where V_D and J_D are the applied bias and diode current density, respectively. Rearrangement of this equation leads to an expression for the overall efficiency of the electrical-to-optical conversion process, the so-called **wall-plug efficiency**:

$$\frac{S_{out}}{P_{in}} = \eta_V \eta_C \eta_{rad} \eta_{ext}. \qquad (8.12)$$

Wall-plug efficiencies for high-brightness LEDs are steadily improving, and a value of 56% has recently been reported [4]. As you will have probably deduced from the foregoing sections, the extraction efficiency is presently limiting performance, and, consequently, it is the subject of intensive research and development. High values of the other efficiencies can be obtained by: operating at a forward-bias voltage close to E_g/q, using a heterostructure to confine the recombination to a well-defined active layer and operating at high current to enhance radiative recombination.

8.6   Luminous efficacy and efficiency

The wall-plug efficiency is an example of a **radiometric** measurement, i.e., one that involves the objective physical properties of power, as would be recorded by a calibrated photodetector and a wattmeter. More commonly, figures-of-merit for LED optical performance are quoted in terms of **photometric** units, i.e., relating to optical power that is perceived by the eye. The eye is most sensitive to the colour green, as can be seen from the plot of the **eye sensitivity function** γ in Fig. 8.7. So, the optical power perceived by the eye can be expressed in watts as $\int_\lambda \gamma S'_{out}(\lambda) d\lambda$, where the prime signifies power, rather than power density, and $S'_{out}(\lambda)$ is the spectral power density in W/m. In practice, the perceived optical power is usually expressed in units of lumens (lm), by applying a conversion factor of 683 lm/W. This factor can be traced back to the old light-intensity unit of candlepower, which was originally defined in terms of the light output of a standard candle. Thus, S' watts of optical power are perceived by the eye as Φ lumens of **luminous flux** according to

$$\Phi = 683 \int_\lambda \gamma S'_{out}(\lambda) d\lambda \quad \text{(lumens)}. \qquad (8.13)$$

The **luminous efficacy** measures the effectiveness of the eye in perceiving optical power:

$$\text{luminous efficacy} = \frac{\Phi}{S'_{out}} \quad \text{(lm/W)}. \qquad (8.14)$$

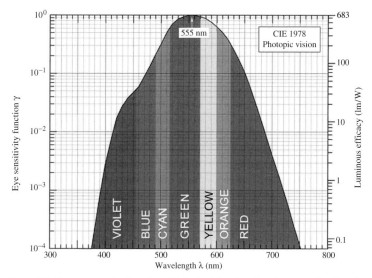

Figure 8.7 Eye sensitivity function. Note that the function is normalized to the peak sensitivity, which occurs in the green part of the spectrum. From Schubert [2, Fig. 16.7], © Cambridge University Press 2006, reproduced with permission.

Thus, for monochromatic emission at 555 nm, the luminous efficacy is 683 lm/W, as indicated on Fig. 8.7. Outside of the visible range, the luminous efficacy is essentially zero.

The final photometric property of interest to us is the **luminous efficiency**: it is also expressed in lm/W, and relates the perceived optical power to the electrical power supplied to the LED. Thus

$$\text{luminous efficiency} = \frac{\Phi}{I_D V_D} \quad (\text{lm/W}). \tag{8.15}$$

To relate our earlier radiometric property of wall-plug efficiency to these photometric properties, note that

$$\text{luminous efficiency} = \text{wall-plug efficiency} \times \text{luminous efficacy}. \tag{8.16}$$

8.7 White-light LEDs

With the advent of blue-green InGaN LEDs, it is now possible to obtain white light by mixing the output from such LEDs with the red light from AlGaInP LEDs. This opens up enormous opportunities for solid-state lighting in general-purpose applications.

The human eye senses colour through rod and cone cells in the retina that are specifically sensitive to red, green, or blue light. Combinations of certain intensities of these colours are perceived by the eye as white light. For standardization and colourimetric reasons, three **colour-matching functions** have been developed. Each is centred on a particular wavelength (red, green, or blue) and has some spectral content such that specific combinations of the intensities of the three sources can produce any desired colour. The

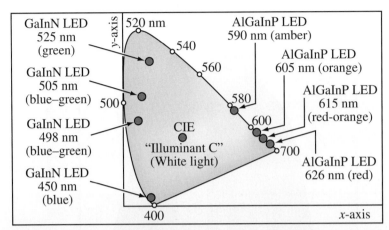

Figure 8.8 Chromaticity diagram, showing the colours of the light from various LEDs. From Schubert [2, Fig. 17.10], © Cambridge University Press 2006, reproduced with permission.

chromaticity coordinates corresponding to the red, green, and blue colour-matching functions are labelled x, y, z, respectively. The x-coordinate, for example, measures the ratio of the stimulus of the red cells in the eye to the total stimulus of all the colour cells to the entire visible spectrum. Thus, $x + y + z = 1$, and it is only necessary to stipulate two chromaticity coordinates when specifying a particular colour. This is the basis of the **chromaticity diagram**, as illustrated in Fig. 8.8. The point at the centre of the white zone has chromaticity coordinates $(x, y) = (1/3, 1/3)$.

Consider the points on the chromaticity diagram of the 450 and 525 nm GaInN LEDs and of the 626 nm AlGaInP LED. Join-up the points to form a triangle. The area within the triangle defines the range of colours that can be obtained by mixing different intensities of the light from each of the three diodes. The range of attainable colours is large; it is called the **colour gamut**, and it includes white light. Now take two LEDs, e.g., the blue-green LED at 498 nm and the red LED at 626 nm. The line between these two passes through the white zone, indicating that some power ratio of these two diodes could produce white light.

Rather than use separate, discrete diodes to obtain the additive colour mixing, it is possible to integrate diodes into a stack of layers, with one sequence of layers (confinement regions and active region). Such monolithic dichromatic structures, and even higher-order stacks of diodes, are presently under development.

At the moment (2009), white-light LEDs usually employ **wavelength conversion** rather than additive colour mixing. In wavelength conversion, some of the diode-generated light is absorbed by a material and re-emitted at longer wavelengths. The phosphorescent light and the residual shorter wavelength light can, collectively, span a sufficiently large wavelength range that the output appears white. Stimulation of a cerium doped yttrium aluminum garnet (YAG) phosphor by 460 nm light from a GaInN diode is one particular, useful combination. A practical example using the high-brightness flip-chip diode discussed earlier under a phosphor dome is shown in Fig. 8.9.

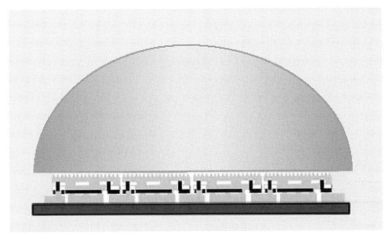

Figure 8.9 The flip-chip InGaN/GaN LED of Fig. 8.6 emits into a YAG phosphor dome, resulting in the emission of white light. From Shchekin and Sun [4], © 2007 Institute of Physics, reproduced with permission of the publishers, the author, and Philips Lumileds.

As white-light LEDs penetrate into general-purpose lighting markets, the issue of colour rendering arises. The ability of an illuminating light source to faithfully render the colours of the object being illuminated is called the **colour rendering index**, CRI. The Sun is characterized by a CRI of 100, and other light sources are measured relative to this. Tungsten-filament and quartz-halogen lamps have a CRI close to 100, but fluorescent lights have a considerably lower index. The fluorescence is from a phosphor, which is excited by a gaseous discharge, and the light emission lacks intensity in the red end of the spectrum, giving a 'cool-light' experience. It follows that for LEDs to obtain high CRI values, the light from several different LEDs will have to be combined. Trichromatic white-light LEDs have been demonstrated with CRI> 90 [2, p. 338], and research in this area is intense.

8.8   Prospects for general-purpose solid-state lighting

In 2001, 22% of the electricity consumed in the USA was used for lighting. The generation of this amount of electricity was responsible for 7% of all the carbon emitted into the atmosphere in the US [6]. These startling figures stem from the poor wall-plug efficiency of conventional lamps, e.g., ≈5% for incandescent lamps, and ≈20% for fluorescent lamps. The former lamps originated in the 1870s, and the latter in the 1930s, so it is perhaps time for something new on the lighting scene. High-intensity discharge lamps are relatively new, and relatively energy efficient, but it is solid-state lighting, with its direct conversion of electrical energy to light, as described in this chapter, that offers most promise for sustainable lighting. This fact has been recognized, at least in the US, and a roadmap to spur the development of white-light LEDs for general-purpose lighting has been drawn-up. This map is shown in Fig. 8.10. Note that 'luminous efficiency', as

Lamp Targets	SSL-LED 2002	SSL-LED 2007	SSL-LED 2012	SSL-LED 2020	Incandescent	Fluorescent	HID
Luminous Efficacy (lm/W)	20	75	150	200	16	85	90
Lifetime (hr)	20,000	20,000	100,000	100,000	1,000	10,000	20,000
Flux (lm/lamp)	25	200	1,000	1,500	1,200	3,400	36,000
Input Power (W/lamp)	1.3	2.7	6.7	7.5	75.0	40.0	400.0
Lamp Cost (in US$/klm)	200.0	20.0	5.0	2.0	0.4	1.5	1.0
Lamp Cost (in US$/lamp)	5.0	4.0	5.0	3.0	0.5	5.0	35.0
Color Rendering Index (CRI)	70	80	80	80	100	75	80
Derived Lamp Costs							
Capital Cost (US$/Mlmh)	12.00	1.25	0.30	0.13	1.25	0.18	0.05
Operating Cost (US$/Mlmh)	3.50	0.93	0.47	0.35	4.38	0.82	0.78
Ownership Cost (US$/Mlmh)	15.50	2.18	0.77	0.48	5.63	1.00	0.83

Figure 8.10 Roadmap for the development of solid-state lighting. Projected performance of LEDs is compared with conventional lighting. Note that 'luminous efficacy' in this table is the same as 'luminous efficiency' as we have defined it. From Tsao [5], © 2004 IEEE, reproduced with permission.

defined in this book, is sometimes confusingly referred to as 'luminous efficacy of the source'. Thus, the first row of the table in Fig. 8.10 is what we have called luminous efficiency.

From the Table it is clear that the luminous efficiency and longevity of LEDs already exceed that of incandescent lamps, but the flux (brightness), cost and CRI are all in need of improvement. Therein lies the challenge for future designers!

Exercises

8.1 The E-k relationships for the conduction bands of two semiconductor materials, A and B, each with spherical constant-energy surfaces, can be expressed as

$$E_A - 0.7 = \alpha k^2 \quad \text{and} \quad E_B - 1.4 = 2\alpha(k - k')^2,$$

respectively, where α is a constant, $k' > 0$, and the energies are in units of eV. Both materials have the same valence-band structure, with the top of the valence band at $E = 0$ and $k = 0$.

Which material would make the better LED?

8.2 Fig. 8.2 shows the band diagram for a P^+pN^+ AlGaAs/GaAs/AlGaAs LED under forward bias. Construct the corresponding band diagram for an $In_{0.49}Ga_{0.51}P$/GaAs/$In_{0.49}Ga_{0.51}P$ LED.

8.3 Explain which of the above two LEDs (the AlGaAs device or the InGaP device) is likely to have the better current efficiency.

8.4 When GaP is co-doped with oxygen and zinc two impurity levels occur at similar spatial coordinates: the O-level is 0.80 eV below the GaP conduction-band edge, and the Zn-level is 0.04 eV above the valence-band edge. LEDs made from GaP:O:Zn emit in the red (≈ 700 nm). Although these diodes are not used commercially nowadays, they are interesting because the emitted wavelength is much shorter than would be expected from a simple O $\rightarrow$ Zn transition.

Suggest a possible reason for this 'shift' in wavelength.

8.5 Fig. 3.7 shows radiative recombination involving an electron and hole separated in energy by the bandgap energy E_g. Radiative recombination can also occur between carriers not at the band extrema but with equal momenta $\hbar k$.

(a) Show that the photon energy dispersion relationship can be written as

$$E_{ph} = E_g + \frac{\hbar^2 k^2}{2m_r^*}, \tag{8.17}$$

where m_r^* is called the **reduced effective mass**, and (8.17) defines the joint dispersion relation.

(b) Show that the reduced effective mass for GaAs is $\approx 0.059m_0$.

8.6 Regarding the energy dependence of the emission intensity of an LED, there are two main factors to consider: (*i*) the **joint density of states** increases as $\sqrt{E - E_g}$, analogously to the density of states in (3.32); (*ii*) the probability of occupancy of these states falls off as $\exp(-E/k_B T)$, at least for a Maxwell-Boltzmann distribution.

Show that these dependencies conspire to give maximum emission at an energy of $E_g + k_B T/2$.

8.7 Two LEDs, A and B, each emit 1 mW of optical power when operating at a current of 1 mA and a forward bias of 2 V.

LED A emits in the ultra-violet part of the spectrum, and LED B emits at 470 nm in the blue part of the spectrum. For each diode evaluate:

(a) the wall-plug efficiency;

(b) the luminous efficacy;

(c) the luminous efficiency.

References

[1] F. Schwierz, Microwave Transistors – the Last 20 Years, *Proc. IEEE 3rd Int. Caracas Conf. on Devices, Circuits and Systems*, pp. D28/1–7, 2000.

[2] E.F. Schubert, *Light-Emitting Diodes*, 2nd Edn., Cambridge University Press, 2006.

[3] O.B. Shchekin, J.E. Epler, T.A. Trottier, T. Margalith, D.A. Steigerwald, M.O. Holcomb, P.S. Martin and M.R. Krames, High Performance Thin-film Flip-chip InGaN-GaN Light-emitting Diodes, *Appl. Phys. Lett.*, vol. 89, 071109, 2006

[4] O.B. Shchekin and D. Sun, Evolutionary New Chip Design Targets Lighting Systems, *Compound Semiconductors*, vol. 13(2), 2007.

[5] J.Y. Tsao, Solid-state Lighting: Lamps, Chips and Materials for Tomorrow, *IEEE Circuits and Devices Magazine*, pp. 28–37, May/June 2004.

[6] *Basic Research Needs of Solid-State-Lighting*, Report of the Basic Energy Sciences Workshop on Solid-State Lighting, May 22–24, 2006. Online [http://www.er.doe.gov/bes/reports/abstracts.html#SSL].

9   HBT basics

The first commercial bipolar junction transistors (BJTs) were made from germanium. Because of the low bandgap of this material (0.67 eV), the intrinsic carrier concentration is high. As n_i increases exponentially with temperature (see (4.19)), these Ge BJTs were unstable, unless operating in a temperature-regulated circuit. Silicon, with its larger bandgap, proved to be a better proposition, and the first Si BJTs appeared in the early 1950s. These transistors ushered in the era of solid-state electronics. They were not challenged until MOSFETs started to appear in the 1960s, and to provide a superior transistor for circuits in which a high input impedance was important. With the advent of CMOS in 1963, the age of large-scale integration began, and the MOSFET became the more ubiquitous transistor. However, as we show elsewhere in this book, the bipolar transistor has inherent advantages in high-frequency performance, due to its superior transconductance, and in high-power applications, due to its favourable geometry. BJTs are also more robust than MOSFETs, which is why many readers will have become familiar with them during their electronics laboratory classes.

Perhaps the biggest event in BJT development in the last 20 years has been the advent of heterojunction bipolar transistors (HBTs). In the single heterojunction version of these transistors, dissimilar semiconducting materials are used for the emitter and the base, whereas the base and collector are made from the same semiconductor. Thus, there is one heterojunction, e.g., an Np-junction as discussed in Section 6.7, and one homojunction, e.g., a pn-junction, as discussed in Chapter 6. For an Npn HBT, the idea is to choose two materials that give a barrier height for holes that is much larger than that for electrons. This allows a much larger doping density to be used in the base, when compared to conventional BJTs, thereby giving a very low base resistance, without excessive base/emitter hole current. The benefit of this is discussed at length in the later chapter on high-frequency transistors. We have already come across heterojunction diode structures in the chapter on LEDs, and have described the formation of such structures by the sequential, epitaxial deposition of thin semiconducting films onto a substrate that would, ideally, be lattice-matched to the deposited films. In particular, the two materials comprising the heterojunction must have very similar lattice constants (see the length a in Fig. 2.1), otherwise there will be defects at the interface, causing unwanted recombination-generation centres (see Section 3.2.2). From the data in Fig. 8.3 it can be seen that materials from the AlAs-GaAs materials system, and from the GaP-InP system, are prime candidates for HBTs constructed on GaAs substrates. Silicon is not such a promising material with which to form heterojunctions as there is not the same

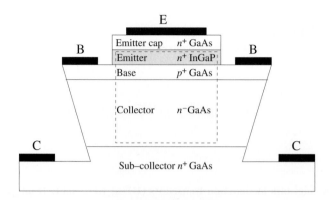

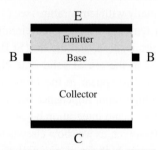

Figure 9.1 Basic structure of a InGaP/GaAs HBT. The n^+ emitter cap layer facilitates ohmic contacting to the emitter metallization, and the n^+ sub-collector reduces the collector resistance. The main part of the transistor is within the region defined by the dashed-line box. This part of the transistor can be represented by the basically 1-D element shown in the lower part of the figure. This reduced structure is valid because the critical dimension is the basewidth (vertical dimension), which is much less than any significant lateral feature size.

variety of lattice-compatible materials to choose from. However, suitably defect-free junctions can be grown between Si and a dilute alloy of $Si_{1-x}Ge_x$, with the Ge mole fraction x not exceeding about 15%. By increasing the Ge mole fraction through the base, the bandgap gets progressively smaller, and a field is created in the base. This field aids the transport of electrons across the base to the collector, thereby improving the high-frequency performance of the transistor, as discussed in Chapter 14. The HBT is a good example of a device designed by **bandgap engineering** [1].

9.1   Basic properties

Fig. 9.1 shows the basic structural arrangement of an HBT using semiconductor materials of the III-V compound variety. The various semiconducting layers would be deposited epitaxially. Note the pairs of contacts for the base and the collector. This arrangement is to reduce both the base spreading resistance (see Section 16.3.1) and the series resistance of the access regions to the main part of the transistor. The latter is, essentially, a vertical slice through the transistor, as shown in the lower part of Fig. 9.1.

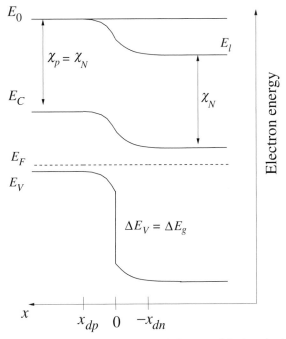

Figure 9.2 Equilibrium energy-band diagram of the base/emitter region of a p-GaAs/N-InGaP HBT. Note the equality of the electron affinities.

The energy band diagram for an HBT is constructed following the procedure detailed in Section 6.1.2. In Fig. 9.2 we show the particular example for the single heterojunction bipolar transistor depicted in Fig. 9.1. It comprises: an emitter of N-In$_{0.49}$Ga$_{0.51}$P, a base of p^+-GaAs and a collector of n^--GaAs. The particular mole fractions for In and Ga in the InGaP layer are chosen to give a good lattice match to GaAs (see Fig. 8.3). The bandgap for this material is 1.86 eV. The electron affinities of GaAs and In$_{0.49}$Ga$_{0.51}$P are essentially the same, so the difference in bandgaps ($\approx$0.4 eV) is taken up entirely by the band-edge offset in the valence band. This illustrates the defining feature of HBTs: the creation of different energy barriers for electrons and holes at the emitter/base junction. As we have done elsewhere in this book, we follow the common practice of denoting the region of higher bandgap by an uppercase symbol.

By having a higher hole barrier, the doping density in the base can be increased without compromising the current gain.[1] The high base doping density also allows a very narrow base to be used without causing the base access resistance to become excessive. A narrow base is advantageous because of two reasons: it enables a steep profile for the minority carrier concentration to be maintained, which leads to a high collector current; it means that there is less minority carrier storage and, consequently, less base-storage capacitance. Both these factors help in the attainment of a high f_T. The

---

[1] Base doping densities of 10^{19}–10^{20} cm^{-3} are common, and are 1–2 orders of magnitude higher than in the emitter.

reduced base resistance helps in realizing a high f_{max}. Both of these frequency metrics are discussed in Chapter 14 on high-frequency devices.

From (6.39) it follows that the built-in voltage for our sample device is

$$q V_{bi} = k_B T \ln \left[\frac{N_{Cp} \, n_{0N}}{N_{CN} \, n_{0p}} \right]. \tag{9.1}$$

Assuming similar effective densities of states for InGaP and GaAs, the built-in voltage for an emitter doping density of $N_D = 3 \times 10^{17} \, \text{cm}^{-3}$ and a base doping density of $N_A = 6 \times 10^{19} \, \text{cm}^{-3}$, for example, is 1.46 V.

In preparation for deriving an expression for the current in the next section, we recognize a very important difference between the ideal diode discussed in Section 6.6 and a modern HBT: the width of the base. In an ideal diode it is infinite, so any carriers injected from the emitter have ample opportunity to make collisions, with the result that their distribution at the edge of the emitter/base space-charge region is of a near-equilibrium form. It is this characteristic that allowed us to use Shockley's Law of the Junction to get $n(x_{dp})$, for example.

In a modern HBT, the basewidth can be ≈ 30 nm. To estimate the mean free path for our example from (5.45) we need the electron mobility in GaAs for $N_A = 6 \times 10^{19} \, \text{cm}^{-3}$, and the mean unidirectional velocity for injection from an emitter with $N_D = 3 \times 10^{17} \, \text{cm}^{-3}$. From Fig. 5.3 we find $\mu_e \approx 0.075 \, \text{m}^2 (\text{V s})^{-1}$, and from Table 4.2 we get $v_R \approx 10^5 \, \text{m s}^{-1}$. Thus, the mean-free-path length is ≈ 10 nm. This is insufficient to assume that all carriers injected into the base are 'thermalized' by collisions. However, it is reasonable to assume that the electrons injected from the emitter maintain a hemi-Maxwellian distribution through the emitter/base space-charge region. We denote the part of this positive-going (collector-directed) electron distribution that overcomes the barrier as $n_E^*/2$. Further, we recognize this as one-half of the boundary condition $n(x_{dp})$ given by Shockley's Law of the Junction. Thus,

$$\frac{n_E^*}{2} = \frac{n_{0E}}{2} e^{-(V_{bi} - V_{BE})/V_{th}} \equiv \frac{n_{0B}}{2} e^{V_{BE}/V_{th}}, \tag{9.2}$$

where n_{0E} and n_{0B} are the equilibrium electron concentrations in the emitter and base, respectively.

9.2   Collector current

Our intention is to obtain an expression for the collector current I_C in terms of the transistor's main controlling voltage $V_{aj} \approx V_{BE}$. We do this by considering electron flow in the base. The portion of the base between the junction space-charge regions is called the **quasi-neutral base**: it is essentially field-free, so we can assume reasonably safely that any minority-carrier currents therein are due to diffusion. This transport mechanism is fuelled by carrier concentration gradients, so we need to know the electron profile in the quasi-neutral base. This can be done by combining the equations for electron transport and electron continuity from our master set of equations (5.24). Under the

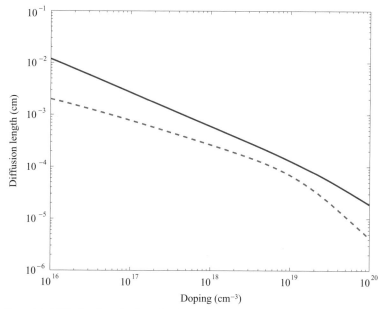

Figure 9.3 Diffusion lengths for minority carrier electrons (solid line) and holes (dashed line) in gallium arsenide. Computed from data on the doping density dependence of μ and τ from (5.31) and (3.22), respectively.

drift-diffusion approximation and in steady-state, these yield

$$\frac{d^2n}{dx^2} - \frac{n - n_{0B}}{L_e^2} = 0, \tag{9.3}$$

where L_e is the electron minority-carrier diffusion length. This equation is the same as appeared in our analysis of the ideal diode (6.31), but its solution here will be different because of the different boundary conditions. The minority-carrier diffusion length can be thought of as a measure of the distance a minority carrier diffuses before it recombines with a majority carrier. Typical values for GaAs are shown in Fig. 9.3, and can be evaluated using the diffusivity from (5.31) and the Einstein Relation, and the minority-carrier lifetime from (3.22).

If the quasi-neutral basewidth W_B in an Npn HBT is much less than L_e, then essentially no recombination takes place in the base. Under these circumstances, (9.3) reduces to

$$\frac{d^2n}{dx^2} = 0, \tag{9.4}$$

i.e., the electron diffusion current $J_e = q D_e \, dn/dx$ is a constant. Having a short base is important practically, as it leads to a high I_C and, as explained in the high-frequency chapter, to good high-frequency performance. Here, we consider this case: it allows us to compute J_e if we can find the electron concentration at any two points in the base. We will attempt to do this for the points at either end of the quasi-neutral base, x_{dp} and $x_{dp} + W_B$ in Fig. 9.4.

The forward bias lowers the energy barrier at the emitter/base junction, facilitating injection of electrons from the emitter into the base. These electrons are drawn from the positive-going hemi-Maxwellian at $x = -x_{dn}$, which is the other end of the

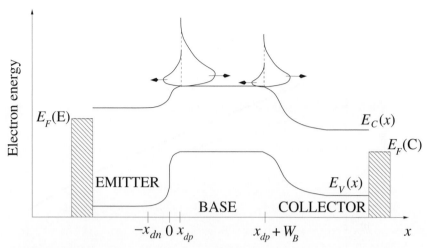

Figure 9.4 HBT in the active mode of operation ($V_{BE} > 0$, $V_{BC} < 0$).

space-charge region in the depletion approximation. The total width of the depletion region is $W = x_{dn} + x_{dp}$. Let us assume that W is less than a mean-free path for electrons.[2] Under these conditions, the right-going part of the electron distribution at $x = x_{dp}$, as illustrated in Fig. 9.4, is given by (9.2). As these electrons enter into the base they will scatter, and some will be re-directed towards the emitter, thus contributing to the left-going concentration at $x = x_{dp}$. Another contribution to this left-going distribution will come from electrons injected into the base from the collector. There will not be many of these electrons because the reverse bias at the collector/base junction increases the barrier therein. However, we will allow for some to be present, and some of them, after scattering and reaching $x = x_{dp}$ will be moving to the left. Let n_L represent the total, left-going distribution of electrons at $x = x_{dp}$. The situation is illustrated in Fig. 9.4. Thus

$$n(x_{dp}) = \frac{n_E^*}{2} + n_L .$$ (9.5)

Further, let us assume that the scattering events have reduced n_L to a hemi-Maxwellian distribution. Therefore, the electron current density at $x = x_{dp}$ is

$$J_e(x_{dp}) = -q\frac{n_E^*}{2}2v_R - (-qn_L 2v_R).$$ (9.6)

Equation (9.6) can be used to eliminate n_L from (9.5), yielding the boundary condition

$$n(x_{dp}) = n_E^* + \frac{J_e(x_{dp})}{q2v_R} .$$ (9.7)

Note that this expression reduces to Shockley's Law of the Junction (6.29) for either low currents or very high v_R. In a completely analogous way, the boundary condition at the

---

[2] This is a good assumption, at least for forward bias, as can be verified by doing Exercise 9.1.

other end of the quasi-neutral base can be written

$$n(x_{dp} + W_B) = n_C^* - \frac{J_e(x_{dp} + W_B)}{q2v_R} ,$$

(9.8)

where $n_C^*/2$ is the concentration of electrons in the collector with enough left-directed kinetic energy to surmount the collector/base barrier, i.e.,

$$\frac{n_C^*}{2} = \frac{n_{0B}}{2} e^{V_{BC}/V_{th}} .$$

(9.9)

We now have two carrier concentrations, at points separated by a known distance, thus

$$J_e = q D_e \left[n(x_{dp} + W_B) - n(x_{dp}) \right] / W_B.$$

(9.10)

Substituting for the carrier concentrations from (9.7) and (9.8), we obtain

$$J_e = -q n_{0B} \left[e^{V_{BE}/V_{th}} - e^{V_{BC}/V_{th}} \right] \cdot \frac{1}{\frac{W_B}{D_e} + \frac{1}{v_R}} \cdot$$

(9.11)

This equation has been ordered to highlight the charge and velocity terms. The reciprocal of the latter is the sum of two reciprocal velocities, or 'slownesses'. If the 'diffusion velocity' D_e/W_B is much less than the 'injection' velocity v_R, then the process of diffusion limits the current. This used to be the case in practical bipolar transistors, but fabrication techniques have improved to the point where basewidths of < 50 nm are routinely employed, in which cases it becomes necessary to include the $1/v_R$ term, otherwise the current would be overestimated, and would not be asymptotic to its ballistic limit of $-q(n_E^*/2)2v_R$. This is illustrated in Fig. 9.5. The electron current exiting the collector can be viewed as a positive charge flow into the collector from the external circuit: IEEE convention deems this to be a positive current, therefore

$$I_C \equiv -J_e A \equiv I_S \left[e^{V_{BE}/V_{th}} - e^{V_{BC}/V_{th}} \right] ,$$

(9.12)

where A is the cross-sectional area, and I_S collects together the non-exponential terms from (9.11) and is the transistor equivalent to the ideal diode saturation current of (6.37).

Transistor action in the HBT is exemplified by the plot of J_C shown in Fig. 9.6. The controlling voltage of the base/emitter junction is V_{BE}, and the voltage across the other junction V_{BC} is embedded in $V_{CE} = V_{CB} + V_{BE} \equiv -V_{BC} + V_{BE}$. Note the similarity of this set of curves, which constitute the **collector current characteristic**, with the drain current characteristic shown in the MOSFET chapter. There are some subtle differences, though. Firstly, there is not really a linear region at low collector/emitter bias. This is because conduction is never resistive, but is always diffusive (except in the ballistic case). Secondly, we have produced a characteristic without recourse to a threshold voltage, as is invoked in some MOSFET models to define the ON condition. In bipolar transistors, I_C depends exponentially on the controlling voltage V_{BE}, so it is arbitrary where the ON

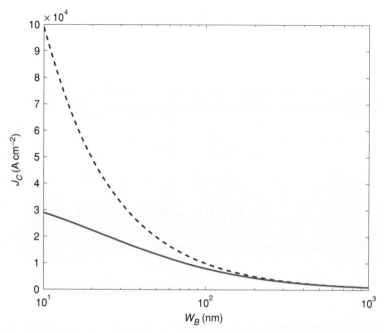

Figure 9.5 Collector current density vs. basewidth, showing the effect of including the velocity-boundary condition v_R (solid line), and of neglecting to include v_R (dashed line). The parameters for the InGaP/GaAs device are: $N_E = 3 \times 10^{17}$ cm^{-3}, $N_B = 6 \times 10^{19}$ cm^{-3}, $V_{BE} = 1.4$ V, $V_{CE} = 3$ V.

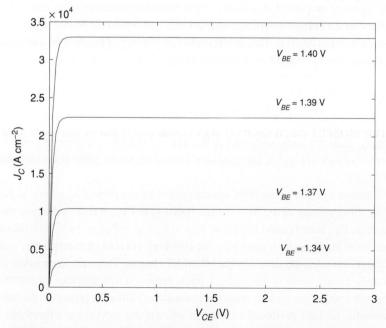

Figure 9.6 Collector current characteristic with V_{BE} as a parameter. The parameters for the InGaP/GaAs device are: $N_E = 3 \times 10^{17}$ cm^{-3}, $N_B = 6 \times 10^{19}$ cm^{-3}, $W_B = 30$ nm.

condition is deemed to begin.[3] Often in books on electronic circuits a **turn-on** voltage is specified for a bipolar transistor, and is usually taken to be $V_{BE} = 0.7\,\text{V}$ for a Si BJT. For HBTs made from wider bandgap materials, a larger value is appropriate.

In the description of the derivation of (9.11) the base/collector was mentioned as being reverse biased. However, the mathematics is quite general, and the equation applies for both forward- and reverse-bias of the base/collector junction and, indeed, to both bias modes for the base/emitter junction. To appreciate the effect of forward biasing the collector/base junction, pick a curve on Fig. 9.6 and begin at large V_{CE}, i.e., when V_{CB} is positive, meaning that the collector/base n/p junction is reverse biased. Now, stay on the curve (fixed V_{BE}) and note that I_C does not change as the reverse bias on the collector/base junction is reduced. This is because the electrons injected into the base from the emitter are easily collected by the favourable electric field in the base/collector space-charge region. However, there comes a point as V_{CE} is reduced when the collector/base junction becomes forward biased; this happens when $V_{CB} < 0$, i.e., when $V_{CE} < V_{BE}$. The collector now starts to inject electrons into the base, and this flow counters the flow from the emitter. The flow from the collector increases exponentially with V_{BC}, so I_C decreases drastically. Confusingly, this region where both junctions are forward biased is called the **saturation region**, and the region where I_C is actually constant, i.e., when the base/emitter junction is forward biased and the collector/base junction is reverse biased, is called the **active region**.[4]

To complete the description of the operating modes: when both junctions are reverse biased the HBT is said to be **cut-off**. Our treatment has focused on the **normal mode** of operation. If the transistor were to be operated with the actual collector being used as the emitter, then that would be the **inverse mode** of operation. Such a distinction is not made in the case of the MOSFET because in that device the source and drain are usually identical physically.

9.3   Base current

In the MOSFET the presence of the gate oxide means that we don't have to be concerned with the DC current at the controlling electrode. In a bipolar junction transistor, however, there is a DC current at the controlling base electrode (see Fig. 9.7). Holes flow into the base to: (i) replenish holes lost to recombination with electrons in the base; (ii) replenish holes lost to recombination with electrons in the base-emitter space-charge region; (iii) supply any hole current across the reverse-biased base/collector junction; (iv) supply the hole current due to injection of holes into the emitter across the forward-biased base-emitter junction. We have already given reasons for ignoring recombination in the base, *but only for the purpose of computing the collector current.* In other words, we have said

---

[3] In a MOSFET, I_D depends exponentially on the surface potential ψ_s, but the latter is not always linearly related to the controlling voltage V_{GS}, as we describe in Chap. 13.

[4] The reason why 'saturation' is used to describe a region in which I_C is not constant is explained in Section 13.2.

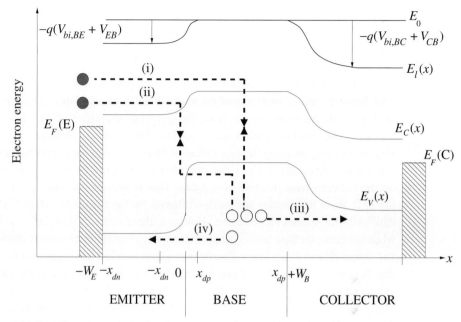

Figure 9.7 The various mechanisms of hole current in an HBT in the active mode of operation.

that any recombination current in the base is small compared to the collector current. This does not mean that any recombination current is small compared to the base current, which, in a well-designed transistor is much lower than I_C. We calculate this portion of I_B in the next subsection.

Regarding mechanism (ii), there will be recombination in the forward-biased base/emitter space-charge region because it is a region where there are large numbers of both electrons and holes. Any current due to this recombination will show up in the emitter and base leads at low V_{BE}, when the base/emitter barrier is high and there is little net hole injection into the quasi-neutral emitter. However, at moderate forward bias, this injection current, which is mechanism (iv), becomes large and dominates over the space-charge region recombination current. Here, we will ignore the latter. Also, the reverse-bias current of mechanism (iii) can be safely neglected. The remaining mechanism (iv), the injection of holes into the emitter, is treated in Section 9.3.2.

9.3.1   Recombination in the base

Mechanism (i) in Fig. 9.7 is due to recombination in the base, and can be computed from the electron continuity equation in the master set of equations (5.24)

$$J_{e,\mathrm{rec}} = \int_{J_e(x_{dp})}^{J_e(x_{dp}+W_B)} dJ_e = q \int_{x_{dp}}^{x_{dp}+W_B} \frac{n_B(x) - n_{0B}}{\tau_e} dx \,, \qquad (9.13)$$

where n_B is the spatially varying electron concentration in the base. Assuming the linear profile used in deriving the earlier expression for I_C, we have[5]

$$n_B(x) = n(x_{dp}) - \left[n(x_{dp}) - n(x_{dp} + W_B)\right](x/W_B)$$

$$= \left(n_E^* + \frac{J_e(x_{dp})}{q2v_R}\right)\left(1 - \frac{x}{W_B}\right) + \left(n_C^* - \frac{J_e(x_{dp} + W_B)}{q2v_R}\right)\frac{x}{W_B}. \quad (9.14)$$

Substituting into (9.13), integrating, re-arranging, and converting to current, yields

$$I_{B,\text{rec}} = Aqn_{0B}\left[(e^{V_{BE}/V_{th}} - 1) + (e^{V_{BC}/V_{th}} - 1)\right]\left\{\frac{1}{\frac{2\tau_e}{W_B} + \frac{1}{2v_R}}\right\}, \quad (9.15)$$

where A is the cross-sectional area of the emitter: in our one-dimensional model, it is the same area as used for evaluating the collector current from (9.11). The effective velocity in this case is the term within the curly brackets: it reduces to the usual expression of $W_B/2\tau_e$ when v_R is large.

9.3.2   Hole injection into the emitter

The region of interest is that of the quasi-neutral emitter, $(-W_E - x_{dn}) < x < -x_{dn}$, as shown in Fig. 9.7, and we are concerned with the hole current J_h in the emitter due to hole injection over the base/emitter barrier into the quasi-neutral emitter. The emitter is usually much wider than the base so recombination of the injected holes must be considered. Thus, from the equations for the hole diffusion current and the hole charge continuity in the drift-diffusion approximation we get

$$\frac{d^2p}{dx^2} - \frac{p - p_{0E}}{L_h^2} = 0, \quad (9.16)$$

where p_{0E} is the equilibrium hole concentration in the emitter, and L_h is the hole minority carrier diffusion length in the emitter. A convenient form of the general solution to (9.16) is

$$p(x) - p_{0E} = B\cosh\frac{x}{L_h} + C\sinh\frac{x}{L_h}. \quad (9.17)$$

Because the emitter length may be greater than L_h, we cannot ignore recombination of the injected holes in the quasi-neutral emitter. However, the relatively long length of the emitter means that there is no need to bound the carrier velocity as we did for J_e in the base by including the thermal velocity $2v_R$ in the boundary conditions. Thus, the boundary condition for the hole concentration at the emitter-edge of the depletion region follows without modification from Shockley's Law of the Junction

$$p(-x_{dn}) = p_{0E}e^{V_{BE}/V_{th}}. \quad (9.18)$$

[5] Assuming a linear profile, which results from having no recombination, and then using that profile to estimate recombination by allowing for a finite value of τ_B is permissible so long as there is not much recombination, i.e., $L_e > W_B$ (see Exercise 9.6).

At the other end of the emitter we assume an ohmic contact, thus

$$p(-x_{dn} - W_E) = p_{0E} .$$ (9.19)

To simplify the algebra, let us use the length variable x' in place of x, where $x' = x + x_{dn}$. Thus, the boundary conditions yield values for the constants

$$B = p_{0E}(e^{V_{BE}/V_{th}} - 1)$$

$$C = -p_{0E}(e^{V_{BE}/V_{th}} - 1)\coth\frac{-W_E}{L_h} .$$ (9.20)

The hole current density is

$$J_h(x') = -q\,D_h\frac{dp}{dx'} = -q\frac{D_h}{L_h}\left(B\sinh\frac{x'}{L_h} + C\cosh\frac{x'}{L_h}\right).$$ (9.21)

Substituting for the constants B and C, the hole current in the emitter at the edge of the depletion region is

$$J_h(-x_{dn})A = -q\,A\frac{D_h}{L_h}p_{0E}(e^{V_{BE}/V_{th}} - 1)\coth\frac{W_E}{L_h} ,$$ (9.22)

where A is the emitter area.[6]

Assuming no recombination in the depletion region, then $J_h(-x_{dn})A$ is also the hole current on the base side of the depletion region. We label this as $I_{B,\mathrm{inj}}$; thus, it can be added to $I_{B,\mathrm{rec}}$ to get the total base current.

The two components of I_B are plotted in Fig. 9.8 for the case of a InGaP/GaAs HBT with $W_E = 159\,\mathrm{nm}$ and $W_B = 30\,\mathrm{nm}$. Even for such a short-base device, the recombination component of base current is much more significant than the so-called **back-injection** hole current. This is a testament to the formidable reflecting action of the large hole barrier, which is engineered by choosing materials for the heterojunction with a large valence-band offset. Operation of the transistor in the saturation regime would increase the importance of $I_{B,\mathrm{rec}}$, due to the influx of electrons from the forward-biased collector/base junction. Also shown in the figure is the collector current; plotted in this way it shows the **transfer characteristic** of the device. A semi-logarithmic plot of collector and base currents vs. V_{BE} is known as a **Gummel plot**, after H.K. Gummel, a pioneer in the field of semiconductor device modelling.

9.4   DC equivalent-circuit model

The DC equivalent-circuit model of the HBT must represent the collector current, (9.11) and (9.12), and the base current, (9.15) and (9.22). The circuit of Fig. 9.9 does this, and also includes the base current due to injection of holes into the collector. The latter is mechanism (iii) from Section 9.3, and it is described by an expression similar to (9.22). Note that, in the equivalent circuit, the base recombination term has been separated into

---

[6] The minus sign in this case arises because the hole flow we are considering is in the negative x direction.

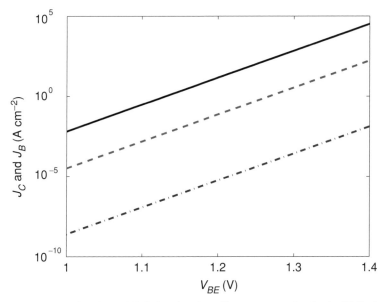

Figure 9.8 *J-V* plots for an HBT showing the collector current density (solid line), and two components of the base current density: $J_{B,\text{rec}}$ (dashed line) and $J_{B,\text{inj}}$ (stippled line). The parameters for this InGaP/GaAs device are: $N_E = 3 \times 10^{17}$ cm^{-3}, $N_B = 6 \times 10^{19}$ cm^{-3}, $W_E = 150$ nm, $W_B = 30$ nm, $V_{CE} = 3$ V, and the minority carrier properties are as per Fig. 9.3.

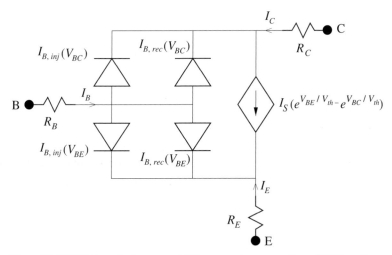

Figure 9.9 HBT DC equivalent circuit. The current source is from (9.11), with I_S representing the non-exponent part of that equation. Hole injection from the base into the emitter is given by (9.22) and is represented by the diode $I_{B,\text{inj}}(V_{BE})$. Hole injection into the collector is given by a similar expression (replace E with C and use the correct, doping-density-dependent values for D_h and L_h), and is represented by the diode $I_{B,\text{inj}}(V_{BC})$. The other two diodes represent recombination in the base of electrons injected from the emitter and the collector, as described by (9.15).

two diode-like terms, one associated with each junction, thereby emphasizing the origin of the electrons in the base, with which the holes recombine. Resistors are also shown to represent the resistances of the various quasi-neutral and access regions.[7]

An often-used metric for a bipolar transistor is the **DC, common-emitter current gain** β_0. In this configuration, the emitter lead is common to the input- and output-pairs of terminals, so that the controlling parameter is the base current. Thus

$$\beta_0 = I_C / I_B . \tag{9.23}$$

In the active mode of operation the upper two diodes in the equivalent circuit can be neglected, and the current gain can be simply expressed in a constant, bias-independent form, using (9.11), (9.22) and the relevant part of (9.15). Thus, if the parameter in Fig. 9.6 were I_B rather than V_{BE}, then a family of curves with equal base steps would show equally spaced collector currents in the active region.

In the **common-base connection** the input current is the emitter current and the **DC, common-base current gain** is given by

$$\alpha_0 = I_C / |I_E| . \tag{9.24}$$

A typical value might be $\alpha_0 = 0.99$, which would yield a transistor with a β_0 that was 100 times larger.

Exercises

9.1   Consider an Np^+n, $Al_{0.3}Ga_{0.7}As/GaAs$ HBT, for which the emitter and base doping densities are 5×10^{17} cm^{-3} and 1×10^{19} cm^{-3}, respectively. The HBT is operating in the active mode with a forward bias of 1.4 V.

   (a) Calculate the width of the space-charge region W at the emitter/base junction using the Depletion Approximation.

   (b) Estimate the mean-free-path length $\bar{l}$ for electrons, and determine if it is reasonable to assume that there is no scattering in the junction space-charge region.

9.2   Consider an npn GaAs BJT operating in the active mode with $V_{BE} = 1.25$ V and $V_{BC} = -3.0$ V. The emitter doping density is 10^{18} cm^{-3} and the width of the emitter quasi-neutral region is 100 nm. The corresponding values for the base are 10^{19} cm^{-3} and 25 nm.

   Estimate β_0, and show your calculations of the relevant currents.

9.3   Make the above BJT into an Npn HBT by changing the emitter to lattice-matched InGaP, i.e., $In_{0.49}Ga_{0.51}P$ (see Fig. 8.3). The minority carrier properties of the InGaP can be taken to be the same as for correspondingly doped GaAs.

   Estimate β_0.

---

[7] See Section 14.6.1 for details on how to estimate R_B, which is particularly important in high-frequency applications.

9.4 Compare the values of β_0 from the previous two questions and state the principal reason(s) for their large difference.

9.5 Besides the improvement in current gain, the HBT offers a much reduced base resistivity, which is very important in high-frequency devices.

Re-visit Exercise 9.2 and adjust the base doping density so that this BJT yields a similar β_0 to the HBT of Exercise 9.3.

Compare the base resistivities of the BJT and HBT.

9.6 In Section 9.3.1, the base recombination current was computed from the seemingly inconsistent approximation of a linear profile for the minority carrier electrons. If this approximation is not made, but Shockley boundary conditions are assumed, i.e., $v_R \to \infty$, then the base current is easily shown to be [2]

$$I_{B,\text{rec}} = Aqn_{0B}\left[(e^{V_{BE}/V_{\text{th}}} - 1) + (e^{V_{BC}/V_{\text{th}}} - 1)\right]$$

$$\cdot \frac{D_e}{L_e}\left\{\coth(W_B/L_e) - \frac{1}{\sinh(W_B/L_e)}\right\}. \qquad (9.25)$$

(a) How short does the quasi-neutral base have to be, relative to the electron diffusion length, for this equation to be adequately approximated by (9.15)?

(b) Is this condition satisfied in the HBT of Exercise 9.3?

9.7 Consider an Np^+n, $Al_{0.3}Ga_{0.7}As/GaAs$ HBT under forward bias.

(a) Using Fig. 6.9a as a guide, sketch the energy-band diagram for the emitter/base part of this Type I HBT. Note that the electron energy barrier is parabolic, and that tunnelling from the emitter to the base is possible.

(b) To derive an expression for the transmission probability, start with (5.57), which is an approximate expression for $T(E)$ for a non-rectangular barrier. Apply this equation to the parabolic barrier of height qV_N on the N-side of the junction, and show that

$$T(E) = e^{-\gamma}, \qquad (9.26)$$

where

$$\gamma = \frac{qV_N}{E_0}\left[\sqrt{1-X} - X\ln\left(\frac{1+\sqrt{1-X}}{\sqrt{X}}\right)\right],$$

$$E_0 = (\hbar q/2)\sqrt{N_D/(m_1\epsilon_1)},$$

and

$$X = E/qV_N.$$

9.8 Consider an Np^+n, $Al_{0.3}Ga_{0.7}As/GaAs$ HBT, for which the emitter and base doping densities are 5×10^{17} cm^{-3} and 1×10^{19} cm^{-3}, respectively. The HBT is operating in the active mode with a forward bias of 1.4 V.

(a) Use (9.26) to obtain a plot of $T(E)$ vs. E/qV_N within the energy range for tunnelling, i.e., from Fig. 6.9a, $-\chi_p \le E \le -\chi_N - qV_{bi} + qV_N$.

(b) Now estimate the spectral carrier density $n(E) = g(E)f(E)$ of electrons available for tunnelling. Use the Maxwell-Boltzmann distribution function, and display your result graphically.

(c) The spectral tunnelling flux is $n(E)T(E)$: plot this against the normalized energy E/qV_N.

The result should show that the peak tunnelling flux occurs at an energy of $\approx 0.75qV_N$.

References

[1] D.L. Pulfrey, *Heterojunction Bipolar Transistor*, Wiley Encyclopedia of Electrical and Electronics Engineering, J.G. Webster, Ed., John Wiley & Sons, Inc., vol. 8, 690–706, 1999.

[2] D.L. Pulfrey and N.G. Tarr, *Introduction to Microelectronic Devices*, p. 348, Prentice-Hall, 1989.

10 MOSFET basics

The MOSFET[1] was the subject of a patent in 1933 [1], but did not reach commercial maturity until about thirty years later. The delay was principally due to a lack of understanding of the importance of the oxide/semiconductor interface, and to the time taken to develop suitable fabrication procedures, notably for the growth of the thin gate oxide. Now, in the early 21st century, the science of silicon, and the art and technology of its processing into electronic devices have reached such a state of maturity that billions of Si MOSFETs are made weekly. The claim that the Si MOSFET is the most abundant object made by mankind is difficult to refute [2].

In this chapter the so-called 'long-channel' FET is considered. The basic electrostatics of the device is developed, and the DC current-voltage characteristics are derived using two models that are very widely used in the simulation of Si MOSFET integrated circuits: PSP and SPICE. PSP stands for 'Penn-State Philips', after the two organizations that have been largely instrumental in bringing this surface-potential model to a state of commercial viability [3]. It is the Compact Model Council's new, industrial-standard, MOSFET model.[2] SPICE stands for 'Simulation Program with Integrated Circuit Emphasis'. It was originally developed by Lawrence Nagel at the University of California at Berkeley in the mid-1970s, and has evolved extensively since then [4]. PSP is surface-potential based, whereas SPICE is threshold-voltage based. Here, we start with the more general and accurate surface-potential model, and then we show how the threshold-voltage model is derived from it.

The material in this chapter provides the basis for understanding how the MOSFET has developed into today's dominant digital transistor, as described in Chapter 13. Our treatment draws heavily on the classic work of Tsividis [5].

10.1   Transfer characteristic

A typical Si MOSFET is shown in Fig. 10.1. The particular transistor shown has two np-junctions, with the highly doped (n^+) regions being called the **drain** and the **source**.

---

[1] Metal-Oxide Field-Effect Transistor. In this acronym 'M' literally stands for metal, but it is also used for any highly conductive gate material, e.g., heavily doped polycrystalline silicon, as is still used for most Si MOSFETs, although there is now a return to metal gates in very high-performance devices (see Chapter 13).
[2] Compact Model Council: http://www.geia.org/Standard-Models-and-Downloads

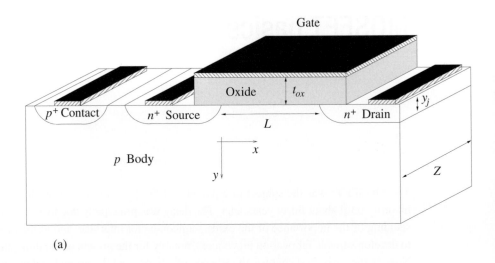

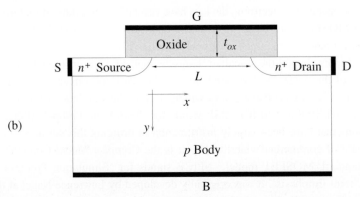

Figure 10.1 Basic MOSFET structure. Note the orientation of the axes x and y. (a) Actual arrangement, showing the body contact on the top surface, and defining the length L and width Z of the channel, the thickness t_{ox} of the oxide, and the depth y_j of the source and drain np-junctions. (b) 2-D representation with the body contact on the bottom.

The p-region is called the **body** or the **substrate**. The separation between the source and drain regions is the defining physical metric; values less than 50 nm are achievable today.[3] We will show how the drain current I_D is due to the flow of electrons from the source to the drain. The conductivity type of the carriers leads to this device being called an **n-channel MOSFET**, or an **N-FET**. A complementary **P-FET** can be realized by reversing the doping type of each region. Only the N-FET will be discussed here.

Strictly speaking, the MOSFET is a four-terminal device. For the moment, let us assume that no voltage is applied between the source and the body, i.e., $V_{SB} = 0$. Application of voltage between the gate and the body V_{GB} ($\equiv V_{GS}$ in this case), for a

---

[3] To attempt to comprehend such a small size, realize that the diameter of a human hair is about 2000 times greater.

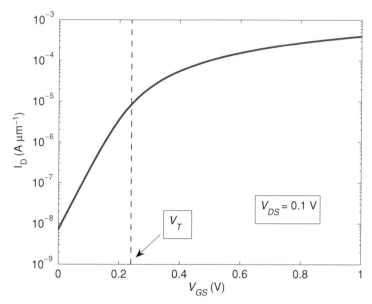

Figure 10.2 Transfer (or gate) characteristic at $V_{DS} = 0.1$ V for a CMOS90 NFET with the properties listed in Appendix C. The threshold voltage is 0.24 V. MEDICI (Synopsys) simulation using the DDE version of (5.24).

given drain-source voltage $V_{DS} > 0$, leads to the **transfer-** or **gate-characteristic** shown in Fig. 10.2. Note how the drain current I_D increases exponentially at first, and then less strongly as V_{GB} is increased. The sequence of events that causes this behaviour is:

- the positive V_{GB} repels holes from the substrate close to the oxide/semiconductor interface (also known as **the surface**). This creates a space-charge layer, across which, in the y-direction, some of V_{GB} is dropped;
- this potential in the top part of the body also causes a voltage drop across the depleted body/source pn-junction, i.e., in the x-direction. This forward biases the source/body diode and electrons are injected into the body;
- the electron injection is heaviest at the surface because this is where the potential in the body is highest. Thus, the electrons form a thin **channel** at the surface. In accordance with standard pn-junction theory (6.29), this charge increases exponentially with applied bias;
- however, eventually, this electron charge becomes so dense that it electrostatically screens the body from the gate. Further increases in V_{GB} are absorbed almost entirely in the oxide, and the exponential relationship between channel charge and gate bias is lost.

Don't be misled by the above into thinking that the basic, exponential, I-V relationship of a forward-biased diode no longer applies at high gate bias. It is just that V_{GB} is no longer the operative bias. The relevant bias for the forward-biased source/channel diode is $(\psi_s - V_S)$, where ψ_s is the **surface potential** in the body, and V_S is the source potential, which we'll take to be zero here. I_D is still related exponentially to ψ_s, but the

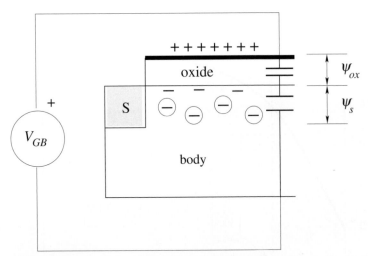

Figure 10.3 Representation of the oxide and the electrons and ions in the semiconductor as capacitors, for the purpose of determining the surface potential.

latter is no longer nearly equal to V_{GB}. This can be appreciated from Fig. 10.3, which shows the potential-divider action of the series arrangement of the oxide capacitance and the semiconductor capacitance, i.e.,

$$\Delta\psi_s = \Delta V_{GB} \frac{1}{1 + C_s/C_{ox}}. \tag{10.1}$$

C_{ox} can easily be appreciated by its geometric, parallel-plate, nature; C_s is due to the gate-potential-induced changes in charge in the channel ΔQ_n, and in the body ΔQ_b. Writing the sum of these charges as ΔQ_s, the capacitances per unit area, as defined and discussed in Chapter 12, are given here by

$$C_{ox} = -\frac{\Delta Q_s}{\Delta\psi_{ox}} \equiv \frac{\epsilon_{ox}}{t_{ox}} \quad \text{and} \quad C_s = -\frac{\Delta Q_s}{\Delta\psi_s}, \tag{10.2}$$

where ψ_{ox} is the voltage drop across the oxide, ϵ_{ox} and t_{ox} are the permittivity and thickness of the oxide, respectively.

Equation (10.1) supports our earlier description of the gate characteristic: at low gate bias, the charge in the semiconductor is small, so $C_s \ll C_{ox}$, and ψ_s changes almost exactly as V_{GB}, giving an exponential relationship between I_D and gate bias; at higher bias, C_s becomes appreciable due to the increased charge in the channel, and, consequently, the change in ψ_s with V_{GB} is much diminished. The electron charge Q_n, the movement of which constitutes the current I_D, still increases exponentially with ψ_s, but no longer exponentially with V_{GB}. The boundary between the exponential and non-exponential regions is gradual, but it is usually associated with the **threshold voltage** V_T. The region $V_{GS} < V_T$ is the **sub-threshold region**, and the drain current at $V_{GS} = 0$ is called the **OFF current**. For $V_{GS} > V_T$ we have the **super-threshold region**, and the drain current in this region is often called the **ON current**.

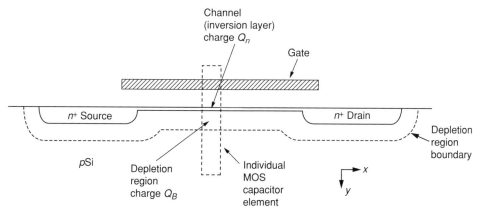

Figure 10.4 MOS capacitor resulting from taking a slice through the MOSFET and ignoring the source and drain. From Pulfrey and Tarr [6, Fig. 7.9].

10.2   Electrostatics

10.2.1   MOS capacitor

Consider a slice of the MOS structure from gate to body (see Fig. 10.4); ignore the lateral connections to the source and drain, and imagine a metallic contact to the body on the bottom. Such a two-terminal structure is called a MOS capacitor. The three components of the structure – gate, insulating oxide, semiconducting body – are represented energetically in Fig. 10.5(a). For specificity, the gate is taken to be heavily doped, n-type polysilicon, for which the Fermi level is coincident with the conduction-band edge, the insulator is silicon dioxide, and the semiconducting body is p-type silicon. To equilibrate the system, the Fermi level in the body must be raised; this is accomplished by transfer of electrons from the gate. The transferred electrons recombine with holes in the p-type body, creating a space-charge region of ionized acceptors near to the interface with the oxide. Thus, there is band-bending in this region. The gate acquires a net positive charge due to its loss of electrons. The charge difference across the oxide creates an electric field therein. The total potential differences across the oxide and semiconductor are ψ_{ox} and ψ_s, respectively (see Fig. 10.5b). The sum of these two gives the built-in voltage V_{bi} for the MOS capacitor. Inspection of Fig. 10.5b reveals

$$E_0 - qV_{bi} - \Phi_G = E_0 - \Phi_S$$

$$V_{bi} = \frac{\Phi_S - \Phi_G}{q}, \tag{10.3}$$

where Φ_G and Φ_S are the **work functions** of the gate and the semiconducting body, respectively.

Note that the difference in work functions in the case of Fig. 10.5b is such that the semiconductor surface region is effectively less p-type than the rest of the body, i.e., it has assumed a more n-type character. This phenomenon is known as **inversion**. As the intention is to create a high density of electrons at the surface, it follows that this will be

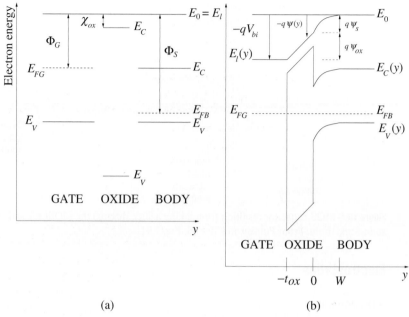

Figure 10.5 Band diagrams for the electron-energy variation in the y-direction for the MOS capacitor in Fig. 10.4. For the gate, $\tilde{E}_{FG}$ is taken to be coincident with the conduction-band edge, as would be the case for n^+ polysilicon. (a) The separated components. (b) The completed band diagram at equilibrium. χ_{ox} is the electron affinity of the oxide, Φ_G and Φ_S are the work functions of the gate and the semiconductor, respectively.

achieved at a lower positive applied gate/body voltage than would be the case if there were no band-bending at equilibrium, i.e., if V_{bi} were zero. The no-band-bending condition is called the **flat-band condition**; it can be achieved in our example by applying a negative potential to the gate (see Fig. 10.6a). Evidently, $V_{fb} = -V_{bi}$. The electron concentration (per unit volume) in the semiconductor is given by

$$n(y) = n_i e^{(E_{FB} - E_{Fi}(y))/kT} \equiv n_i e^{(\psi(y) - \phi_B)/V_{th}} .$$ (10.4)

Deep in the body of the device we have

$$n(B) = n_i e^{-\phi_B/V_{th}} = \frac{n_i^2}{N_A} ,$$ (10.5)

where a uniform doping density of acceptors has been assumed and ϕ_B is defined as

$$\phi_B = \frac{1}{q} [E_{Fi}(B) - E_{FB}] .$$ (10.6)

These equations can be used to express n in terms of N_A:

$$n(y) = N_A e^{(\psi(y) - 2\phi_B)/V_{th}} ,$$ (10.7)

where $\psi_s \equiv \psi(0)$ is the potential at the semiconductor surface. Evidently, when ψ_s equals $2\phi_B$, then $n(0) = N_A$. This condition which is illustrated in Fig. 10.6b, is defined

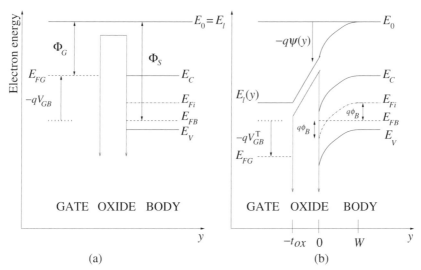

Figure 10.6 (a) Energy-band diagram at flatband: $V_{GB} = V_{fb} \equiv -V_{bi}$. (b) Energy-band diagram at the onset of strong inversion: $\psi_s = 2\phi_B$, $V_{GB} = V_{GB}^T$.

as the **onset of strong inversion** in the surface channel of a MOS capacitor. The electron charge per unit area is given by

$$Q_n = \int_0^W -qn(y)\,dy,\qquad(10.8)$$

where W is the depth of the space-charge layer that extends from the surface into the body.

10.2.2   MOSFET

Now let us consider the effect of contacting the inversion layer via the drain and source regions, between which we apply a voltage V_{DS}. This voltage is dropped along the channel and is the driving force for electron flow from the source to the drain. To recognize the non-equilibrium nature of the situation now that charge flow is permissible, a separate electron quasi-Fermi level E_{Fn} is introduced (see Fig. 10.7). The difference in quasi-Fermi levels for electrons at the surface and for holes deep in the body defines the channel/body voltage V_{CB}:

$$-qV_{CB}(x) = E_{Fn}(x) - E_{FB}.\qquad(10.9)$$

The presence of a positive V_{CB} reduces the field in the oxide, and increases the potential drop across the depletion layer, as shown in Fig. 10.7. The effect of the former is to reduce the charge on the gate Q_G; and the effect of the latter is to make the body charge Q_b more negative. Overall charge neutrality demands, therefore, that the channel charge Q_n becomes more positive. Thus, with reference to the two band diagrams in Fig. 10.7, the reduction in channel charge in (b) means that the applied gate/body voltage is no longer sufficient to put the channel in inversion at the point x

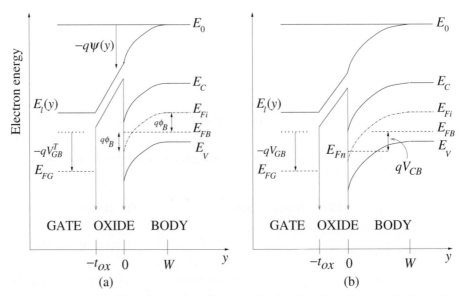

Figure 10.7 Energy-band diagrams in the y-direction, showing the effect of a channel voltage V_{CB} arising from the application of a drain-source voltage. (a) The threshold condition with $V_{DS} = 0$. (b) With $V_{DS} > 0$ and V_{GB} unchanged.

under consideration. As $x \to 0$, i.e., as the source end of the channel is approached, the effect diminishes. Thus, the electron concentration is a function of x and y, and (10.7) becomes

$$n(x, y) = N_A e^{[\psi(x,y) - 2\phi_B - V_{CB}(x)]/V_{\text{th}}} . \tag{10.10}$$

If we can determine $n(x, y)$, then the drain current should easily follow, using the Drift-Diffusion Equation, for example. The following Sections 10.3 and 10.4 describe two approaches to achieve this goal.

10.3   MOSFET *I-V* characteristics from the surface-potential model

10.3.1   Surface potential

The MOSFET we are considering is assumed to be invariant in the z-direction, so Poisson's Equation for the body and channel regions is, neglecting hole charge in the body's space-charge region,

$$\frac{\partial^2 \psi}{\partial x^2} + \frac{\partial^2 \psi}{\partial y^2} = \frac{q N_A}{\epsilon_s} \left[1 + e^{(\psi - 2\phi_B - V_{CB})/V_{\text{th}}} \right] . \tag{10.11}$$

The controlling field for the charge in a well-designed FET is $\mathcal{E}_y$. Recognizing this, and simultaneously allowing a tractable solution to (10.11), we make the **Gradual Channel**

Approximation:

$$\frac{\partial^2 \psi}{\partial x^2} \ll \frac{\partial^2 \psi}{\partial y^2}. \tag{10.12}$$

To solve (10.11), multiply both sides of it by $(\frac{\partial \psi}{\partial y})\partial y$ and then integrate from deep in the body, where $\psi = 0$ (assuming V_B, the potential applied to the body terminal, is zero) and $\partial \psi / \partial y = 0$, to the surface, where $\psi = \psi_s$ and $\partial \psi / \partial y = \mathcal{E}_y(0)$. The last limit is the field in the semiconductor at the oxide/semiconductor interface. Using this and Gauss's Law ($\mathcal{E}_y(0) = -Q_s/\epsilon_s$), we can then get an expression for the charge per unit area in the semiconductor:

$$Q_s = -\sqrt{2q\epsilon_s N_A}\sqrt{\psi_s + \frac{k_B T}{q}\left[e^{\psi_s/V_{th}} - 1\right]e^{-(2\phi_B+V_{CB})/V_{th}}}, \tag{10.13}$$

where Q_s is a function of x through $\psi_s(x)$. As we are working towards an expression for the terminal characteristics of the device, we need to bring in the applied voltages. For V_{GB} we can do this by realizing from Fig. 10.5 and Fig. 10.6 that

$$V_{GB} - V_{fb} = \psi_{ox}(x) + \psi_s(x), \tag{10.14}$$

and by applying Gauss's Law once more to obtain

$$Q_s(x) = -Q_G(x) = -C_{ox}\psi_{ox} = -C_{ox}\left[V_{GB} - V_{fb} - \psi_s(x)\right], \tag{10.15}$$

where Q_G is the charge per unit area on the gate electrode.

(10.13) and (10.15) provide an implicit relation between ψ_s and V_{GB}:

$$\psi_s = V_{GB} - V_{fb} - \gamma\sqrt{\psi_s + \frac{k_B T}{q}\left[e^{\psi_s/V_{th}} - 1\right]e^{-(2\phi_B+V_{CB})/V_{th}}}, \tag{10.16}$$

where $\gamma = \sqrt{2q\epsilon_s N_A}/C_{ox}$ is called the **body factor**.[4] The x-dependence of ψ_s enters through the channel/body voltage $V_{CB}(x)$.

Before examining how ψ_s changes with V_{GB}, let us use the surface potential to define the state of inversion in the channel of the MOSFET. Here, we take inversion to mean the presence of electrons at the surface, as is required for there to be a drain current.

A point x in the channel is in **weak inversion** when

$$0 < \psi_s(x) < \phi_B + V_{CB}(x); \tag{10.17}$$

a point x in the channel is in **moderate inversion** when

$$\phi_B + V_{CB}(x) < \psi_s(x) < 2\phi_B + V_{CB}(x); \tag{10.18}$$

a point x in the channel is in **strong inversion** when

$$2\phi_B + V_{CB}(x) < \psi_s(x). \tag{10.19}$$

---

[4] The name refers to the fact that the term $\sqrt{2\epsilon_s N_A}$ comes from consideration of the charge in the body.

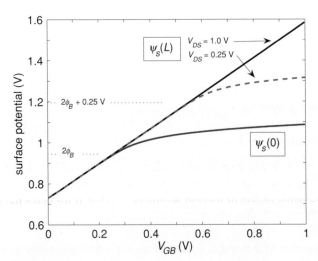

Figure 10.8 Surface potential vs. V_{GB} for a CMOS90 N-FET with parameters listed in Appendix C.

Thus, at the **onset of moderate inversion** $n(x, 0) = n_i$, and at the **onset of strong inversion** $n(x, 0) = N_A$. These important markers are shown in Fig. 10.8, which is a plot of ψ_s vs. V_{GB} at the two ends of the channel: $\psi_s(y) = \psi_s(0)$ at the source end and $\psi_s(L)$ at the drain end. A drain-source voltage is applied, i.e., $V_{SB} = 0$ and $V_{DB} > 0$.

Note that, for this particular example, V_{fb} is sufficiently negative that the source is in moderate inversion even at $V_{GB} = 0$. As V_{GB} increases, ψ_s follows with a linear correspondence. This is the situation described in Section 10.1 when Q_s is so small that $C_s \ll C_{ox}$, and the capacitor-divider action allows ψ_s to track V_{GB} almost perfectly. When ψ_s at the source end rises to $2\phi_B$, the electron concentration per unit volume reaches N_A at the surface, and the strong-inversion region is entered. Shortly after this, the charge per unit area in the semiconductor $(Q_n + Q_b)$ is sufficiently large (negatively) that C_s becomes comparable to C_{ox}. As the surface concentration of electrons increases exponentially with ψ_s (due to injection across the source/body np-junction), it is not long before C_s becomes much larger than C_{ox}, leading to near-saturation of $\psi_s(0)$. Electrostatically, the near-saturation of $\psi_s(0)$ occurs because the body becomes shielded, or screened, from the gate by the large negative charge Q_n in the channel, i.e., field lines from any new charge on the gate, due to any increase in V_{GB}, terminate almost entirely on the electron charge in the channel. Thus, there is barely any further change in the depletion layer at the source-end of the channel.

Turning now to the drain-end of the channel, Fig. 10.8 shows the situation for two values of V_{DS}. When $V_{DS} = 0.25$ V, it can be seen clearly that strong inversion at the drain is not reached until $\psi_s(L) > 2\phi_B + V_{DS}$, i.e., until the effect of $V_{CB}(L) = V_{DS}$ has been overcome by V_{GB}. The drain-induced potential on the channel reduces the y-directed field in the oxide, and consequently requires that less negative charge be present in the channel. It follows that strong inversion can only be re-established by increasing V_{GB}. Note that, in our particular example, at $V_{DS} = 1$ V, if V_{GB} is limited to 1 V also, then strong inversion will not be achieved at the drain end of the channel!

10.3.2   Drain current

The charge per unit area in the semiconductor comprises electrons near the surface Q_n, and ionized acceptors in the depleted region of the body Q_b:

$$Q_s(x) = Q_n(x) + Q_b(x).$$ (10.20)

To proceed towards finding the current we need to know Q_n, the density per unit area of the mobile charge. To do this we make the **Charge Sheet Approximation**, i.e., we assume that all the electrons reside in a sheet at the semiconductor surface. The benefit of this is that there would be no voltage drop through such a sheet *in the y-direction*, so the voltage $\psi_s - V_B$, where V_B is the potential applied to the body terminal, is dropped entirely across the space-charge region of ionized acceptors. Further, if we make the Depletion Approximation for this region, i.e., neglect the charge due to holes, then Q_b is simply

$$Q_b(x) = -qN_AW(x) = -\sqrt{2\epsilon_s qN_A\psi_s(x)},$$ (10.21)

where $W(x)$ is the depletion-region width. The charge density per unit area in the channel then follows from (10.15) and (10.20):

$$Q_n(x) = -C_{ox}\left[V_{GB} - V_{fb} - \psi_s(x) - \gamma\sqrt{\psi_s(x)}\right].$$ (10.22)

The electron current density in a 2-D sheet, using the Drift-Diffusion Equation is

$$\vec{J}_e = Q_n\vec{v}_d - D_e\frac{dQ_n}{dx},$$ (10.23)

and the electron current is

$$I_e = ZQ_n\mu_e\frac{d\psi_s}{dx} - Z\frac{k_BT}{q}\mu_e\frac{dQ_n}{dx},$$ (10.24)

where Z is the width of the transistor, as noted in Fig. 10.1, and we have invoked the non-degenerate form of the Einstein Relation. The flow of electrons in our case is in the positive x-direction inside the device, so the current at the drain is positive, by IEEE convention.[5] Thus

$$\int_0^L I_D\,dx \equiv -\int_0^L I_e\,dx = -Z\int_{\psi_s(0)}^{\psi_s(L)}\mu_eQ_n\,d\psi_s + Z\frac{k_BT}{q}\int_{Q_n(0)}^{Q_n(L)}\mu_e\,dQ_n.$$ (10.25)

Because we have assumed that there is no hole conduction, and that there is current leakage neither to the gate nor to the substrate, I_e is constant throughout the device. Thus, the left-hand side of the above equation is simply $I_D L$. On the right-hand side, we recognize that the mobility will be affected by the y-directed electric field, causing scattering at the semiconductor surface where it meets the oxide. Because of this, we call the mobility in the channel the **effective mobility**, and expect it to be less than the mobility in the bulk of a material with near-perfect crystallinity. At this stage of our model development, we also assume that μ_{eff} does not depend upon the x-directed

---

[5] This convention defines a device current as positive when positive charge enters a terminal of the device.

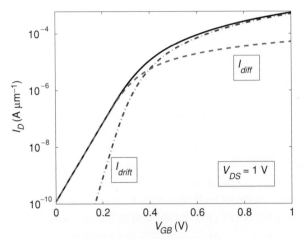

Figure 10.9 MOSFET gate characteristic from the surface potential model. Model parameters as given in Appendix C for a CMOS90 N-FET.

electric field, so the drain current can be expressed as

$$I_D = \frac{Z}{L}\mu_{\text{eff}}\left[\int_{\psi_s(0)}^{\psi_s(L)} -Q_n \, d\psi_s + \frac{k_B T}{q}\int_{Q_n(0)}^{Q_n(L)} dQ_n\right]. \qquad (10.26)$$

To perform the first integration, use the dependence of Q_n on ψ_s from (10.22). The second integration is trivial, and, after doing it, substitute for $Q_n(0)$ and $Q_n(L)$ from (10.22). Both the drift and diffusion currents now become functions of ψ_s at the two ends of the channel:

$$I_{D,\text{drift}} = \frac{Z}{L}\mu_{\text{eff}}C_{ox}\left[(V_{GB} - V_{fb})(\psi_s(L) - \psi_s(0))\right.$$

$$\left. -\frac{1}{2}(\psi_s(L)^2 - \psi_s(0)^2) - \frac{2}{3}\gamma\left(\psi_s(L)^{3/2} - \psi_s(0)^{3/2}\right)\right]$$

$$I_{D,\text{diff}} = \frac{Z}{L}\mu_{\text{eff}}C_{ox}\frac{k_B T}{q}\left[(\psi_s(L) - \psi_s(0)) + \gamma\left(\psi_s(L)^{1/2} - \psi_s(0)^{1/2}\right)\right], \qquad (10.27)$$

where $\psi_s(L)$ and $\psi_s(0)$ are obtained from (10.16), with $V_{CB} = V_{DB}$ and V_{SB}, respectively.

Note how each current component would merely change sign if $\psi_s(L)$ and $\psi_s(0)$ were reversed. This symmetry is to be expected because of the geometrical symmetry of the FET.

Examples of the current from this model are shown in Fig. 10.9 for the case of varying V_{GB}, which produces the **gate characteristic**, and in Fig. 10.10 for the case of varying V_{DB}, which produces the **drain characteristic**. The gate characteristic is like the one shown in Fig. 10.2. The initial exponential form of the curve, and then the 'flattening out', were explained in Section 10.1. The new information in Fig. 10.9 is the dominance of the diffusion current in the 'exponential' regime, and the dominance of drift at higher V_{GB}. Drift gains importance over diffusion as the field $\mathcal{E}_y$ strengthens with V_{GB}, and makes for a more uniform electron distribution along the channel, i.e., in the x-direction.

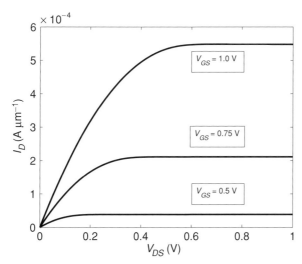

Figure 10.10 MOSFET drain characteristic from the surface potential model. Note: $V_{SB} = 0$, so that $V_{GS} \equiv V_{GB}$ and $V_{DS} \equiv V_{DB}$. Model parameters as given in Appendix C for a CMOS90 N-FET.

The drain characteristic is shown on a linear-linear plot for three values of V_{GS} corresponding to the non-exponential part of the gate characteristic. Initially, the $I_D - V_{DS}$ relation is essentially linear: V_{CB} is small everywhere, the channel is in strong inversion along its entire length, Q_n does not vary much with x. In other words, the channel behaves like a resistor. As V_{DB} increases, strong inversion is lost progressively towards the drain, until the drain-end becomes moderately or weakly inverted (see Fig. 10.8). Under these conditions, Q_n near $x = L$ is so small that the field due to the gate potential terminates almost entirely on the ionic charge in the body, i.e., V_{CB} in this region no longer affects the surface potential. Thus, the latter becomes solely determined by V_{GB}. This region of strong correlation between ψ_s and V_{GB} is evident in the near-linear portion of Fig. 10.8. An estimate of the relation can be obtained by setting $Q_n = 0$ in (10.22)

$$\psi(L) \equiv \psi_s(Q_n = 0) = \left[-\frac{\gamma}{2} + \sqrt{\frac{\gamma^2}{4} + V_{GB} - V_{fb}} \right]^2 \approx V_{GB} - V_{fb}, \quad (10.28)$$

where the approximate form applies if $(V_{GB} - V_{fb}) \gg \gamma^2/4$. Under these circumstances, the overall potential drop along the channel in the x-direction reaches its maximum value, i.e., $\psi_s(L) - \psi_s(0) \approx V_{GB} - V_{fb} - 2\phi_B - V_{SB}$, assuming that the source-end of the channel is in strong inversion. Thus, the driving force for the drain current reaches its maximum value and, consequently, I_D saturates. To put it another way, the drain is 'sinking' all the charge that the source can supply for a given V_{GB}.[6]

---

[6] This behaviour is characteristic of a long-channel device. In Chapter 13 we consider short-channel devices in which L is so small that V_{DB} also affects $\psi_s(0)$; this influences I_D and is undesirable because it is the drain's job merely to sink the charge, not to control its flow: that is the gate's job.

10.3.3   Pinch-off and channel-length modulation

Two cautions are issued about the above description of saturation of I_D.

Firstly, note that Q_n can never equal zero if current is to be maintained, because some electrons are needed to carry the charge. The condition for saturation of I_D is not really $Q_n(L) = 0$, but rather $|Q_n(L)| \ll |Q_b(L)|$. Thus, the channel never **pinches off** completely.

Secondly, as V_{DB} increases beyond the value at which $|Q_n(L)| \ll |Q_b(L)|$, then the condition $|Q_n(x)| \ll |Q_b(x)|$ occurs at x values closer to the source. This is as though the effective channel length of the device were decreasing. This has the effect of increasing the drain current; the phenomenon is known as **channel-length modulation**. In modern devices, e.g., in which $L \leq 100$ nm, the effect is masked by the **short-channel effect**, which we will discuss at length in Chapter 13.

10.4   MOSFET *I-V* characteristics from the strong-inversion, source-referenced model

The next MOSFET model to be described is used in SPICE, and is the one that is most frequently featured in textbooks. It is not as accurate as the surface-potential model, nor does it preserve the symmetry of the FET's characteristics, nor does it provide a single expression that describes the drain current in all regimes of operation. However, it is completely analytical in nature, and yields useful equations for circuit-simulation purposes. However, as noted earlier, the surface-potential model has recently been adopted by CMC as the industry standard.[7] The computational burden of the surface-potential model has been lessened by the advent of faster computers, and any disadvantage it still has in this regard is considered by some to be outweighed by its greater accuracy. In this section we derive the strong-inversion, source-referenced model as a rather gross simplification of the surface-potential model. We also use it to introduce the effect of a field-dependent mobility on the drain current.

10.4.1   Basic assumptions of the model

The model has 'strong inversion' in its title because it assumes that the surface potential is everywhere at its strong-inversion value, i.e.,

$$\psi_s(x) = 2\phi_B + V_{CB}(x). \tag{10.29}$$

Obviously, this assumption is going to cause the model to breakdown at high V_{DS}, when we have already seen that strong inversion is lost at the drain.

The model has 'source-referenced' in its title because it refers all potentials to the source, rather than to the body, e.g.,

$$V_{CS}(x) = V_{CB}(x) - V_{SB}. \tag{10.30}$$

---

[7] Compact Model Council, http://www.geia.org/index.asp?bid=597

This change in reference causes the source/drain symmetry of the device to be lost, but it opens the way for the introduction of a very useful parameter, the **threshold voltage**.

The second assumption of the model is that

$$V_{CS} \ll 2\phi_B + V_{SB} \,. \tag{10.31}$$

This just reinforces the notion of strong inversion everywhere in the channel, i.e., the potential in the channel should not be disturbed too much from its strong-inversion value at the source by the application of a voltage V_{DS}.

10.4.2 Drain current for constant mobility

The surface-potential model indicates that the drain current is mainly due to drift when the source is in strong inversion. Accordingly, we start by taking the drift portion of the expression for I_D from the surface-potential model (10.27). This expression contains the surface potentials at the source- and at the drain-ends of the channel. The implication of the first assumption of the model is that

$$\psi_s(0) = 2\phi_B + V_{SB}$$

$$\psi_s(L) = 2\phi_B + V_{DB}$$

$$\psi_s(L) - \psi_s(0) = V_{DS}$$

$$\psi_s(L) = 2\phi_B + V_{SB} + V_{DS} \,. \tag{10.32}$$

The implication of the second assumption of the model is that

$$V_{DS} \ll 2\phi_B + V_{SB} \,. \tag{10.33}$$

We make use of this inequality to perform a binomial expansion to the second order of the term $\psi_s^{3/2}$ in the drift current part of (10.27). The result is a compact expression for the drain current:

$$I_D = ZC_{ox}\left[V_{GS} - V_T - m\frac{V_{DS}}{2}\right] \cdot \mu_{\text{eff}}\frac{V_{DS}}{L} \,, \tag{10.34}$$

where various terms have been collected together to define two useful parameters: the **threshold voltage**

$$V_T = V_{fb} + 2\phi_B + \gamma\sqrt{2\phi_B + V_{SB}} \,, \tag{10.35}$$

and the **body-effect coefficient**

$$m = 1 + \frac{\gamma}{2\sqrt{2\phi_B + V_{SB}}} \,. \tag{10.36}$$

Note that the expression for the threshold voltage contains no reference to the channel length L, nor to the drain/source bias V_{DS}, i.e., these properties do not influence conditions at the source. For this reason, (10.35) is appropriate for FETs that exhibit so-called

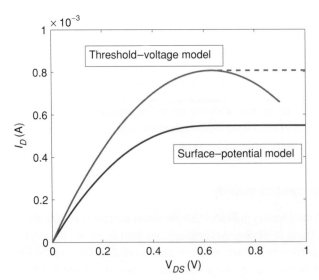

Figure 10.11 Drain characteristic: comparison of SPICE Level 1 model and the PSP model. Model parameters as given in Appendix C for a CMOS90 N-FET at $V_{GS} = 1$ V.

'long-channel' behaviour. 'Short-channel' effects on the threshold voltage are discussed in Chapter 13.

We'll return to these model parameters later, but now let us focus on Fig. 10.11, in which (10.34) is plotted. Note that for low V_{DS}, when $V_{DS} \ll 2(V_{GS} - V_T)/m$, then (10.34) reduces to the linear relationship of a resistor. At higher V_{DS} the current starts to decrease! This strange behaviour is indicative of the breakdown of the model, and, as could have been anticipated from our earlier discussion, this occurs when strong inversion is lost at the drain. In fact, the value of V_{DS} at which the model breaks down can be determined from Fig. 10.8. Note from this figure that, for a supply voltage of 1 V, the maximum value of $\psi_s(L)$ is about 1.58 V. As $2\phi_B = 0.95$ V in our example, strong inversion is lost at the drain when $V_{DS} = 0.63$ V. In other words, ψ_s in Fig. 10.8 would start to flatten-out at $V_{GB} = 1$ V when $V_{DS} = 0.63$ V. This is the voltage at which the drain current in Fig. 10.11 reaches its peak.

The way that the model copes with its deficiency is to assume that I_D stays constant at the value at the top of the curve. Differentiation of (10.34) with respect to V_{DS} indicates that the maximum in current occurs at

$$V_{DS} \equiv V_{DSsat} = \frac{V_{GS} - V_T}{m}, \tag{10.37}$$

i.e., at $V_{DS} = 0.63$ V in our example. V_{DSsat} is the **saturation voltage**. Substituting V_{DSsat} into (10.34) gives an expression for the **saturation current**

$$I_{Dsat} = \frac{Z}{L} C_{ox} \mu_{\text{eff}} \frac{(V_{GS} - V_T)^2}{2m}. \tag{10.38}$$

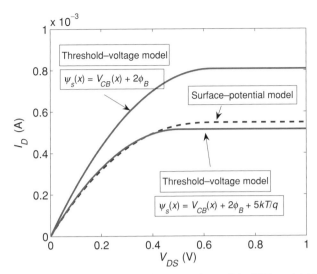

Figure 10.12 Drain characteristic: comparison of the PSP model (dashed line) with the SPICE Level 1 model, for both the case of $\psi_s(0) = 2\phi_B$ and for the case of $\psi_s(0)$ increased above this value by $5V_{\mathrm{th}}$. Model parameters as given in Appendix C for a CMOS90 N-FET at $V_{GS} = 1$ V.

This constant-mobility model is often referred to as **the square-law model**. This equation for the saturation current and (10.34) for the below-saturation region constitute the **Level 1** model in SPICE.

10.4.3 Comparison of the surface-potential and SPICE models

The drain characteristics for the two models are compared in Fig. 10.12. It can be seen, and also from Fig. 10.11, that the agreement is not very good. This leads us to call into question the validity of the first assumption (10.29). Consider the source-end of the channel, for example. The assumption there is that $\psi_s(0) = 2\phi_B + V_{SB}$ at the onset of strong inversion. Fig. 10.8 shows that this is, indeed, the value at which ψ_s starts to increase less strongly with V_{GB}, but it is clearly not the value at which ψ_s becomes relatively flat, indicating the screening of the body from the gate voltage by the large charge density in the channel. This does not happen until about 0.1 V later, i.e., until the volumetric charge density of electrons at the surface has become much greater than N_A. This observation was made many years ago [7], but seems to have been largely ignored. Tsividis suggested that the surface potential at strong inversion should be raised by about $6V_{\mathrm{th}}$. Here, in Fig. 10.12, we find that augmenting $2\phi_B$ by $5V_{\mathrm{th}}$, brings the SPICE model into much better agreement with the surface-potential model.

10.4.4 Threshold voltage, body-effect coefficient and channel charge density

We now discuss the threshold voltage and the body-effect coefficient, first introduced in Section 10.4.2, and relate them to the charge density in the channel Q_n.

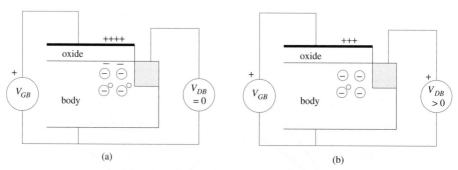

(a)   (b)

Figure 10.13 Illustration of the body effect at the drain end of the channel. The application of $V_{DS} > 0$ in (b) changes the charge distributions from their values at $V_{DS} = 0$, as shown in (a). $m(L) \equiv |\Delta Q_n(L)/\Delta Q_G(L)| = 2$ in this case.

The surface-potential model expresses Q_n in (10.22), where the term $\sqrt{\psi_s}$ appears. Applying the second assumption of the SPICE model to this term and expanding it to first-order, the result is

$$Q_n(x) = -C_{ox}\left[V_{GS} - V_{fb} - 2\phi_B - \gamma\sqrt{2\phi_B + V_{SB}}\right.$$

$$\left. -V_{CS}(x)\left\{1 + \frac{\gamma}{2\sqrt{2\phi_B + V_{SB}}}\right\}\right], \qquad (10.39)$$

where V_G has been referenced to the source, using $V_{GS} = V_{GB} - V_{SB}$. This equation can then be written compactly as

$$Q_n(x) = -C_{ox}\left[V_{GS} - V_T - mV_{CS}(x)\right]. \qquad (10.40)$$

At the source-end of the channel

$$Q_n(0) = -C_{ox}\left[V_{GS} - V_T\right]. \qquad (10.41)$$

This leads to the useful interpretation of the threshold voltage as the gate/source voltage at which the channel becomes strongly inverted at the source-end. In light of the discussion at the end of the previous subsection, V_T would be even more useful if the occurrences of $2\phi_B$ in its expression (10.35) were augmented by $5V_{th}$.

Turning now to the body-effect coefficient m, recall that it entered into the strong-inversion model during the expansion of the term involving $\psi_s(0)^{3/2}$ and $\psi_s(L)^{3/2}$ in the equation for the drift current in the PSP model (10.27). Its appearance in the strong-inversion model is an attempt to include in this model the effect of the body in contributing to the reduction in channel charge Q_n when V_{DS} is increased. To appreciate this, consider Fig. 10.13. The left-hand sketch illustrates conditions at the drain end of a FET with $V_{DS} = 0$, and with the channel in strong inversion everywhere. The portion of the gate that is shown has 4 positive charges; these are complemented by 2 negative charges in the channel and 2 negative charges in the body. The right-hand sketch shows the charge distribution after application of some $V_{DS} > 0$. The positive potential on the drain has removed the electrons from the drain-end of the channel: one electron flowed around the drain/body circuit and ionized another acceptor; the other flowed around the drain/gate circuit and annihilated one positive charge on the gate.

In a two-terminal structure, such as a parallel-plate capacitor, there is a one-to-one correspondence between the change in charges on the two plates, which would be represented by Q_n and Q_G in this case. However, as Fig. 10.13b illustrates, the presence of a third region, the body, causes $|Q_n| > Q_G$, i.e., a loss of channel charge in excess of the corresponding loss of gate charge. This adversely impacts the drain current, and warrants the inclusion of m in the expressions for I_D in the strong-inversion model.[8]

Mathematically, considering a change in V_{DS} that changes the channel/source voltage from 0 to $V_{CS}(x)$ at some point x in the region shown in Fig. 10.13b, the changes in charge are

$$\Delta Q_G(x) = C_{ox}\Delta\psi_{ox}(x) \qquad = -C_{ox}V_{CS}(x)$$

$$\Delta Q_b(x) = \Delta(C_b(x)V_{CB}(x)) \approx -C_b(x)V_{CS}(x)$$

$$\Delta Q_n(x) = -\Delta Q_G - \Delta Q_b = (C_{ox} + C_b(x))V_{CS}(x)$$

$$\equiv C_{ox}\,m(x)\,V_{CS}(x), \qquad (10.42)$$

where an alternative to (10.36) for the definition of m has emerged:[9]

$$m(x) = 1 + \frac{C_b(x)}{C_{ox}}. \qquad (10.43)$$

The 'bottom line' here is that $m(x)$ is the ratio of the change in charge in the channel to the change in charge on the gate:

$$m(x) = \left|\frac{\Delta Q_n(x)}{\Delta Q_G(x)}\right|. \qquad (10.44)$$

Generally, $m(x) > 1$ because some of ΔQ_n goes to change the charge in the body, e.g., $m(L) = 2$ in Fig. 10.13. In practice, m is closer to 1 than to 2.

10.4.5 I_D when mobility is field-dependent

We now remove one of our earlier assumptions by allowing for the dependence of the effective mobility on the longitudinal field in the channel $\mathcal{E}_x$. The relationship between the drift velocity and $\mathcal{E}_x$ is shown in Fig. 5.2, and can be expressed as

$$\frac{1}{v_d} = \frac{1}{\mu_{eff}|\mathcal{E}_x|} + \frac{1}{v_{sat}}. \qquad (10.45)$$

The drift current can be simply written as the charge times the drift velocity

$$I_D = ZC_{ox}\,[V_{GS} - V_T - mV_{CS}(x)]\cdot v_d(x). \qquad (10.46)$$

---

[8] Note that m is not explicit in the PSP model because that model takes into account the effect of V_{DS} on Q_n via its more exact treatment of the surface potential.

[9] It is left as an exercise for the reader to show that the two expressions for m are equivalent. See Exercise 10.3.

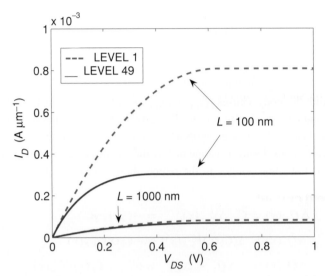

Figure 10.14 Comparison of the SPICE Level 1 and Level 49 models. The device parameters other than the gate length, which is the parameter here, are listed in Appendix C for a CMOS90 N-FET at $V_{GS} = 1$ V.

Substituting for v_d, and using $|\mathcal{E}_x| = dV_{CS}(x)/dx$, integration of (10.46) gives

$$I_D = ZC_{ox}\left[V_{GS} - V_T - m\frac{V_{DS}}{2}\right] \cdot \mu_{\text{eff}}\frac{V_{DS}}{L + (\mu_{\text{eff}}V_{DS}/v_{\text{sat}})} . \quad (10.47)$$

As in (10.34), the charge appears as if it is evaluated where $V_{CS} = V_{DS}/2$, but the velocity is modified. Clearly, if v_{sat} were infinite, the two equations would be the same.

This equation also exhibits a maximum value when the basic assumption of strong inversion breaks down at the drain-end of the channel (see Exercise 10.6). As in the constant-mobility case, the maximum value of I_D can be taken as the value for the drain saturation current. This new value of I_{Dsat} is

$$I_{Dsat} = ZC_{ox}(V_{GS} - V_T) \cdot v_{\text{sat}}\frac{\sqrt{1 + 2\mu_{\text{eff}}(V_{GS} - V_T)/(mv_{\text{sat}}L)} - 1}{\sqrt{1 + 2\mu_{\text{eff}}(V_{GS} - V_T)/(mv_{\text{sat}}L)} + 1} . \quad (10.48)$$

With respect to I_{Dsat} for the Level 1 model, it can be seen that the charge is still evaluated at $x = 0$, but the velocity is now limited to a maximum value of v_{sat}. The corresponding new expression for the drain saturation voltage is

$$V_{DSsat} = \frac{2(V_{GS} - V_T)/m}{\sqrt{1 + 2\mu_{\text{eff}}(V_{GS} - V_T)/(mv_{\text{sat}}L)} + 1} . \quad (10.49)$$

The set of equations, (10.47), (10.48) and (10.49), constitute the **Level 49 model** in SPICE.

The two models, Level 1 and Level 49, are compared in Fig. 10.14 for two MOSFETs, which differ only in their channel length: 1000 nm vs. 100 nm. The models agree well in the linear region, where both reduce to (10.50)

$$I_D = ZC_{ox}(V_{GS} - V_T) \cdot \mu_{\text{eff}} \frac{V_{DS}}{L}. \tag{10.50}$$

In the saturation regime there is good agreement in the long-channel case, but very poor agreement in the shorter-channel case. The disagreement is due to the fact that the Level 1 model uses a linear relationship between velocity and field ($v = \mu_{\text{eff}}\mathcal{E}_x$). This means that at the high values of $\mathcal{E}_x$ that can arise in the channel of a short-length device, v_d can exceed v_{sat}, which is the limiting velocity imposed by Level 49.

10.5   Sub-threshold current

In the sub-threshold regime of operation, $V_{GS} < V_T$ and the channel is everywhere in weak or moderate inversion. The aim of this section is to link the sub-threshold current to the threshold voltage and to V_{GS}. The resulting expression is used in Chapter 13 to examine two important properties of transistors intended for high-speed logic applications: the ratio of the onset of the ON-current to the OFF-current, and the inverse sub-threshold slope. From (10.13), (10.20) and (10.21) we have

$$Q_n = -\sqrt{2q\epsilon_s N_A} \left[\sqrt{\psi_s + \frac{k_B T}{q} e^{(\psi_s - 2\phi_B - V_{CB})/V_{\text{th}}}} - \sqrt{\psi_s} \right]. \tag{10.51}$$

In moderate inversion $\psi_s < (2\phi_B + V_{CB})$, so the first square-root term in the square brackets can be expanded using the binomial theorem. Keeping terms to first-order yields

$$Q_n = -\frac{\sqrt{2q\epsilon_s N_A}}{2\sqrt{\psi_s}} \frac{k_B T}{q} e^{(\psi_s - 2\phi_B - V_{CB})/V_{\text{th}}}. \tag{10.52}$$

In our discussion of the body-effect coefficient we emphasized that, in weak inversion, the surface potential is barely influenced by V_{CB}. This means that there is essentially no field $\mathcal{E}_x$, so that the drain current is due to diffusion. Therefore, taking the diffusive part of (10.27), and substituting (10.52), with $Q_n(0)$ evaluated at $V_{CB} = V_{SB}$, and $Q_n(L)$ at $V_{CB} = V_{DB}$ gives

$$I_D = \frac{Z}{L} \mu_{\text{eff}} \left(\frac{k_B T}{q} \right)^2 \frac{\sqrt{2q\epsilon_s N_A}}{2\sqrt{\psi_s}} e^{(\psi_s - 2\phi_B)/V_{\text{th}}} e^{-V_{SB}/V_{\text{th}}} \left\{ 1 - e^{-V_{DS}/V_{\text{th}}} \right\}. \tag{10.53}$$

Because the dominant effect of ψ_s on the current occurs via the exponential term in the numerator, no great loss of accuracy would result from choosing some constant value for $\sqrt{\psi_s}$ in the denominator. Here, we choose the value at the extreme end of the weak-inversion regime ($=2\phi_B + V_{SB}$), because this then allows the term $\sqrt{2q\epsilon_s N_A}/(2\sqrt{\psi_s})$ to be written succinctly as $C_{ox}(m - 1)$. To bring in the threshold voltage, consider the linear portion of Fig. 10.8, which shows the gate-bias dependence of ψ_s for a particular V_{SB}. Identify a particular surface potential ψ_s' with a particular gate-source voltage V_{GS}', with it being understood that $V_{GS} < V_T$ in this weak-inversion condition; and identify

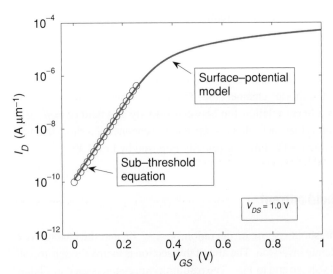

Figure 10.15 Sub-threshold current comparison: (SPICE Level 49 vs. PSP). Model parameters as given in Appendix C for a CMOS90 N-FET.

the surface potential $2\phi_B + V_{SB}$ with V_T. In sub-threshold $|Q_n| \ll |Q_b|$, so from (10.43) for the body-effect coefficient and (10.1) for the potential-divider action of the oxide and body capacitances, we have

$$\frac{d\psi_s}{dV_{GS}} = \frac{1}{m} = \frac{2\phi_B + V_{SB} - \psi_s'}{V_T - V_{GS}'}. \tag{10.54}$$

Putting this into (10.53) and dropping the prime symbol gives the desired form for the sub-threshold current

$$I_D = \frac{Z}{L}\mu_{\text{eff}}\left(\frac{k_B T}{q}\right)^2 C_{ox}(m-1)e^{(V_{GS}-V_T)/mV_{\text{th}}}\left\{1 - e^{-V_{DS}/V_{\text{th}}}\right\}. \tag{10.55}$$

This expression brings out the exponential dependence of I_D on V_{GS} in the sub-threshold region. This relationship is evident in Fig. 10.2, and also in Fig. 10.15. The latter compares the result from (10.55) with the result from the surface-potential model (10.27): evidently the two equations give excellent agreement.

10.6   Applying the long-channel models

The models presented in this chapter are appropriate for use with MOSFETs in which **long-channel behaviour** applies. This does not mean that the transistor is necessarily physically long: it means that the charge and potential at the source end of the channel, from which the current issues, are not influenced by conditions at the drain, i.e., by the drain potential V_D.

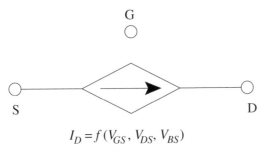

$$I_D = f(V_{GS}, V_{DS}, V_{BS})$$

Figure 10.16 MOSFET DC equivalent circuit.

The charge at the source end of the channel in the strong-inversion, source-referenced model is

$$Q_n(0) = -C_{ox}(V_{GS} - V_T),$$ (10.56)

where the threshold voltage is given by (10.35), and is independent of V_{DS}. In the surface-potential method, the surface potential at the source-end of the channel is given by (10.16)

$$\psi_s(0) = V_{GB} - V_{fb} - \gamma \sqrt{\psi_s(0) + \frac{k_B T}{q} \left[\exp\left(\frac{\psi_s(0)}{V_{th}}\right) - 1 \right] \exp\frac{-(2\phi_B + V_{SB})}{V_{th}}},$$

(10.57)

which is independent of V_{DB}.

These equations, and the models from which they are derived, emphasize the control of the critical source conditions by the gate potential V_G. In FETs with physically short channels, the proximity of the drain to the source leads to the possibility of some of this control being surrendered to V_D. This is undesirable because the threshold voltage, for example, would not then be determined solely by pre-designed physical properties: it would become bias dependent, i.e., it would depend on the operating conditions. To maintain gate control of the channel in FETs that are physically short is one of the challenges faced by designers of modern Si FETs, for which the channel length is <50 nm. The structural and physical changes to FETs that have been introduced to maintain long-channel behaviour in physically short devices are discussed in Chapter 13. The success that has been achieved in this regard means that the models of this chapter, with some modifications, can still be applied to truly short-channel FETs.

10.7 DC equivalent-circuit model

The DC equivalent circuit for a MOSFET is very simple if the possibilities of charge flow from the channel to the gate and to the substrate are ignored. These are usually good assumptions to make because the gate oxide, even though it is very thin, is a good insulator, and because the source/substrate n^+/p barrier is much higher than the source/channel n^+/n barrier, i.e., the source-to-substrate current path, and the source-to-drain current path through the substrate, are both shorted out when there is an inversion layer present. Thus, the equivalent circuit for the ON condition can be represented as a voltage-controlled current generator (see Fig. 10.16), with the particular $I_D(V_{GS}, V_{DS}, V_{BS})$ relationship being given by whichever of the expressions for I_D and I_{Dsat} given in this chapter is appropriate.

Exercises

10.1 A process engineer incorrectly sets the flow rate of phosphorus that is used to dope the poly-silicon gates of N-MOSFETs. Instead of the usual very high doping, the gates are only doped at $1 \times 10^{17}\,\text{cm}^{-3}$.

By how much does this change the threshold voltage of the transistors?

10.2 One of the major assumptions in deriving the 'strong-inversion, source-referenced' models from the 'surface-potential' model is that

$$V_{CS}(x) \ll (2\phi_B + V_{SB}). \tag{10.58}$$

Justify the making of this assumption for CMOS90 N-MOSFETs.

10.3 Show that (10.36) for the body-effect coefficient m is equivalent to the definition for $m(x)$ in (10.43), when the latter is evaluated at the source end of the channel.

10.4 Compute the magnitude of the body-effect coefficient for the CMOS90 process.

10.5 Perform the derivation of (10.47) for I_D following the procedure stated in the text.

10.6 Show that the reason why the Level 49 model breaks down is because it predicts $\mathcal{E}(x) \to \infty$ at $V_{CS}(L) = V_{DSsat}$.

10.7 A widely used expression for the saturation voltage in long transistors is $V_{DSsat} = (V_{GS} - V_T)/m$.

At what gate length does this expression start to become inaccurate?

10.8 A certain semiconductor material has an energy band structure in which the curvature around the extrema is greater for the valence band than it is for the conduction band.

If MOSFETs of a given channel-length L were made from this semiconducting material to supply a given current at a given overdrive voltage $(V_{GS} - V_T)$, which type of FET (n-channel or p-channel) would have the smaller footprint?

10.9 (a) Employ the 'long-channel' threshold voltage expression (10.35) to evaluate the threshold voltage for N-FETs from the CMOS90 and CMOS65 technologies.

(b) Of all the parameters that were changed in going from the 90-nm technology node to the 65-nm technology node, which is the one most responsible for the large difference in long-channel V_T?

10.10 (a) For an N-FET with the CMOS90 parameters listed in Appendix C, obtain a $\log_{10} I_D$-V_{GS} plot for the sub-threshold regime, with $V_{DS} = 1.0\,\text{V}$.

(b) Add to the plot from (a) the corresponding data for an N-FET from the CMOS65 process.

(c) If the 65-nm technology was designed to give the same OFF current as for CMOS90 devices, what is the threshold-voltage shift in 65-nm FETs due to short-channel effects? Assume that there are no short-channel effects in CMOS90.

(d) Imagine that the microprocessor in your laptop computer uses CMOS90 FETs. Estimate the temperature of the CPU chip and add the appropriate curve to your sub-threshold plot.

Is the temperature rise desirable?

10.11 In the previous two questions the trend for the 'long-channel threshold voltage' to increase as devices scale down was noted. The reason for this is to accommodate the reduction in actual threshold voltage due to increasing short-channel effects.

The parameters for CMOS45 are not yet well known, but one might expect a thicker high-k dielectric for the gate oxide, e.g., $\epsilon_r = 16$ and $t_{ox} = 8\,\text{nm}$, a metal gate, and perhaps a slightly reduced doping density in the body, e.g., $1 \times 10^{18}\,\text{cm}^{-3}$.

If the target value for the long-channel V_T is 0.45 V, what must be the work-function of the metal gate?

10.12 From Question 10.10 it can be taken that real N-FETs in the CMOS90 and CMOS65 technologies have the same OFF current. If performance can be evaluated as the ratio of maximum ON current (I_{Dsat}) to OFF current, did the change from CMOS90 to CMOS65 follow Moore's Law?

10.13 Evaluate the saturation current if ballistic transport were to occur in the channel of an N-FET. Explain your choice of electron effective mass.

If you were R&D Manager of a large semiconductor company, would the margin between the values of I_{Dsat} for dissipative and ballistic transport lead you to try to develop ballistic Si MOSFETs?

10.14 By transferring a large amount of charge to the gate of a FET it is possible to cause the gate oxide to break down. This is why engineers wear a well-grounded wrist band when testing FETs.

If an engineer neglected to wear a ground band, and tested FETs from the following two batches, which FETs would be more likely to be destroyed?

Batch A: $L = Z = 200\,\text{nm}$ and $t_{ox} = 40\,\text{nm}$;
Batch B: $L = Z = 2000\,\text{nm}$ and $t_{ox} = 4\,\text{nm}$.

References

[1] J.E. Lilienfeld, *Device for Controlling Electric Current*, US patent, 1,900,018, March 7, 1933.
[2] C.T. Sah, *Fundamentals of Solid-State Electronics*, p. 521, World Scientific Publishing Co., Singapore, 1991.
[3] G. Gildenblat, X. Li, W. Wu, H. Wang, A. Jha, R. van Langevelde, G.D.J. Smit, A.J. Scholten and D.B.M. Klaassen, PSP: An Advanced Surface-Potential-Based MOSFET Model for Circuit Simulation, *IEEE Trans. Electron Dev.*, vol. 53, 1979–1993, 2006.

[4] W. Liu, *MOSFET Models for SPICE Simulation, Including BSIM3V3 and BSIM4*, John Wiley & Sons, Inc., 2001.

[5] Y. Tsividis, *Operation and Modeling of the MOS Transistor*, 2nd Edn., Chap. 4, Oxford University Press, 1999.

[6] D.L. Pulfrey and N.G. Tarr, *Introduction to Microelectronic Devices*, Prentice-Hall, 1989.

[7] Y. Tsividis, *Operation and Modeling of the MOS Transistor*, 1st Edn., p. 57, McGraw-Hill, 1987.

11 HJFET basics

Originally, the most significant difference between a MOSFET and a BJT was the high input impedance afforded by the gate insulation of the FET. The high quality of the SiO_2/Si system in Si MOSFETs has not proved possible to replicate in other semiconductor systems, particularly those involving III-V compound semiconductors. Therefore, to capitalize on the advantages that the latter semiconductors may have over silicon, such as mobility (see Fig. 11.1), and still realize a FET device, some other way of implementing the field-effect is necessary. This has been done by using a metal/semiconductor junction, rather than a metal/insulator/semiconductor combination, to create a vertical field to control the charge in the channel. The two main devices that exploit this are shown in Fig. 11.2: the MESFET (metal-semiconductor FET), and the HEMT (high-electron-mobility transistor). The latter device is also sometimes called a MODFET, where MOD refers to modulation doping, and relates to the fact that the doping in the top barrier layer plays a key role in controlling (modulating) the channel charge. All these transistors are HJFETs (heterojunction FETs) on account of the presence of a metal/semiconductor heterojunction in their structures.

The use of high-mobility semiconductors in MESFETs and HEMTs enables attainment of a high transconductance, which is a prerequisite for good high-frequency perfomance (see Section 14.4). The mobility is further enhanced in a HEMT by reducing the doping density in the barrier layer proximal to the channel, and by exploiting the two-dimensional nature of the channel. Quantum-mechanical effects due to the confinement of electrons in the channel not only improve the mobility, but also lead to improved noise characteristics, as explained in Chapter 17.

11.1 Schottky barrier

Metal/semiconductor junctions that exhibit rectifying characteristics are called **Schottky barriers**.[1] The simplest situation is when the electrostatics of the junction is determined by the work functions Φ of the two components, and by the electron affinity χ_S of the semiconductor.[2] Such a situation is shown in Fig. 11.3 for the case of $\Phi_M > \Phi_S$. On

---

[1] After Walter Schottky, who performed pioneering studies on these contacts in the 1930s.
[2] Charges at the interface, in the form of dipoles and surface states, often due to the contamination and imperfection of the interface, can also affect the electrostatics, but these factors are not considered here.

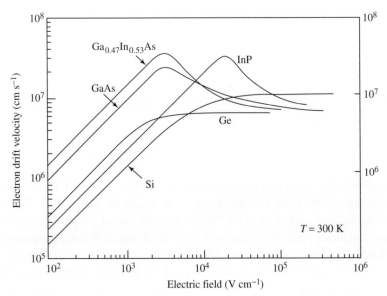

Figure 11.1 Drift velocity vs. field for various semiconductors, illustrating the relative inferiority of silicon. From Sze [1], © John Wiley & Sons Inc. 2002, reproduced with permission.

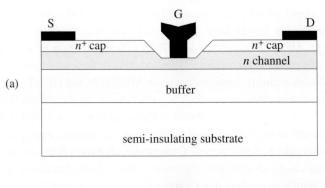

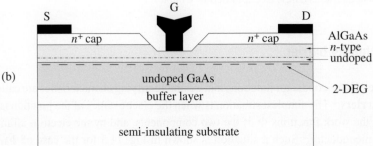

Figure 11.2 The epitaxial-layered structures of two HJFETs. Top: the MESFET. Bottom: the HEMT. Both are grown on a semi-insulating substrate.

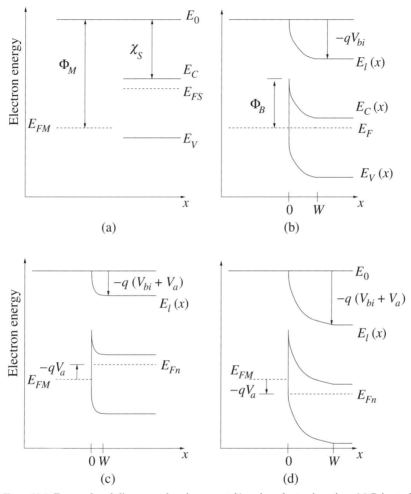

Figure 11.3 Energy-band diagrams showing a metal/semiconductor junction. (a) Prior to joining the two components. (b) At equilibrium. (c) Forward bias. (d) Reverse bias.

bringing together the metal and the semiconductor, electrons transfer from the latter to the former to equilibrate the system. Thus, a depletion layer forms (see Fig. 11.3b), the width of which can be modulated by applying a voltage to the diode. For an n-type semiconductor, forward bias means a negative potential on the semiconductor with respect to that on the metal (see Fig. 11.3c).

The applied bias also changes the height of the barrier for electrons crossing the junction from the semiconductor to the metal. In this respect, the behaviour is no different from that of an np-junction. However, in the latter, the net flow across the junction is determined to some extent by the subsequent diffusion of the injected electrons in the p-region. This is usually the bottleneck to transport, and does not lead to much disturbance from equilibrium in the junction region. Thus, the quasi-Fermi levels can be regarded as being almost constant across the junction (see Section 6.5). Contrast this to the situation shown in Fig. 11.3c, where there is an abrupt change in E_{Fn} at the interface. As discussed in Section 6.7.3, this indicates a severe departure from equilibrium. Such a situation arises because the electrons injected into the metal are not throttled by diffusion. They join the multitude of electrons already in the conduction band of the metal ($\approx 10^{23}$ cm^{-3}), and a large current can be maintained by an infinitesimal field. Thus, the bottleneck to charge transport is the junction itself, and the current across the junction has to be driven by a change in E_{Fn}.

The situation is akin to that in an Np heterojunction with a very short base. From (9.11) the current density due to electrons injected into the metal can be written as

$$J_e\,(\mathrm{S} \rightarrow \mathrm{M}) = q n_0 \frac{e^{-V_a/V_{\mathrm{th}}}}{\frac{1}{v_R}}. \tag{11.1}$$

where $V_a < 0$ in forward bias. Note that this current density is positive if the x-direction is defined as x increasing to the right. From Fig. 11.3b, and from (4.14)

$$n_0 = N_C e^{-\Phi_B/k_B T}, \tag{11.2}$$

so $J_e\,(\mathrm{S} \rightarrow \mathrm{M})$ can be written as

$$J_e\,(\mathrm{S} \rightarrow \mathrm{M}) = q N_C e^{-\Phi_B/k_B T} e^{-V_a/V_{\mathrm{th}}}\, v_R, \tag{11.3}$$

where $\Phi_B = (\Phi_M - \chi_S)$ is the **Schottky barrier height**. At zero-bias, $J_e\,(\mathrm{S} \rightarrow \mathrm{M})$ must be balanced by an equal flow of electrons from the metal to the semiconductor $J_e\,(\mathrm{M} \rightarrow \mathrm{S})$. This flow from the metal is not affected by the applied bias as Φ_B is fixed by the material constants. Thus, the full expression for the current is

$$J_e \equiv J_e\,(\mathrm{S} \rightarrow \mathrm{M}) - J_e\,(\mathrm{M} \rightarrow \mathrm{S}) = q N_C e^{-\Phi_B/kT}(e^{-q V_a/kT} - 1)v_R. \tag{11.4}$$

11.1.1   Thermionic emission and tunnelling

Despite the discontinuity in the quasi-Fermi level at the interface, quasi-equilibrium can still be considered to apply in the semiconductor itself. The situation is illustrated in Fig. 11.4a, which depicts thermionic emission, i.e., the current is due to a relatively few energetic electrons drawn from a much larger pool of electrons that remain in an equilibrium distribution. It is guaranteed that the number of thermionically emitted electrons is relatively small if the depletion-region potential drops by at least $k_B T$ eV within one mean-free-path length $\bar{l}$ of the interface [2].

If the doping density is very large, then the bands will bend very steeply in the semiconductor (see Fig. 11.4b). In this case, if the barrier becomes very thin (less than ≈ 5 nm), then tunnelling of electrons can take place. Electrons can pass through such a barrier in either direction, so the rectifying nature of the contact is lost. The I-V characteristic becomes linear about the origin, and the contact is said to be **ohmic**. Such contacts are desirable when the metal is required merely to connect the semiconductor to external circuitry with minimal voltage drop. Both ohmic and rectifying contacts are used in the HJFETs described in the following sections.

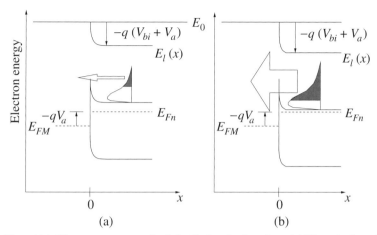

(a)   (b)

Figure 11.4 Charge transport at the Schottky barrier interface. (a) Thermionic emission of electrons in the shaded portion of the distribution over a barrier with a thick space-charge region. (b) Tunnelling of electrons in the shaded portion of the distribution through a barrier with a thin space-charge region.

11.2   MESFET

The basic MESFET structure is shown in the top part of Fig. 11.2. The n^+ 'cap' layer assures ohmic contacts for the source and drain, while the gate metallization is chosen to make a rectifying contact to the less heavily doped n-type 'active layer'. These n-regions are grown epitaxially either on a weakly doped p-region or on a semi-insulating substrate.[3]

11.2.1   Channel formation and threshold voltage

In the top part of Fig. 11.5 the space-charge region is shown in a MESFET with $V_{DS} > 0$. This bias makes the semiconductor potential at the drain end of the device more positive than at the source end. Thus, with respect to the gate, the drain end of the gate/semiconductor Schottky barrier is more reverse biased, so the space-charge region is wider there. The band-bending in this layer, and at the n/semi-insulating junction, define a channel, through which charge must pass en-route to the drain from the source, as illustrated in the lower part of the figure. Overall control of the space-charge-region width $W(x)$ is due to the bias V_{GS} applied to the Schottky diode. If, at zero bias, W is less than the thickness of the active layer, then the channel is 'open'. To 'close' it the Schottky diode must be reverse biased sufficiently for the depletion-region edge to reach the interface between the active and semi-insulating regions. The value of V_{GS} required to do this is the threshold voltage V_T: it is negative for n-type semiconductors. FETs which are ON when $V_{GS} = 0$ are termed **depletion-mode** FETs.

---

[3] Semi-insulating refers to a semiconductor in which the conductivity can be reduced to near-intrinsic proportions, usually by charge compensation involving defects in the material [3].

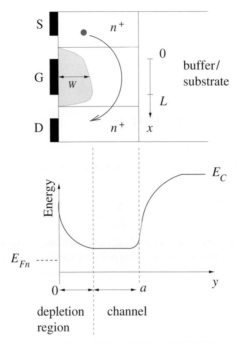

Figure 11.5 Channel formation in a MESFET.

The closure of the channel must not be taken literally. It is to be taken in the same context as 'pinch-off' of the channel is used in MOSFETs. It marks the onset of drain-current saturation in FET models of the 'Level 1' variety. These models predict that the channel charge goes to zero at the drain-end of the channel, whereas, in fact, the charge is small, but finite, and the current is sustained at a high value by virtue of the fact that the electrons are moving very quickly in this part of the channel.

A case in which there would be truly no channel would be if the semiconductor active layer were so thin that the depletion region at zero-bias reached right through the semiconductor. To open up a channel in this case it would be necessary to forward bias the Schottky junction. Recall from Section 11.1 that the current grows exponentially with forward bias, so there is a limit to the forward bias that can be applied: it is set by the 'leakage' current to the gate that can be tolerated, and is only a few tenths of a volt. A MESFET working in this mode would have a positive threshold voltage and would be called an **enhancement-mode** device.

11.2.2   Drain current

To obtain a rudimentary expression for the ON current in a MESFET, consider that transport is due only to drift:

$$\vec{J}_e = -qn\vec{v}_{de} , \tag{11.5}$$

with $\vec{v}_{de}$ being the electron drift velocity in the channel. We use a 'Level 1' model for the drift velocity-field relation

$$v_{de} = \mu_e \mathcal{E}_x = \mu_e \frac{\partial V_{CS}}{\partial x}, \qquad (11.6)$$

where V_C is the channel potential. The drain current is

$$I_D = -J_e Z \left(a - W(x)\right), \qquad (11.7)$$

where a is the thickness of the active layer, Z is the width of the transistor, and the leading minus sign supports the IEEE convention, which assigns a positive value to a positive charge flow entering a device from an external terminal. From np-junction theory, (6.22) yields

$$W(x) = \sqrt{\frac{2\epsilon_s}{q N_D} \left(V_{bi} - [V_{GS} - V_{CS}(x)]\right)}. \qquad (11.8)$$

Substituting in (11.7) for W and for v_{de}, and integrating, yields

$$I_D = G_0 \left\{ V_{DS} - \frac{2}{3V_P^{1/2}} \left[(V_{DS} + V_{bi} - V_{GS})^{3/2} - (V_{bi} - V_{GS})^{3/2} \right] \right\}, \qquad (11.9)$$

where G_0 is the channel conductance when there is no depletion region:

$$G_0 = q \mu_e N_D a \frac{Z}{L}, \qquad (11.10)$$

and V_P is the so-called **pinch-off** voltage: it is the potential drop across the depletion region at threshold, i.e.,

$$V_P = \frac{a^2 q N_D}{2\epsilon_s} = V_{bi} - V_T, \qquad (11.11)$$

where N_D is the doping density of the active layer and ϵ_s is the semiconductor permittivity. In a depletion-mode device $V_P > V_{bi}$, so V_T is negative.

If (11.9) is examined at various V_{DS}, it can be seen that the drain characteristic is linear at low V_{DS}, then it reaches a peak, before becoming smaller at higher V_{DS}. This is the same behaviour as observed for the 'strong-inversion, source-referenced' models for the MOSFET in Section 10.4 (see Fig. 10.11), and indicates that our models break down at some point. Here, the MESFET model breaks down when

$$V_{DS} \equiv V_{DSsat} = V_{GS} - V_T. \qquad (11.12)$$

This is the same condition at which the Level 1 model of the MOSFET was found to break down (assuming a body-effect coefficient of unity). It is associated with the channel pinching-off at the drain end, i.e., $W(L) = a$. As before, the model is extended into the saturation region by allowing I_D to remain constant at the peak value given by (11.9), i.e.,

$$I_{Dsat} = G_0 \left\{ V_{GS} - V_T - \frac{2}{3V_P^{1/2}} [(V_P)^{3/2} - (V_{bi} - V_{GS})^{3/2}] \right\}. \qquad (11.13)$$

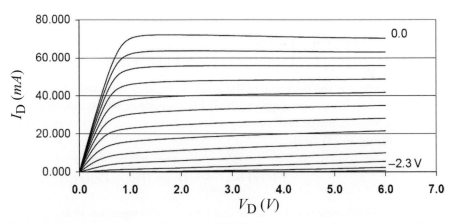

Figure 11.6 Drain characteristic of the TQHiP MESFET from Triquint. The parameter is V_{GS}, which varies from 0 to -2.4 V in increments of -0.2 V. This depletion-mode FET has $L = 500$ nm, $Z = 300$ µm, $V_T = -2.3$ V, and is intended for use in RF power amplifiers and switches.

As an example of the drain characteristic of a MESFET, Fig. 11.6 is offered. It is for a power MESFET from Triquint, and the relevant feature here is the need to apply a negative V_{GS} in order to turn the device OFF, i.e., it is a depletion-mode FET.

11.3   HEMT

In a HEMT, the channel is not contained within the n-type material that forms the Schottky barrier, as occurs in a MESFET. Instead, the channel exists in an underlying, undoped, semiconducting layer (see Fig. 11.2). The absence of doping means no ionized impurity scattering, which is one of the factors that leads to the exceptionally high mobility of this transistor. The barrier semiconductor, besides forming a heterojunction with the gate metal, also forms a heterojunction with the channel layer, as can be seen in Fig. 11.7. The latter heterojunction is constructed in a similar fashion to that illustrated for the InGaP/GaAs HBT in Fig. 9.2. However, in this AlGaAs/GaAs case, the electron affinities of the two semiconductors are not equal, so there is a discontinuity in the conduction-band edge. This discontinuity defines one side of the channel, within which are confined the electrons induced into the undoped semiconductor during the charge transfer that accompanies equilibration of the system. The large band-bending in the undoped semiconductor forms the other side of the confining 'potential well'. The confinement in the y-direction is such that the electrons are free to move only in the two, mutually perpendicular directions: they form a **two-dimensional electron gas** (2-DEG) in the channel in the plane parallel to the interface.

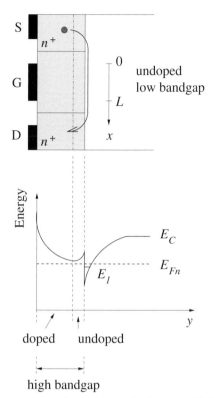

Figure 11.7 Channel formation in a HEMT. The bottom figure is a slice through the HEMT under the gate, showing the Schottky barrier and the conduction-band edge in the underlying layers in an AlGaAs/GaAs HEMT. The band diagram is drawn for equilibrium conditions. E_1 is the energy level of the first conduction sub-band arising from electron confinement in the narrow potential notch.

11.3.1   The 2-DEG

To appreciate the significance of the 2-DEG, let us represent the 'triangular' well at the AlGaAs/GaAs interface by a 'rectangular' well, as illustrated in Fig. 11.8. For the moment, as a further simplification, let us approximate this asymmetrical, finite barrier, by a symmetrical barrier stretching to infinite energy. Within such a well the Schrödinger Wave Equation (2.3) reduces to

$$\frac{d^2 \psi_y}{dy^2} + k_y^2 \psi_y = 0 \,, \tag{11.14}$$

where the wavevector in the y-direction is

$$k_y = \frac{1}{\hbar} \sqrt{2m^* E_y} \,. \tag{11.15}$$

The boundary conditions for this case of an infinitely high barrier are

$$\psi_y(0) = \psi_y(a) = 0 \,. \tag{11.16}$$

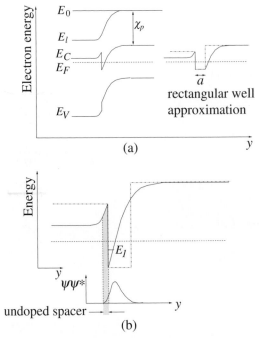

Figure 11.8 Representing the asymmetric, finite, triangular potential well at the AlGaAs/GaAs interface by an asymmetric, finite, rectangular well.

So the solution is

$$\psi_y(y) = A \sin(k_y y),$$ (11.17)

where A is some constant. It follows that k_y is quantized:

$$k_y = \frac{n\pi}{a}, \quad n = 1, 2, 3 \cdots.$$ (11.18)

Therefore, and this is the important point, the y-directed energy E_y is also quantized

$$E_{y,n} = n^2 \frac{\hbar^2 \pi^2}{2m^* a^2}.$$ (11.19)

Of course, as we saw in Chapter 2, the allowed energy levels are always quantized because of their relation to k, which is quantized according to the reciprocal of the length of the region of interest. Here, because the length a is small, the allowed energy levels are widely separated, and cannot be viewed as a continuum, as we have done elsewhere in talking about the conduction band.

However, there are bands in this case too, but only in the directions perpendicular to that of the channel thickness, e.g., associated with each of the allowed energy levels $E_{y,n}$ are allowed bands of energy in the unconstrained perpendicular directions x and z. If we suppose that electrons stay near the bottom of these bands, then a parabolic E-k dispersion relationship can be used to characterize these bands. Further, if we assume

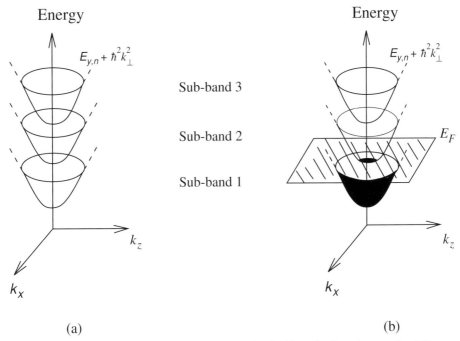

Figure 11.9 Two-dimensional energy sub-bands associated with each allowed energy level $E_{y,n}$. (a) Unfilled bands. (b) Some band-filling, as determined by the position of the Fermi energy. From Roblin and Rohdin [4], © Cambridge University Press 2002, reproduced with permission, and courtesy of Patrick Roblin, Ohio State University.

an isotropic effective mass, the total energy can be written as

$$E = E_{y,n} + \frac{\hbar^2}{2m^*}k_\perp^2, \tag{11.20}$$

where $k_\perp^2 = k_x^2 + k_z^2$. Thus, associated with each allowed energy $E_{y,n}$, there is a 2-D sub-band (see Fig. 11.9).

11.3.2   The finite well

Electrons cannot escape from an infinite well, as we implied above by setting $\psi = 0$ at $y = 0, a$ in formulating our expression for the discrete energy levels. The wavefunctions are sinusoidal and the probability density function $\psi\psi^*$ for the first energy level, for example, indicates the greatest probability of finding the electrons as being at the centre of the well. This is highly desirable for a FET because it keeps the electrons away from the interface, at which there may be crystalline imperfections, which would cause increased scattering and a reduction in mobility.

In real devices, as illustrated in Fig. 11.8, the well is not finite, so electrons can escape through the sides of the potential well. In other words, the probability density function for channel electrons is not zero in the barrier.[4] The situation is illustrated in the lower

---

[4] For a derivation of the wavefunctions in a finite potential well, see Griffiths [5].

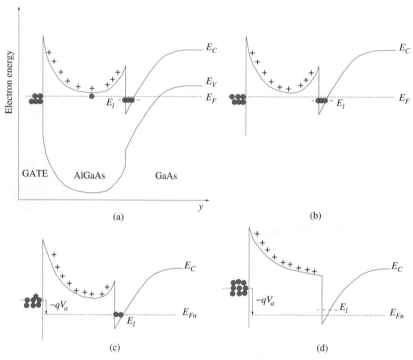

Figure 11.10 Conduction-band diagrams illustrating the control of the electron concentration in the channel by the gate bias. (a) Equilibrium, wide barrier layer. (b) Equilibrium, barrier layer just wide enough for the two depletion regions to just not overlap. (c) Reverse bias. (d) Reverse bias at the threshold condition.

panel of Fig. 11.8. Whereas the penetration into the barrier cannot be prevented, its effect on the mobility can be mitigated by incorporating a thin, undoped region of AlGaAs barrier material, a so-called **spacer layer**, next to the interface. This removes a source of bulk scattering, namely, ionized-impurity scattering.

11.3.3   Electron concentration in the 2-DEG

To find the number of electrons in each sub-band it is first necessary to find the density of states in $k_\perp$-space. Recall that in 3-D the corresponding problem was one of counting states in the volume of a sphere (see Fig. 3.10). Here, it is a case of counting states in a circle of area $\pi k_\perp^2$. The 'skin' of this circle has an area of $2\pi k_\perp \partial k_\perp$. In 1-D a state occupies $2\pi/L$ of k-space, where L is the real-space length in that direction (see (2.12)). Therefore, the number of states in our skin, or annulus, is

$$N_{2D} = 2\,\frac{2\pi k_\perp\,\partial k_\perp}{(2\pi/L_x)(2\pi/L_z)}, \tag{11.21}$$

where L_x and L_z are the side-lengths of the channel, and the leading 2 in the numerator accounts for spin.

Dividing by the real-space area $L_x L_z$, and converting $\partial k_\perp$ to ∂E via (11.20), and finally dividing by ∂E, the 2-D density of states per unit area and per unit energy, for each sub-band, is

$$g_{2D} = \frac{m^*}{\pi \hbar^2}. \tag{11.22}$$

Note that this density of states for each sub-band is independent of energy. This makes for an easy solution for the electron concentration in each sub-band, i.e., from (3.33)

$$n_{2D,n} = \int_{\text{sub-band}} \frac{m^*}{\pi \hbar^2} f(E) \, dE. \tag{11.23}$$

Each sub-band starts at an energy $E = E_{y,n}$, and if we assume that the top of each sub-band is at infinity, then, on using the Fermi-Dirac distribution function and integrating, we obtain

$$n_{2D,n} = \frac{m^* kT}{\pi \hbar^2} \ln \left[\exp(\frac{E_F - E_{y,n}}{kT}) + 1 \right]. \tag{11.24}$$

Thus, the total concentration of electrons in the channel is

$$n_s = \sum_{n=1}^{\infty} n_{2D,n}, \tag{11.25}$$

where the subscript s reminds that this is a surface concentration: an upper limit would be around $(10^{23})^{2/3}, \approx 10^{15} \text{ cm}^{-2}$.

11.3.4 Controlling the channel charge by the gate potential

Having considered the metal/semiconductor and semiconductor/semiconductor junctions separately, let us now bring them both together to form the structure of the HEMT under the gate, as first illustrated in Fig. 11.7. To equilibrate this system the donors in the AlGaAs barrier layer supply electrons to both the gate metal and to the 2-DEG in the channel. Depletion regions result at both ends of the barrier layer (see Fig. 11.10a). If the barrier layer is sufficiently thick that the two depletion regions do not overlap, then n_s will reach its highest value, labelled n_{s0}.[5] Fig. 11.10b shows the case where the two depletion regions just meet at equilibrium, but do not overlap. This is the equilibrium situation that is sought in HEMTs.

If a reverse bias is now applied to the gate/barrier Schottky diode, the gate demands more electrons. These must come from the electrons already donated to the channel (see Fig. 11.10c). Thus, n_s decreases. This is illustrated in Fig. 11.10c by the electron quasi-Fermi level E_{Fn} dropping towards the first allowed energy level E_1 ($\equiv E_{y,1}$). Eventually, as the reverse bias is further increased, a situation will be reached where all the donors are called upon to donate electrons to the gate. In this case $n_s \to 0$, and the applied voltage $V_a = V_T$ (see Fig. 11.10d).

---

[5] In modern HEMTs n_{s0} may be as high as 10^{13} cm^{-2}.

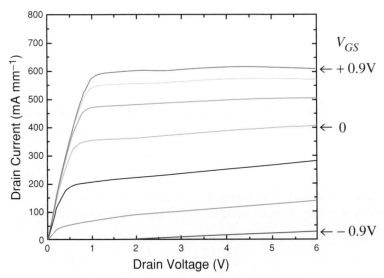

Figure 11.11 Drain I-V characteristic for an AlGaAs/GaAs power HJFET from Triquint (0.15 μm Power pHEMT 3MI). Gate length $= 150$ nm, gate width $= 100$ μm. The V_{GS} increments are 0.3 V.

11.3.5   The drain *I-V* characteristic

As we have just seen, n_s is controlled by $V_a = V_{GS}$. The drain current is drawn from this charge, and the familiar I_D-V_{DS} characteristic of a FET results, with saturation being due to either pinch-off or velocity saturation. However, the device we have been discussing is a depletion device, so it will be ON at $V_{GS} = 0$, and the saturation current will be progressively reduced as V_{GS} is made more negative. It is possible to further increase I_{Dsat} by applying $V_{GS} > 0$, i.e., by operating the FET in the enhancement mode. An example of a commercial device that allows this is shown in Fig. 11.11. Note that a positive V_{GS} forward biases the gate/AlGaAs Schottky diode, so the enhancement mode of operation is limited by the source-gate leakage current that can be tolerated.

Exercises

11.1 Consider two diodes: one is a metal/n-Si Schottky diode, and the other is a p^-/n-Si homojunction. The n-type doping is quite high and is the same for both diodes. Which diode is more likely to deliver a forward-bias electron current at the hemi-Maxwellian velocity limit? Give reasons for your answer.

11.2 Consider a GaAs MESFET with a body thickness of 500 nm, a doping density of $N_D = 10^{16}$ cm^{-3} and a Schottky barrier height of 0.8 eV.
   (a) Is this a depletion-mode or an enhancement-mode FET?
   (b) Calculate the threshold voltage.

11.3 Consider a metal/X/Y HEMT with no spacer layer; X and Y are lattice-matched semiconductors. The work functions of the metal, X and Y can be taken as 3.5, 2, and 4 eV, respectively. The electron affinities of X and Y are 1.5 and 3 eV, respectively. The bandgaps of the X and Y are 4 and 2 eV, respectively.

(a) Draw energy-band diagrams, showing to approximate scale in energy (e.g., $1\,\text{cm} \equiv 1\,\text{eV}$), the zero-field and local-vacuum levels, the conduction bands and valence bands for the two semiconductors, and the Fermi levels for the metal and the semiconductors under the following conditions:

  (i) for the three, separated components of the device;

  (ii) for the joined components at equilibrium.

(b) What is the height of the Schottky barrier?

(c) What is the height of the potential well at the X/Y interface?

11.4 HEMT A and HEMT B are identical, except that $\Phi_{\text{gate,A}} < \Phi_{\text{gate,B}}$, where Φ is the metal work function.

Which transistor has the more negative threshold voltage, and why?

11.5 Consider Fig. 11.10b, c and d. This sequence shows an HJFET at equilibrium, with a small negative gate voltage applied, and with a gate voltage equal to the threshold voltage, respectively. Call this HJFET A.

Consider Fig. 11.10a, which shows the equilibrium band diagram for an HJFET with a thicker barrier layer. Call this HJFET B.

Which of the two HJFETs has the more negative threshold voltage?

11.6 Familiarize yourself with (10.41).

Imagine that an equation of similar form applies to a HEMT over the full length of the channel, and for $V_T \leq V_{GS} \leq 0$.

An $Al_{0.5}Ga_{0.5}As/GaAs$ HEMT has a barrier layer of thickness 30 nm and a channel that can be taken to be an infinite square well of thickness 8 nm.

At equilibrium, E_F lies mid-way between the first and second quantized energy levels in the channel.

(a) Evaluate the electron charge density in the channel from (11.25).

(b) Evaluate the threshold voltage for this transistor.

11.7 For the case of conduction electrons, plot the energy dependence of the 2-D density of states and of the 3-D density of states on the same graph.

The plot should reveal that the 2-D DOS at each sub-band energy level equals the 3-D DOS at the corresponding energies.

References

[1] S.M. Sze, *Semiconductor Devices, Physics and Technology*, 2nd Edn., Fig. 7.15, John Wiley & Sons, Inc., 2002.

[2] F. Berz, The Bethe Condition for Thermionic Emission near an Absorbing Boundary, *Solid State Electronics*, vol. 28, 1007–1013, 1985.

[3] D. Kabiraj, R. Grötzschel and S. Ghosh, Modification of Charge Compenstation in Semi-insulating Semiconductors by High Energy Light Ion Irradiation, *J. Appl. Phys.*, vol. 103, 053703, 2008.

[4] P. Roblin and H. Rohdin, *High-speed Heterostructure Devices*, Fig. 4.7, Cambridge University Press, 2002.

[5] D.J. Griffiths, *Introduction to Quantum Mechanics*, Sec. 2.6, Prentice-Hall, 1995.

12 Transistor capacitances

The capacitance of a transistor is a crucial consideration when designing devices for applications in the commercially and societally important areas of digital logic, high-frequency signal processing, and memory. Accordingly, as a pre-cursor to the subsequent chapters on transistors suited to these applications, transistor capacitance is given a thorough treatment in this chapter of its own.

The approach taken presents capacitance in a general way that can be applied to all transistors. The usual practice is to treat capacitance in an ad hoc manner, sometimes involving charges of opposite polarity, as in junction capacitance, and sometimes considering just one polarity of charge, as in storage capacitance, for example. In fact, the origin of these two capacitances is the same: in the case of the emitter/base capacitance, for example, it is the change in charge within the device due to electrons that have entered from the emitter to set-up a new steady-state charge profile in the transistor in response to a change in base potential. This fact is recognized here, and provides a view of capacitance that is both physically based and intuitively appealing. Our approach is based on that of Tsividis for MOSFETs [1], and leads naturally to a double-subscripted specification of capacitance in all types of transistor, e.g., C_{EB} or C_{SG}. Thus, a 3×3 capacitance matrix captures all the capacitive elements of a three-terminal device, and allows for non-reciprocity.

The capacitances described here are used without further ado in subsequent chapters on transistors for applications in digital logic (Chapter 13), high-frequency circuitry (Chapter 14), and memory (Chapter 15).

12.1 Defining capacitance

Capacitance is related to charge, so let's start with a general version of the equation for the continuity of charge density from (5.22):

$$\frac{\partial Q}{\partial t} + \frac{\partial J}{\partial x} = 0 , \qquad (12.1)$$

where J is a conduction current of positively charged carriers that have a volumetric concentration Q. Recall that this equation was derived from the Boltzmann Transport Equation. To emphasize its fundamental nature, note that it can also be derived from

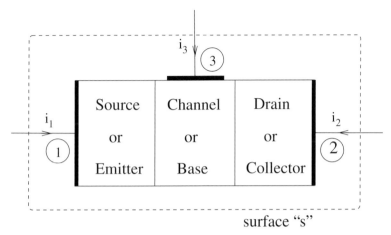

Figure 12.1 Generic transistor, enclosed by a surface s, which is penetrated by three labelled leads, along which charge can enter or leave the transistor.

Electromagnetic Theory via Maxwell's equations, i.e., by applying the divergence operator to Ampere's Law, integrating over some volume, changing to a surface integral, and using Gauss's Law to bring in the charge density Q. Equation (12.1) expresses the universal truth of the conservation of charge, e.g., if there is a decrease in charge within a certain volume, then the removed charge must be carried to somewhere else by a current, i.e., charge cannot be created nor destroyed.

For the generic transistor shown in Fig. 12.1, under steady-state conditions,

$$\frac{\partial Q_1}{\partial t} + \frac{\partial Q_2}{\partial t} + \frac{\partial Q_3}{\partial t} = 0,\qquad(12.2)$$

where Q_j is the charge that enters or leaves the transistor through the lead attached to region j.

It is very important to understand that by invoking steady-state conditions we have not seriously limited the applicabilty of (12.2). For example, imagine that when an N-MOSFET is OFF there is no electronic charge at the semiconductor surface. Now consider what happens when the gate/source voltage is suddenly raised above the threshold voltage. Electrons are injected into the channel and start to migrate towards the drain. For the short period it takes for the electrons to travel to the drain (the so-called **transit time**), there is a source current but no drain current. As (12.1) demands, the charge builds up within the channel during this time. After the elapse of the transit time, a current issues from the drain and we have steady-state conditions. For a 45-nm Si MOSFET and electrons moving at $v_{\mathrm{sat}} = 10^5 \mathrm{\,m\,s^{-1}}$, the transit time is 450 fs. So, providing we don't need to know anything about this device's performance during time-frames of this order, then the steady-state condition can be applied with impunity.

Each of the charges relevant to Fig. 12.1 is a function of the potentials applied to the three leads: V_1, V_2, V_3. Let's use V_1 as a reference; so the Q's are now functions of the

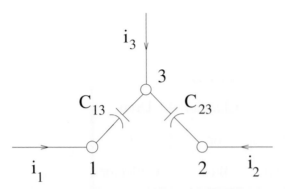

Figure 12.2 Equivalent-circuit representation of (12.7).

voltages V_{21}, V_{31}. Thus

$$\frac{\partial Q_1}{\partial t} = \frac{\partial V_{21}}{\partial t}\frac{\partial Q_1}{\partial V_{21}} + \frac{\partial V_{31}}{\partial t}\frac{\partial Q_1}{\partial V_{31}}$$

$$\frac{\partial Q_2}{\partial t} = \frac{\partial V_{21}}{\partial t}\frac{\partial Q_2}{\partial V_{21}} + \frac{\partial V_{31}}{\partial t}\frac{\partial Q_2}{\partial V_{31}}$$

$$\frac{\partial Q_3}{\partial t} = \frac{\partial V_{21}}{\partial t}\frac{\partial Q_3}{\partial V_{21}} + \frac{\partial V_{31}}{\partial t}\frac{\partial Q_3}{\partial V_{31}}. \tag{12.3}$$

For a specific example, consider $\partial V_{21}/\partial t = 0$. This is the situation when V_{DS} or V_{CE} is held constant in a FET or HBT, respectively. From (12.2) and (12.3)

$$\frac{\partial V_{31}}{\partial t}\left[\frac{\partial Q_1}{\partial V_{31}} + \frac{\partial Q_2}{\partial V_{31}} + \frac{\partial Q_3}{\partial V_{31}}\right] = 0. \tag{12.4}$$

For this to be true for any $\partial V_{31}/\partial t$, we must have

$$\frac{\partial Q_3}{\partial V_{31}} = -\frac{\partial Q_1}{\partial V_{31}} - \frac{\partial Q_2}{\partial V_{31}}, \tag{12.5}$$

from which capacitances are defined:

$$C_{33} = C_{13} + C_{23}. \tag{12.6}$$

Equation (12.5) is to be interpreted as: in response to a change in voltage ∂V_{31}, the change in charge that flows in through electrode 3 equals the negative of the changes in charge that flow in through the other two electrodes. Capacitance allows this relationship to be neatly represented in an equivalent circuit. As an illustration of this, use (12.3) and (12.6), to recognize that

$$0 = -C_{13}\frac{\partial V_{31}}{\partial t} - C_{23}\frac{\partial V_{31}}{\partial t} + C_{33}\frac{\partial V_{31}}{\partial t}$$

$$\equiv i_1 + i_2 + i_3, \tag{12.7}$$

where the currents are often called **charging currents**, i.e., they exist only when the charge is changing. This equation can be represented by the simple equivalent circuit of Fig. 12.2. Usually, the capacitances are positive, as would be expected naturally, but

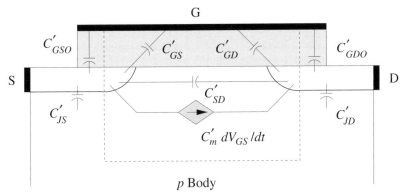

Figure 12.3 Illustration of the principal intrinsic and extrinsic capacitance in a MOSFET. The capacitors in the intrinsic case are within the dashed box; those without represent extrinsic capacitance. The prime indicates total capacitance to distinguish it from capacitance per unit area.

there is at least one exception, as we mention in the next section. Our notation gives

$$C_{jk} = -\frac{\partial Q_j}{\partial V_k} \qquad \text{if } k \neq j$$

$$C_{jk} = +\frac{\partial Q_j}{\partial V_k} \qquad \text{if } k = j$$

$$C_{jj} = \sum_{k \neq j} C_{jk} = \sum_{k \neq j} C_{kj}. \qquad (12.8)$$

For a proof of the last equation see Exercise 12.1.

12.2 MOSFET capacitance

The principal sources of capacitance in a MOSFET are shown in Fig. 12.3. Those capacitors within the dashed border represent intrinsic capacitance, and those outside represent extrinsic capacitance. All the capacitors are labelled with a prime, which indicates that they represent total capacitances, unlike C_{ox}, which represents capacitance per unit area.

12.2.1 Intrinsic MOSFET capacitances

For the MOSFET, $j = 1 \equiv S$, $j = 2 \equiv D$, $j = 3 \equiv G$; the terminal currents can be written as[1]

$$\begin{pmatrix} i_S \\ i_D \\ i_G \end{pmatrix} = \begin{pmatrix} C'_{SS} & -C'_{SD} & -C'_{SG} \\ -C'_{DS} & C'_{DD} & -C_{DG} \\ -C'_{GS} & -C'_{GD} & C'_{GG} \end{pmatrix} \begin{pmatrix} \partial V_S/\partial t \\ \partial V_D/\partial t \\ \partial V_G/\partial t \end{pmatrix}. \qquad (12.9)$$

Because of (12.8) only 6 of these 9 capacitances are independent.

---

[1] Here we've neglected the body of the device: including it would lead to a 4×4 capacitance matrix.

Note that C_{jk} is not necessarily equal to C_{kj}, i.e., the capacitances are not necessarily reciprocal. For example, consider C'_{GD} and C'_{DG} when the FET is operating in the saturation regime. Changing V_{DS} brings about essentially no change in drain current, which means there is no change in the total channel charge Q'_n, which, in turn, means there is no change in Q'_G. Hence, in saturation, $C'_{GD} \approx 0$. However, on increasing V_{GS}, the charge all along the channel $Q_n(x)$ increases negatively (see (10.40)). Some of this charge is supplied by the drain, so $\partial Q'_D < 0$ and $C'_{DG} > 0$.

Because of (12.7), which is Kirchoff's Current Law, only two of the three equations in (12.9) are necessary to specify the currents. Taking the drain and gate currents, and using V_S as the reference potential:

$$i_D = (C'_{GD} + C'_{SD})\frac{\partial V_{DS}}{\partial t} - C'_{DG}\frac{\partial V_{GS}}{\partial t}$$

$$i_G = -C'_{GD}\frac{\partial V_{DS}}{\partial t} + (C'_{GD} + C'_{GS})\frac{\partial V_{GS}}{\partial t}, \tag{12.10}$$

where use has been made of the summation expression in (12.8) to expand the C_{jj} terms. Recognizing that $V_{DS} = (V_{GS} - V_{GD})$, and doing a bit of algebra, leads to

$$i_D = C'_{SD}\frac{\partial V_{DS}}{\partial t} - (C'_{DG} - C'_{GD})\frac{\partial V_{GS}}{\partial t} - C'_{GD}\frac{\partial V_{GD}}{\partial t}$$

$$i_G = C'_{GS}\frac{\partial V_{GS}}{\partial t} + C'_{GD}\frac{\partial V_{GD}}{\partial t}. \tag{12.11}$$

These equations are represented in the schematic of Fig. 12.4. This is an extension of Fig. 12.2, and is a connected version of the intrinsic part of Fig. 12.3. Figure 12.4b clearly shows that C'_{GD} represents the effect of the drain on the gate. In the case of Fig. 12.4c, recognize that $\partial V_{GS} = \partial V_{GD}$, and define the **transcapacitance** as $C'_m = (C'_{DG} - C'_{GD})$. After doing this, it is clear that

$$i_D = -C'_{DG}\frac{\partial V_{GD}}{\partial t}. \tag{12.12}$$

This shows that C'_{DG} represents the effect of the gate on the drain. The transcapacitance is, therefore, the circuit element that allows non-reciprocity to be realized in the capacitance between any two terminals.

An unusual intrinsic capacitance is C'_{SD} for a FET operating in the non-saturation regime. An increase in V_{DS} widens the space-charge region, causing Q_b to increase negatively. This causes a positive charge change in the channel charge Q_n, as can be seen from Fig. 10.13. Because Q_n is intimately related to Q_S and Q_D, the changes in theses charges are positive too. Thus, C'_{SD} is negative, via the first equation in (12.8).

The intrinsic capacitances are computed automatically in numerical circuit simulators such as SPICE.[2] Analytically, it is not easy to derive expressions for the capacitances, but it is worthwhile trying to do so in order to get a 'feel' for the factors that influence them. Here, we consider the case of operation in the triode mode, and make the simplifying assumptions of the Level 1 model (see Section 10.4.1), one of which is $Q_b(x) = Q_b(0)$.

---

[2] The capacitances are computed from the currents and voltages in the circuit via (12.7).

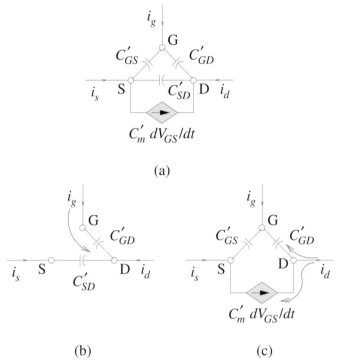

(a)

(b)   (c)

Figure 12.4 Schematics of intrinsic MOSFET capacitance. (a) Complete representation. (b) Determination of the effect of ∂V_{DS} on i_G. (c) Determination of the effect of ∂V_{GS} on i_D. In (b) and (c) only those elements of (a) that are operative under the stated conditions are shown.

Further, we assume that the body-effect coefficient is unity. Thus, from (10.40), and recognizing that, in the intrinsic device,

$$Q_G + Q_n + Q_b = 0,\tag{12.13}$$

we have

$$Q_G(x) = C_{ox}(V_{GS} - V_T - V_{CS}(x)) - Q_b(0).\tag{12.14}$$

Recall that these charges are per unit area. The total gate charge is

$$Q'_G = C_{ox} Z \int_0^L (V_{GS} - V_T - V_{CS}(x))\, dx - ZL Q_b(0),\tag{12.15}$$

where Z and L are the width and length, respectively, of the gate. Now, turn this integral over dx into one over dV_{CS} by using the expression for current (10.46), and integrate over the range of V_{CS} from 0 to V_{DS}, after eliminating I_D from the equation by using (10.34). The answer is, after some rearranging [2, pp. 242–244]

$$Q'_G = \frac{2C_{ox} ZL}{3} \frac{(V_{GS} - V_T)^3 - (V_{GD} - V_T)^3}{(V_{GS} - V_T)^2 - (V_{GD} - V_T)^2} - ZL Q_b(0).\tag{12.16}$$

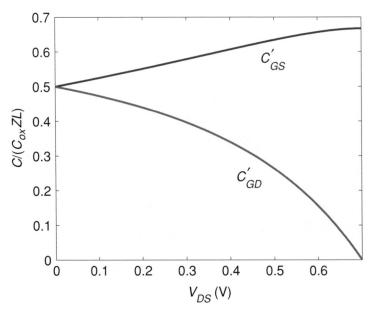

Figure 12.5 Bias dependence in the triode regime ($V_{DS} < V_{GS} - V_T$) of the intrinsic gate-source and gate-drain capacitances. $V_T = 0.3$ V, $V_{GS} = 1.0$ V.

Differentiating this, with respect to V_{SG} ($= -V_{GS}$) to get C'_{GS}, and with respect to V_{SD} to get C'_{GD}, gives

$$C'_{GS} = \frac{2}{3}C_{ox}ZL\left[1 - \frac{(V_{GD} - V_T)^2}{(V_{GS} - V_T + V_{GD} - V_T)^2}\right]$$

$$C'_{GD} = \frac{2}{3}C_{ox}ZL\left[1 - \frac{(V_{GS} - V_T)^2}{(V_{GS} - V_T + V_{GD} - V_T)^2}\right]. \qquad (12.17)$$

Recall that these expressions are valid for the triode regime only (and under Level 1 conditions), but they do serve the useful purpose of illustrating (see Fig. 12.5) the following features: the capacitances are of the order of $C_{ox}ZL$; the capacitances are bias dependent; as saturation is approached ($V_{DS} \rightarrow (V_{GS} - V_T)$ in Level 1), $C'_{GD} \rightarrow 0$, and $C'_{GS} \rightarrow \frac{2}{3}C'_{ox}$.

To briefly explain the trends in the figure, consider first C'_{GS}, which represents the effect of V_S on Q'_G. In the context of increasing V_{DS} with V_D held constant, V_S has to change negatively. Thus, V_{GS} increases, which means more charge on the gate. Hence, C'_{GS} increases with drain/source bias. Regarding C'_{GD}, we know from our earlier discussion that it is approximately zero in saturation. Figure 12.5 shows the trend towards this condition as the triode regime is traversed. Recall that I_D increases with V_{DS} in the triode regime, even when V_{GS} is held constant. This means that the channel charge density at the source end of the channel, $Q_n(0) \approx C_{ox}(V_{GS} - V_T)$, is fixed. The drain current at the source end of the channel is given by $I_D = Q_n(0)v(0)$, where v is the electron velocity. At the drain end of the channel the current is unchanged, but v will be higher because of the increased field $\mathcal{E}_x$. This means that $|Q_n(L)|$ decreases with

V_{DS}, and we know it tends to zero when saturation via pinch-off is reached. Thus the total channel charge $|Q'_n|$ decreases with drain/source bias, as illustrated in Fig. 10.13. Because we are assuming no change in the body charge Q'_b, the gate charge must decrease, and C'_{GD} too.

The fact that the intrinsic capacitances are proportional to the gate length has meant that their absolute magnitudes have decreased with scaling. The extrinsic capacitances described below have also diminished, but not necessarily by as much as the intrinsic capacitances because they are not directly dependent on the gate length. However, in the highest performance Si CMOS circuitry, transistors are becoming so small that their total capacitance is of less significance than the capacitance of the connecting wires that join together the transistor logic gates.

12.2.2   Extrinsic MOSFET capacitances

In addition to intrinsic capacitance, there is also extrinsic capacitance in a MOSFET (see Fig. 12.3). This results from the overlap of the gate and the heavily doped source and drain regions, and from the np-junctions between each of these regions and the body, as indicated in Fig. 12.3.

For the overlap capacitance at the drain end, for example,

$$C'_{GDO} = -\frac{\partial Q'_{GO}}{\partial V_D},$$ (12.18)

where the subscript O refers to 'overlap'. From Gauss's Law

$$Q'_{GO} = \epsilon_{ox} A_O \mathcal{E}_O,$$ (12.19)

where A_O is the area of overlap, and $\mathcal{E}_O$ is the magnitude of the y-directed field in the overlap region. If we assume that there is no field-fringing, $\mathcal{E}_O = -V_{DG}/t_{ox}$, where t_{ox} is the oxide thickness. Thus, with this simplification, the overlap capacitance is just that of a parallel-plate capacitor

$$C'_{GDO} = \frac{\epsilon_{ox} A_O}{t_{ox}}.$$ (12.20)

For the junction capacitances, see the treatment of bipolar transistors in the next section. Other extrinsic capacitances in a MOSFET, such as the gate-body overlap capacitance and the junction sidewall capacitance, are treated elsewhere [2, pp. 248–249].

12.3   HBT capacitance

The following capacitances are relevant to all bipolar junction transistors, but we will refer to them as 'HBT capacitances' because the HBT is the bipolar transistor we examine in the later chapters on digital switching and high-frequency performance.

In HBTs the base is the controlling electrode, so we will start by expressing some intrinsic capacitances due to changes in the base potential V_B, or to the voltages V_{BE} and

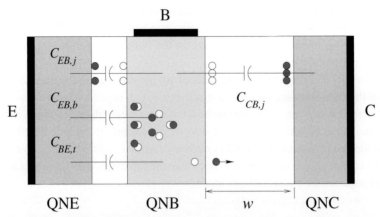

Figure 12.6 Concept of intrinsic capacitance in an HBT. The space-charge regions at each junction are unshaded, and the various capacitors representing different intrinsic processes, as described in the text, are shown within these regions.

V_{BC}. The cross-sectional areas of the emitter and the collector are usually larger than that of the gate of modern FETs, so intrinsic capacitances tend to dominate in HBTs. As usual, we consider only *Npn* transistors.

12.3.1   Emitter-base junction capacitance

When V_B changes, such that $\partial V_{BE} > 0$, for example, the emitter-base potential barrier is lowered, and the space-charge region shrinks. The shrinkage on the n-side, for example, is due to electrons flowing in from the emitter lead and nullifying the charge of some of the ionized donors that constituted the previous depletion region (see Fig. 12.6). The emitter-base junction capacitance is

$$C'_{EB,j} = -\frac{\partial Q'_{E,j}}{\partial V_{BE}}, \tag{12.21}$$

where $Q'_{E,j}$ refers to the total number of electrons that have entered through the emitter contact and come to reside at the edge of the junction. Note from Fig. 12.6 that holes also flow in from the base contact to effect a corresponding shortening of the depletion region on the base side of the junction.

If we wish to consider the hole charge explicitly, we examine

$$C'_{BE,j} = -\frac{\partial Q'_{B,j}}{\partial V_{EB}}. \tag{12.22}$$

Evidently, this junction capacitance is reciprocal.

Inasmuch as the base-emitter junction is planar, the junction capacitance can be evaluated as we did earlier for C'_{GDO}, resulting in the 'parallel-plate' expression

$$C'_{EB,j} = \frac{\epsilon_s A_E}{W}, \tag{12.23}$$

where ϵ_s is the permittivity of the semiconductor, A_E is the area of the emitter-base junction, and W is the depletion region width (6.17).[3]

12.3.2 Base storage capacitance

Another important capacitance is the base storage capacitance

$$C'_{EB,b} = -\frac{\partial Q'_{E,b}}{\partial V_{BE}}, \tag{12.24}$$

where $Q'_{E,b}$ is the electronic charge in the base that is injected from the emitter, in response to ∂V_{BE}, in order to establish a new electron profile in the base, from which the new collector current is derived. For example, if ∂V_{BE} increases the forward bias of the emitter-base junction, electrons will enter the base to form a steeper, steady-state, concentration gradient, from which a larger $I_C(V_{BE})$ will issue. As one electron exits the base to the collector, another enters the base from the emitter, so the electron 'storage' in the base is, in fact, a dynamic process. The charging current that sets up the steady-state concentration gradient is completed by holes flowing into the base from the base lead (see Fig. 12.6). This flow maintains charge neutrality in the quasi-neutral base.

If we choose to look at this capacitance from the hole storage point of view, we write

$$C'_{BE,b} = -\frac{\partial Q'_{B,b}}{\partial V_{EB}}. \tag{12.25}$$

It follows that the base storage capacitance is reciprocal.

To evaluate $C'_{EB,b}$, let us simplify the problem by assuming that the base is so short that there is negligible recombination of electrons and holes.[4] The electron profile, stretching across the quasi-neutral base from $x = 0$ to $x = W_B$ will be linear in this case. Thus:

$$\begin{aligned}
Q'_{E,b} &= -q\frac{W_B A_E}{2}\left[n(0) - n(W_B)\right] - q W_B A_E n(W_B) \\
&= -q\frac{W_B A_E}{2}\left[n(0) + n(W_B)\right] \\
&= -q\frac{W_B A_E}{2}\left[n^*_E + n^*_C\right], \tag{12.26}
\end{aligned}$$

where we have used the boundary conditions (9.7) and (9.8). Recalling the bias-dependence of $n(0)$ (see (9.2)), and differentiating, we obtain

$$C'_{EB,b} = \frac{q^2 W_B A_E}{2k_B T} n_{op} \exp(V_{BE}/V_{\text{th}}). \tag{12.27}$$

The importance of this capacitance can be appreciated by noting its strong bias dependence. However, its significance has decreased over the years as basewidths have shrunk.

---

[3] If the bias change ∂V_{BE} is not small compared to the dc bias V_{BE}, then W in (12.23) should be an average value.

[4] In modern HBTs W_B can be as small as 30 nm, which is much less than the minority carrier diffusion length, so the approximation of no recombination is satisfactory in such cases.

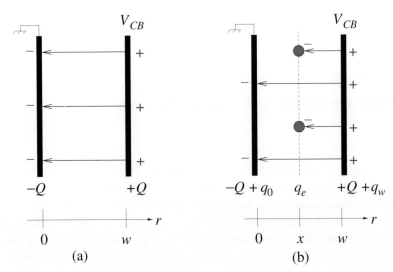

Figure 12.7 Illustration of the change in electric field profile in the base-collector depletion region due to the passage of a charge q_e. In this case $Q = 3q$, and $q_e = -2q$; the latter is located half-way between the edges of the depletion region, across which a constant voltage V_{CB} is maintained.

12.3.3   Emitter storage capacitance

The corresponding emitter storage capacitance $C'_{BE,h}$ is not usually of significance in HBTs because injection of holes into the quasi-neutral emitter from the base is thwarted by a large potential due to the valence band offset, as we demonstrated in Section 9.3.2. Thus

$$C'_{BE,e} = -\frac{\partial Q'_{B,e}}{\partial V_{EB}} \approx 0. \tag{12.28}$$

12.3.4   Base-emitter transit capacitance

We now discuss a rather unusual capacitance that is related to the transit of electrons across the relatively wide space-charge layer at the reverse-biased base-collector junction (active mode of operation). The electrons originate in the emitter lead, but the holes they draw in from the base are different in number, and this turns out to be particularly important in estimating f_T, a high-frequency figure-of-merit (see Chapter 14). We call this capacitance the base-emitter transit capacitance:

$$C'_{BE,t} = -\frac{\partial Q'_{B,h}}{\partial V_{EB}}. \tag{12.29}$$

'Base-emitter' is used in the description, rather than 'base-collector', because, even though the electrons are in the base/collector space-charge region, they originate in the emitter.

Imagine that a uniform field $\mathcal{E}_w$ exists in the base-collector space-charge layer, which we represent here by a depletion region of width w. The field is maintained by a constant voltage V_{CB}, resulting in charges $+Q$ and $-Q$ at the edges of the depletion region (see Fig. 12.7a). Electrons, comprising a charge of q_e, in transit across the region alter the field (see Fig. 12.7b). Physically, this is brought about by positive charges flowing through the external collector-base circuit, and building up at the edges of the depletion region. These charges, q_0 and q_w in Fig. 12.7b, are often called **image charges**. Clearly,

$$q_e + q_0 + q_w = 0 . \tag{12.30}$$

The assumption of constant V_{CB}, and the application of Gauss's Law, yield

$$(Q - q_0)x + (Q + q_w)(w - x) = Qw . \tag{12.31}$$

This leads to expressions for the image charges:

$$q_0 = -q_e \left(1 - \frac{x}{w} \right) \quad \text{and} \quad q_w = -q_e \frac{x}{w} . \tag{12.32}$$

We need now to find the cumulative effect on q_0 of having a beam of electrons stretching across the space-charge region. In each distance element dx let there be q_e / A_E electrons per m^2. Assuming a constant electron drift velocity $\vec{v}_{de}$, the current density due to this moving element is

$$\vec{J}_e = \frac{q_e}{A_E \, dx} \vec{v}_{de} . \tag{12.33}$$

Substituting for q_0, and using the IEEE convention that the collector current is positive for electrons flowing out of the device, yields

$$q_0 = \frac{I_C}{v_{de}} \left(1 - \frac{x}{w} \right) dx . \tag{12.34}$$

Integrating this, and assuming that the drift velocity is constant, gives the hole charge that enters the base in response to a beam of electrons in transit across the collector-base depletion region:

$$Q'_{B,h} \equiv \sum q_0 = \frac{I_C}{v_{de}} \frac{w}{2} . \tag{12.35}$$

Using this in (12.29), we can write

$$C'_{BE,t} \equiv -\frac{\partial I_C}{\partial V_{EB}} \frac{\partial Q'_{B,h}}{\partial I_C} = g_m \frac{w}{2v_{de}} , \tag{12.36}$$

where $g_m = \partial I_C / \partial V_{BE} \equiv \partial I_C / \partial(-V_{EB})$ is the transconductance, which we will encounter in Chapter 14. This capacitance is very important in high-frequency HBTs, on account of the relatively large width of the space-charge region at the base-collector junction.

'Transit capacitance' in FETs

The appearance of $2v_{de}$ in (12.36) rather than simply v_{de} is noteworthy because it implies that the signal velocity exceeds the electron transit velocity! One way of understanding this is to note that charge appears at the collector (in the form of an image charge) before the electron actually reaches that electrode. Recall that the image charges are a necessary feature of the rearrangement of the field due to the electrons in transit. In a MOSFET, electrons in transit along the channel also experience a field $\mathcal{E}_x$ from the applied drain/source voltage. However, there is no corresponding favourable image-charge effect because the image charges appear almost entirely on the gate, and not on the drain [3]. This is because of the two-dimensional geometry of the FET and, in particular, of the gate electrode's close proximity to the channel.

12.3.5   Collector-base junction capacitance

In the active mode of operation, the base-collector junction is reverse biased. Thus, there is negligible injection of minority carriers across the junction, i.e., there are no storage capacitances associated with a change in V_{BC}. Thus, the only significant collector-base capacitance is the junction capacitance

$$C'_{CB} = -\frac{\partial Q'_{C,j}}{\partial V_{BC}} = \frac{\epsilon_s A_C}{w}. \tag{12.37}$$

Even though the depletion region width w at this junction is greater than at the emitter-base junction, because of the reverse bias, C'_{CB} is often significant as the collector area A_C is usually larger than that of the emitter (see Fig. 9.1).

Exercises

12.1 In (12.8) it is claimed that

$$C_{jj} = \sum_{k\neq j} C_{jk} = \sum_{k\neq j} C_{kj}. \tag{12.38}$$

Prove this.

12.2 In Section 12.2.1 the following was quoted as an example of transcapacitance in MOSFETs:

$$C_m = C_{dg} - C_{gd}. \tag{12.39}$$

The inclusion of this element in an equivalent circuit allows for the differences between the effects of the gate on the drain and of the drain on the gate to be properly accounted for.

Does the equivalent circuit need to be modified to allow for differences between C_{gs} and C_{sg}?

12.3 Is it possible to have non-reciprocal capacitance in a two-terminal system?

12.4 In Section 12.2.1 there is some discussion about C_{sd} being negative when a MOSFET is operating in the triode regime. What is C_{sd} when the MOSFET is in saturation?

12.5 Consider a CMOS45 MOSFET of gate length $L = 45\,\text{nm}$ and gate width Z. The edges of the source and drain regions distal from the gate edge can be taken to be $3L$. The lateral intrusion of the source and drain implants under the gate is 3.75 nm. The source and drain doping densities are 2×10^{20} cm^{-3}, and the body doping density is 3.24×10^{18} cm^{-3}. For the oxide, take $\epsilon_r = 15$ and $t_{ox} = 4.8$ nm.

Evaluate the total capacitances C'_{ox}, C'_{GDO} and C'_{JS}.

12.6 Consider the base-emitter transit capacitance $C_{EB,t}$ in an HBT. The benchmark case (case A) is when the electron velocity is constant at v_{sat} throughout the base/collector space-charge layer.

Case B is when the electron velocity overshoots (assumed instantly) to $2v_{\text{sat}}$ in the first half of the space-charge layer, and then continues in the second half of the layer at v_{sat}.

Case C is when the electron velocity changes from v_{sat} to $2v_{\text{sat}}$ at the half-way point.

(a) What are the capacitances in each case?

(b) What is the physical reason for one of the cases being better than the others, i.e. having the lowest capacitance?

12.7 Vertical scaling of HBTs has led to a situation where the base-emitter transit capacitance is now larger than the base-storage capacitance.

Confirm this by evaluating $C_{EB,t}$ and $C_{EB,b}$ for the following HBT, which uses GaAs for the base and collector regions.

$$W_B = 30 \text{ nm}, \ N_B = 5 \times 10^{19} \text{ cm}^{-3}, \ N_C = 5 \times 10^{16} \text{ cm}^{-3}, \ V_{CB} = 3 \text{ V}.$$

12.8 The junction capacitance of a 10^{-3} cm^2 abrupt-junction n^+p diode in reverse bias has the experimental values shown in the table below. Plot these data in such a way that a linear relationship results, from which the p-type doping density and the built-in potential can be ascertained.

C (pF)	3.849	3.288	2.626	2.253	2.008	1.826
V (V)	-0.5	-1	-2	-3	-4	-5

12.9 A p^+n diode is fabricated in n-type silicon of resistivity 10 Ω·cm. For reverse-bias voltages of greater than a few hundred millivolts, the current density in the diode is constant at 100 pA cm^{-2}.

(a) Calculate the base-storage capacitance of this diode when the forward current is 1 mA.

(b) If this diode is now used as a varactor, what must be the diode area if the capacitance is to be 100 pF at a reverse bias of 50 V?

12.10 Two np^+ diodes have p-type GaAs bases of doping density 10^{19} cm^{-3}. Each diode has the same emitter doping density of $N_D = 5 \times 10^{17}$ cm^{-3}, and is operated at a forward bias of 1.25 V. The emitter in Diode A is GaAs, whereas in Diode B it is Al$_{0.3}$Ga$_{0.7}$As.

Which diode has the higher junction capacitance?

References

[1] Y. Tsividis, *Operation and Modeling of the MOS Transistor*, 2nd Edn., Sec. 9.2.1, Oxford University Press, 1999.

[2] D.L. Pulfrey and N.G. Tarr, *Introduction to Microelectronic Devices*, Prentice-Hall, 1989.

[3] D.L. John, Limits to the Signal Delay in Ballistic Nanoscale Transistors, *IEEE Trans. Nanotechnolgy*, vol. 7, 48–55, 2008.

13 Transistors for high-speed logic

Digital logic is a matter of charging and discharging capacitors as quickly as possible. If ΔV is the change in voltage required to signify a change in logic level at a node of constant capacitance C, then the switching time is

$$\tau = \frac{C \Delta V}{I},\qquad (13.1)$$

where I is the average current during the switching cycle. Thus, a successful digital logic technology would demand only small voltage swings, and be capable of supplying large currents to circuits of low capacitance.

This chapter draws on the material on the DC performance of MOSFETs (Chapter 10) and HBTs (Chapter 9), and on the chapter on capacitance (Chapter 12), to explain the features of modern transistors that make them suited to high-speed logic applications. The emphasis is on Si MOSFETs, which, in the form of complementary MOS technology (CMOS),[1] have enabled the ULSI-circuitry[2] that has brought electronics into the lives of so many people. At lower levels of integration, and in applications where speed is more important than static power dissipation, emitter-coupled logic (ECL) using HBTs is a viable technology; it is considered at the end of this chapter.

13.1 Si CMOS

13.1.1 General features of CMOS

In CMOS, logic gates comprise pairs of n- and p-type enhancement-mode MOSFETs. We have only considered the former so far, but the latter can be easily envisaged by changing all doped regions from n-type to p-type, and by reversing the polarity of the applied voltages. In CMOS technology, the threshold voltage of the P-FET is usually made to be opposite, and nearly equal in magnitude, to that of the N-FET. A complementary pair of transistors is shown at the end of the CMOS processing sequence in Fig. 13.1.

---

[1] Invented by Frank Wanlass, Fairchild Semiconductor, 1963.
[2] Ultra-Large-Scale-Integrated, i.e., more than a million devices per chip.

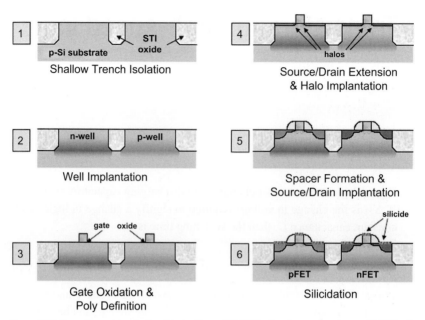

Figure 13.1 Some key steps in the fabrication of Si FETs for deep sub-micron CMOS. Courtesy of Alvin Loke, AMD, Colorado.

In a sub-circuit of one P-FET and one N-FET, the two gates of the two transistors are connected together, and the threshold voltages are such that in either of the logic states (HI or LO) only one of the transistors is ON. Thus there is, in principle, no static power drain. This was the feature that made CMOS an immediate success when it was first introduced in the 1960s. Another attribute is that CMOS logic gates can be made with much fewer transistors than their ECL rival (see Fig. 13.2). Further, in integrated circuits, each transistor has to be contacted at the top surface, and it is difficult to imagine a more compact arrangement than that exhibited by CMOS, in which two of the contacts (source and drain) are so closely aligned with the third contact (gate).

The CMOS industry is huge, and has its own roadmap to chart a path towards ever smaller devices [1]. Technology nodes have been identified, which roughly refer to the minimum line-width or line-spacing that can be achieved by a given CMOS technology. Technologies at 90 nm, 65 nm, and 45 nm have been progressively introduced between 2003 and 2008. Each node number is approximately $\sqrt{2}$-times smaller than the previous one, so, if this shrinking were achieved in the two dimensions of the surface, the size of the object would be halved. This pays homage to Moore's Law, in which the co-founder of Intel observed in 1965 that the number of transistors per square inch on ICs was doubling every year [2]. Transistor channel lengths can be considerably smaller than the technology-node number because of the lateral diffusion of the source and drain implants under the gate, and because of the halo regions (shown in Fig. 13.1 and discussed in Section 13.1.7).

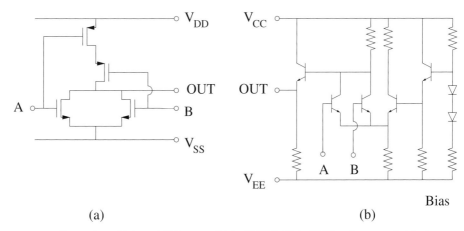

Figure 13.2 Examples of 2-input NOR gates: (a) in CMOS, (b) in HBT emitter-coupled logic (ECL). The latter has a higher transistor count because, in addition to the actual emitter-coupled switching element, level shifters and a current source are required. However, it should also be noted that the ECL gate shown does allow complementary outputs.

The process of systematic shrinking is called **scaling**; it has driven the high-performance end of the digital electronics industry for more than 40 years. In fact, it is the industry's paradigm: make transistors smaller and circuits denser, then new applications will appear and jobs and profits will grow. It is not just the lateral physical dimensions that are scaled, but also the vertical dimensions, the supply voltage, and the doping density. These changes have to be made in concert in order to preserve the long-channel operation of the transistor, i.e., to ensure that the charge at the source end of the channel $Q_n(0)$ is controlled predominantly by the vertical field issuing from the gate. It is becoming increasingly challenging to continue shrinking devices while simultaneously improving device performance, as we shall see in the following sections of this chapter.

13.1.2   The ON-current

In CMOS logic the ON-current is the drain current when $V_{GS} > V_T$, i.e., when the source-end of the channel is in strong inversion. The drain current increases with V_{DS} to reach a maximum value that we have called I_{Dsat}. Over the years, reduction of the channel length L has been more aggressive than reduction of the supply voltage V_{DD}, which sets a maximum limit to the bias voltages V_{DS} and V_{GS}. This is because not only did a shorter L improve I_{Dsat} in physically long devices, (10.38), it also allowed reduction of the area of the device, thereby improving packing density and reducing many of the FET's capacitances (see Chapter 12). The last two attributes still apply, but L no longer has such a strong effect on I_{Dsat} (see (10.48)). This is because L is already sufficiently small for the lateral field $\mathcal{E}_x$ to attain a high-enough value for velocity saturation to occur over a significant part of the channel. The effect of this is illustrated in Fig. 10.14, from which it can be seen that the inclusion of velocity

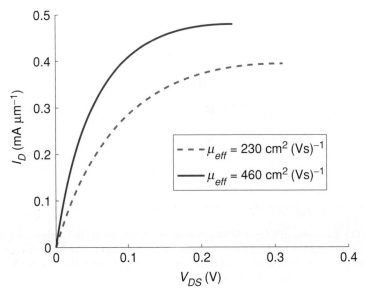

Figure 13.3 Effect of doubling the mobility from 230 to 460 cm²(Vs)⁻¹ in a CMOS90 N-FET with channel length of 50 nm. Other model parameters as given in Appendix C.

saturation (SPICE Level 49 model) gives a much smaller increase in I_{Dsat}, as L is reduced, than is predicted by the SPICE Level 1 model, which does not limit the electron velocity to v_{sat}. However, to attain their saturation velocity, the electrons still have to be accelerated over the source-side of the channel, so a high mobility is still desirable. The effect of doubling the mobility in a short-channel N-FET is shown in Fig. 13.3. Besides the expected increase in I_D in the linear regime, there is some enhancement (about 20%) in I_{Dsat}.

Besides increasing the velocity via μ_{eff}, the ON-current can also be improved by increasing the channel charge. The SPICE models of Chapter 10 inform us that this means increasing C_{ox} and/or the putative **overdrive voltage** ($V_{GS} - V_T$). Ways of increasing C_{ox}, and the implications for leakage of the channel current to the gate, are discussed in Section 13.1.4.

Because CMOS employs only one power supply, V_{DD}, increasing V_{GS} would mean increasing V_{DS}, which, as we have already discussed, is limited by the necessity of keeping $\mathcal{E}_x \ll \mathcal{E}_y$. Additionally, a low value of V_{DD} is desirable for the many portable electronic products that CMOS has enabled. The alternative option for increasing the overdrive voltage is to reduce the threshold voltage, but this would have serious consequences for the sub-threshold current, as discussed in Section 13.1.7. In circuitry as dense as in CMOS microprocessors, where 10^8 transistors per square centimetre is common, one cannot afford to have much of a current per FET when the transistor is OFF ($V_{GS} = 0$ in an N-FET), otherwise the static power drain would be prohibitive (see Section 13.1.10). Unwanted power dissipation also occurs during switching, and is also discussed in Section 13.1.10.

13.1.3   Channel mobility and strain

Starting with the 90-nm generation of MOSFETs, straining of the Si channel has been employed to improve the mobility of both electrons and holes. The strain affects the band structure, and successful strain engineering exploits this to reduce the effective mass in the desired direction of conduction. The strain is produced by stressing the channel region in a variety of ways, as we shall reveal.

Hooke's Law relates the deformation u of a 1-D object to the applied force F by a material-related force constant K:

$$\vec{F} = K\vec{u}\,. \tag{13.2}$$

When generalized to a 3-D object, such as a cube of crystalline semiconductor, we have normal forces and shear forces, and a proliferation of subscripts because of the possibility of deformations of a given side length in a given face arising from all the possible normal- and shear-forces. Each force per unit area is a **stress** and each component of deformation is a **strain**. The force required to produce a given strain is determined by the **elastic stiffness constants** of the material [3]. In cubic crystals such as Si and GaAs, there are just three independent such constants. In Si, their magnitudes range from 64 to 166 GPa, which gives an indication of the immense pressures within a crystalline solid. Conversely, the strain that results from a given stress is determined by the **elastic compliance constants** of the material, which are related to the elastic stiffness constants.

In the simple 1-D model that we used in Chapter 2 to illustrate how band structure is related to crystal structure, we saw the importance of the spacing between atoms. Thus, if this spacing is changed by strain, we can expect changes in band structure to result. If we are looking for strain to improve the mobility of holes in silicon, for example, it would be advantageous if the heavy-hole band could be lowered in energy with respect to the light-hole band.[3] Unfortunately, it is not as simple as this in practice. Both valence bands become so warped by the strain that neither can be considered 'heavy' nor 'light'. Instead, because the strain splits the bands, they are referred to as the 'top' and 'bottom' bands. As the holes preferentially occupy the higher band, it is important that the curvature of the top band be such that the effective mass in the desired direction of conduction be reduced. One way to achieve this for a <110> channel on a {001} Si surface in a sub-100 nm device is to apply a uniaxial compressive stress of about 1 GPa to the p-type channel. This is a huge pressure when one realizes that about 0.8 GPa will break high-strength steel! One way to realize such a high stress is to etch recesses in the Si where the source and drain should be, and then fill-in by epitaxially growing $Si_{1-x}Ge_x$ (see Fig. 13.4). With a Ge mole fraction of $x \approx 30\,\%$, the more expansive SiGe puts the Si channel under a compressive stress of the required magnitude. To date (2009), hole mobility enhancements of up to about four times have been achieved by using this approach.

For n-channel Si MOSFETs on {001} substrates, a <110> tensile stress induces a shear strain that is beneficial for electron conduction in this direction. Three notable

---

[3] Recall that lowering in electron energy is equivalent to raising in hole energy.

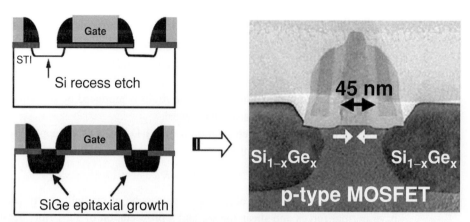

Figure 13.4 Partial process flow and TEM cross-sectional view of a strained-Si *p*-channel MOSFET using embedded SiGe for the source and drain regions. From Thompson *et al.* [4, 5], © 2004 and 2006 IEEE, reproduced with permission.

changes in the conduction-band structure occur: the six-fold symmetry of the conduction-band minima is broken, leading to the two [001] valleys moving to lower energies; the conduction-band minimum moves towards the X-point as the shear strain increases; the degeneracy of the two conduction-band minima at the X-point (see Fig. 2.9) is removed [6]. One manifestation of this is that the prolate spheroid constant-energy surfaces of Fig. 2.12 become scalene spheroids, i.e., the two transverse effective masses are no longer equal. In other words, with reference to Fig. 2.7, the Brillouin zone is no longer symmetrical in directions perpendicular to the new k_x-direction. The situation is obviously quite complicated, and is made more so by quantum confinement effects in the narrow inversion layer. Here, we'll just illustrate the beneficial effect of breaking the six-fold symmetry of the conduction bands extant in unstrained bulk material.

Consider Fig. 13.5a, which loosely represents the *E-k* relationship for Si near the conduction band edge E_C, i.e., the 'valley' is steeper in two of the principal directions than it is in the other, orthogonal, direction. The equivalent constant-energy surfaces in 3-D are shown in Fig. 13.5b. The four spheroids in the horizontal plane are often called the Δ_4 valleys, and the two in the perpendicular direction are called the Δ_2 valleys. Here, we consider the silicon channel to be in the surface plane. The tensile strain breaks the six-fold degeneracy of these valleys, raising the energy of the Δ_4 set, and lowering the energy of the Δ_2 set, i.e., the bottom of the conduction band is raised for the Δ_4 valleys and is lowered for the Δ_2 valleys (see centre and right panels of Fig. 13.5a). For a given energy E with respect to the bottom of the conduction band in the Δ_4 valleys, $E - E_C$ for the Δ_2 valleys is increased, so the spheroids become larger. The electrons naturally seek the lower energy states, so population of the Δ_2 valleys is favoured. If the in-plane effective mass in the Δ_2 valleys remains at m_t^*, which we note from Table 2.1 is considerably less than the longitudinal effective mass m_l^*, then the electron mobility in the channel is increased. As mentioned above, the effective masses are changed by the strain, but to get an idea of the enhancement in μ_e that might be possible, consider the conductivity effective mass in (5.36). Based on the equal probabilities of occupation of the six equivalent conduction bands, a value of $m_{e,\text{COND}}^* = 0.26m_0$ was

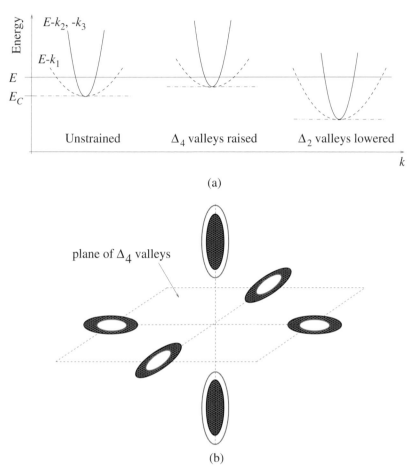

Figure 13.5 (a) Energy-wavevector relations for a material with an effective mass in one direction that is different from the effective mass in the two orthogonal directions. *Left*: all six valleys have the same *E-k* relationship. *Centre*: the Δ_4 valleys (see (b) below) are raised in energy. *Right*: the Δ_2 valleys are lowered in energy. (b) The constant-energy surfaces for the energy $(E - E_C)$ indicated in (a). With respect to the unstrained case (shaded spheroids), the in-plane spheroids are reduced and the out-of-plane spheroids are enlarged.

estimated. If all the electrons could now locate in the Δ_2 valleys, then we would have $m_{e,\text{COND}}^* = m_t^* = 0.19m_0$, i.e., an improvement in mobility of about 37%.

A further bonus is that, whereas electron scattering is equally probable between all six valleys in unstrained silicon, now an energy change is needed to scatter from a Δ_2 valley to a Δ_4 valley. Thus, the in-plane electron mobility may be further increased.

Strain engineering to improve Si MOSFET performance is a very active area of research and development. One **stressor** is a thin layer of silicon nitride, which is deposited by plasma-enhanced CVD[4] over the completed FET. This **stress liner** conforms to the FET as shown in Fig. 13.6, and, depending on the deposition conditions, can be imbued with either a tensile or a compressive stress. For N-FETs a tensile liner is

---

[4] Chemical Vapour Deposition.

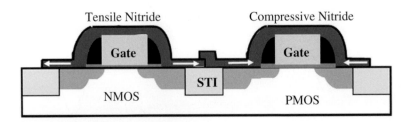

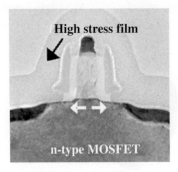

Figure 13.6 Dual stress liner process architecture with tensile and compressive silicon nitride layers deposited by plasma-enhanced CVD. From Thompson *et al.* [4, 5], © 2004 and 2006 IEEE, reproduced with permission.

used, and its adhesion to the source and drain allows its inherent tensile stress to stretch the channel. The contact area in the former regions is much greater than the channel area, so any strain relaxation that takes place occurs mainly outside of the channel. This means that this method of stressing works well if the channel is short, but it also means that it is not easily scaleable. In fact, at the 45-nm node, the source and drain areas may be insufficient to 'anchor' the stress liners. An alternative stressor would then be needed. Creating a trench and filling it with a material that would be compressed by the surrounding silicon is one possibility under investigation at present.

It is amazing that such large and different stresses can be applied to such tiny structures that are in such close proximity. The influence of the tensile stress in an N-FET on the compressive stress of a neighbouring P-FET, for example, is likely to become a necessary consideration in layout design, if, indeed, it is not already so.

13.1.4   Oxide capacitance and high-k dielectrics

Increasing C_{ox} puts more charge in the channel for a given V_{GS} and, consequently, increases I_D. By the 90-nm technology node the oxide thickness had been reduced to ≈ 2 nm, leaving little room for further shrinking as a means of increasing C_{ox}. Apart from the difficulties involved in producing thinner films to acceptable standards of integrity and uniformity, there is the problem of increased leakage to the gate of drain-intended current, as discussed in Section 13.1.6. Traditionally, the gate oxide has been SiO_2: being silicon's native oxide it is obviously compatible with the semiconductor. Further, the interface between the two materials is extremely well understood, and techniques have been developed to effectively passivate the interface [7, Section 6.5.13]. However, a drawback to SiO_2 is its rather low relative permittivity of 3.9. Over the years, nitrogen has been incorporated in the oxide to raise the relative permittivity to 4–5, but, clearly, a higher permittivity dielectric would be helpful. Incorporation of a higher-k dielectric[5] into the CMOS process is another very active area of research and development. At the time of writing (2009), only Intel of the big chip manufacturers has included a new gate-dielectric at the 45-nm technology node, but, undoubtedly, others will follow. The Si/SiO_2 interface is preserved, but the dielectric is thickened by the addition of an oxide derived from hafnium, and probably including some nitrogen. The overall relative permittivity is $\approx$15–20. With respect to an oxide of pure silica, this allows a given oxide capacitance per unit area to be achieved with a thicker oxide, thereby offering the possibility of reducing the gate leakage current (see Section 13.1.6). However, the ON-current would be more directly improved by the new dielectric if C_{ox} was actually made higher than before. This is accomplished by choosing the new dielectric's thickness such that

$$t_{high-k} < t_{silica} \frac{\epsilon_{high-k}}{\epsilon_{silica}} . \tag{13.3}$$

The case for high-k dielectrics is easily stated, but it should be appreciated that the actual replacement of the traditional, tried-and-tested, silica dielectric has posed considerable difficulties for the industry.

13.1.5   Metal gates and poly-silicon capacitance

The industry is moving towards a new **gate stack**, in which a metallic gate electrode is used with the high-k dielectric. This represents another very significant change for the silicon-processing industry, as highly doped polycrystalline silicon gates have been used in CMOS for decades. Originally, in the 1960s, gates were made from aluminium, but this metal's penetration into the underlying silicon dioxide, during thermal treatments later in the device's processing, prevented use of the metal gate as a mask to facilitate the self-alignment of the source and drain regions to the gate. Polycrystalline silicon overcame this problem. However, having a semiconductor for the gate means that there is some depletion at the gate/oxide surface (see Fig. 13.7). This results in an additional potential drop in the device, thus, (10.14) is modified to

$$V_{GB} - V_{fb} = \psi_{poly} + \psi_{ox} + \psi_s . \tag{13.4}$$

In other words, ψ_s, which determines the charge in the channel, experiences less of V_{GB} than it would otherwise. In earlier generations of CMOS this was not a particularly important issue, but now, when material-modification is being increasingly addressed as a means of maintaining the momentum in device-performance enhancement, it needs to be dealt with. Hence, the return of the industry to metal gates, which will need to be less reactive than aluminium, or be deposited later in the fabrication process, perhaps after

---

[5] The 'k' in high-k refers to the symbol often used for dielectric constant, which is an earlier term for relative permittivity.

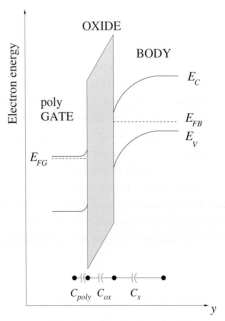

Figure 13.7 Band-bending in the polysilicon gate, and the extra capacitance it causes.

the use of a sacrificial polysilicon gate to allow self-alignment of the source and drain. Nickel and titanium are two possible metals. Co-deposition of metals such as these opens up the possibility of achieving a range of work functions, with which threshold-voltage control could be achieved via variation of V_{fb}. The opportunity of being able to vary V_T across a chip would be welcomed by IC designers.

13.1.6   Gate leakage current

The steady reduction in oxide thickness as part of the scaling process has led to an increase in the flow of electrons from the channel (and from the overlapped part of the drain) in an N-FET to the gate. This tunnelling current detracts from the drain current, hence its name of **leakage current**. The transport mechanism of tunnelling is considered in detail in Section 5.7.3. Expressions for the tunnel current are derived for the cases where the electrons in the channel form a 'classical' 2-D sheet, as we have considered thus far, and where the electrons are confined to quasi-bound states, as discussed in Section 13.1.8. Here we consider the former case, for which the tunnel current density is given by (5.63), which is of the form

$$J_{\text{tunn}} = \int_E \frac{q}{h} n_{2D}(E) T(E) \, dE \, , \tag{13.5}$$

where $n_{2D}(E)$ is the electron density of the 2-D sheet of electrons at the oxide/semiconductor interface, and $T(E)$ is the tunnelling transmission probability. If the tunnelling barrier is approximated as being rectangular in shape, as shown in Fig. 5.8, then there is an analytical expression for $T(E)$ (see (5.54)). This expression contains, in its denominator, hyperbolic functions that involve the thickness of the tunnelling barrier t_{ox}, the electron effective mass in the oxide m^*_{ox}, and the height of the potential barrier.

$$E_{C,ox} - E_C(0) = \chi_{Si} - \chi_{ox}, \tag{13.6}$$

where $E_C(0) \equiv U_1$ is shown in Fig. 5.8, and the χ's are electron affinities. These dependencies are captured in (5.56), which, using the notation of (13.6), is

$$T \approx \exp\left[-\frac{2t_{ox}}{\hbar}\sqrt{2m^*_{ox}(E_{C,ox} - E)}\right]. \tag{13.7}$$

The above equations highlight the fact that tunnelling depends not only on the thickness of the barrier, but also on its height. The latter demands that, for low tunnelling, the oxide must have a low electron affinity. For silica, $\chi = 0.9$ eV, whereas for hafnia $\chi \approx 2.9$ eV. It's reasonable to assume that the electron affinity of present 'high-k' oxides is somewhere between these limits. Let us use χ as a variable and determine the reduction in tunnel current commensurate with a given increase in C_{ox}. For the baseline C_{ox} we assume an oxide of silica with $t_{ox} = 1.75$ nm. We also assume a relative permittivity for the high-k dielectric of four times that of silica.[6] Figure 13.8 shows the ratio of tunnel current in the high-k case to that in the silica case as a function of χ for the high-k dielectric. It is clear from the figure that a 2–3 order-of-magnitude reduction in tunnel current can be achieved simultaneously with a 50% increase in C_{ox}, provided the electron affinity of the high-k dielectric does not exceed about 1.8 eV.[7]

13.1.7   Threshold voltage: the short-channel effect

From the equations developed in Section 10.4.5 for the ON-current, and in Section 10.5 for the sub-threshold current, it is clear that there is a limit to how low V_T can be reduced in practice. For example, if we take $I_D(\text{ONset})$ to be the current at the onset of the ON state, i.e., when $V_{GS} = V_T$; and we take $I_D(\text{OFF})$ as the current in the fully OFF state when $V_{GS} = 0$, then the OFF/ONset current ratio is

$$\frac{I_D(\text{OFF})}{I_D(\text{ONset})} = \exp\frac{-V_T}{m V_{\text{th}}}. \tag{13.8}$$

Thus, if an OFF/ONset current ratio of 10^{-4} is required, then, taking m at its optimum value of unity indicates that V_T must not be less than about 0.24 V. Therefore, the issue of threshold-voltage control is critical, and it brings us to a discussion of factors not previously introduced in Chapter 10, but which can severely affect V_T if they are not dealt with. An illustration of this is seen in Fig. 13.9 and Fig. 13.10. These figures examine the effects on both the ON- and OFF-currents of the depth y_j of the source and drain regions, and of the length L of the channel. To bring y_j into the calculations, the

---

[6] A homogeneous hafnia dielectric would have a relative permittivity of about 5 times that of silica.

[7] It must be noted that our calculation is very sensitive to the values used for the effective masses, which are not well-known. One might even question the use of the effective-mass concept when considering transport through such thin, and non-crystalline, materials. Here, we used 0.91, 0.1, and 0.3 for the effective masses m^*/m_0 of Si, HfO_2 and SiO_2, respectively.

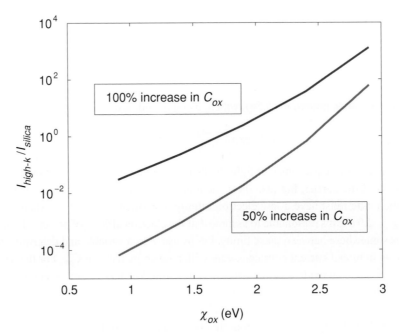

Figure 13.8 Ratio of tunnel current in a high-k dielectric to that in silica as a function of the electron affinity of the high-k dielectric. The high-k dielectric is taken to have a relative permittivity that is four times that of silica. The top curve is for an improvement in C_{ox} of 100%, and the bottom curve is for an improvement of 50%. The parameter values are as given in the caption to Fig. 5.8, unless otherwise stated. $(E_{C,Si} - E_F)$ was taken to be 50 meV.

full set of equations (5.24) in 2-D was solved numerically. The solution for the case of $y_j \equiv 0$ was obtained from the PSP model (10.27).

Consider first the gate characteristic: the bottom three curves show that, for a given channel length, the sub-threshold current increases dramatically as the depth of the source and drain regions is increased. The effect is indicative of a reduction in the threshold voltage. Compare now the two curves at $y_j = 30$ nm, and observe that shrinking L from 100 to 50 nm also effectively reduces the threshold voltage.

Turning now to the drain characteristic, note that the effect of increasing y_j is to enhance the drain ON-current as V_{DS} is increased. This effect is not to be confused with **channel-length modulation**, which occurs in longer devices after V_{DS} is increased beyond the 'pinch-off' value.[8] Current saturation in FETs with sub-micron channel lengths is more due to velocity saturation than to pinch-off (see Section 10.4.5). The phenomenon seen in Fig. 13.10 is indicative of V_T being influenced by V_{DS}. It becomes apparent over the entire range of the drain characteristic when L is reduced to $\approx$50 nm.

The above observations regarding the gate and drain characteristics can be quantified by a reduction in the threshold voltage as either the junction depth is increased, or as

---

[8] In a long-channel device, if V_{DS} is increased beyond the value at which pinch-off occurs, the pinch-off point moves further away from the drain as the depletion region around the drain thickens. This has the effect of shortening the remaining portion of the channel that is still in strong inversion.

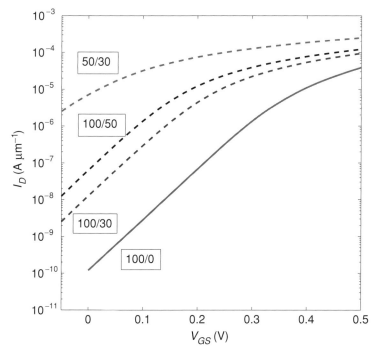

Figure 13.9 Gate characteristics for various combinations of channel length L/source-drain junction depth y_j (both in nm). Unless otherwise stated the parameters are as listed in Appendix C for a CMOS90 N-FET. The bottom curve (solid line) is from the PSP model (10.27) and the other curves are from (5.24), using the Drift-Diffusion Approximation implemented in MEDICI.

the length of an already short channel is decreased. This phenomenon of an increase in drain current due to factors not included in the long-channel model is known as the **short-channel effect**.

Basically, the short-channel effect is an electrostatic effect. Field lines emanate from the positively biased drain, and must terminate on negative charges somewhere in the transistor. The p-type substrate provides suitable sites, leading to a space-charge region extending out from the drain. This region is wider than the space-charge region around the source because of the greater reverse bias at the drain/body np-junction. The drain-related space-charge at some point x' increases the surface potential at x'; this increases $|Q_n(x')|$, and possibly $\mathcal{E}_x(x')$. Both effects lead to an increase in I_D. This particular aspect of the short-channel effect is often called **charge sharing**. The name comes from the fact that the total space charge (depletion) at some point x' is now determined by both the gate and drain potentials. A given $\psi_s(x')$ can now be achieved with a lower V_{GS} than would be needed if there were no depletion due to V_{DS}.

As L decreases, the space-charge region from the drain progressively encroaches on the source. Eventually, $\psi_s(0)$ is affected by V_{DS}. The effect on I_D is direct because $\psi_s(0)$ sets the height of the potential barrier at the source/channel junction, i.e., it controls the electron flow into the channel. This aspect of the short-channel effect is called **drain-induced barrier lowering** (DIBL).

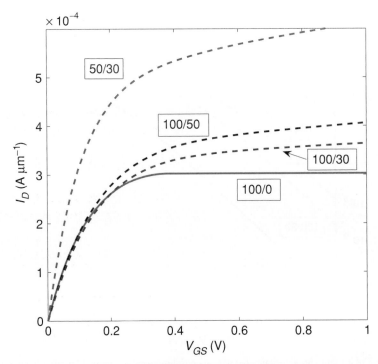

Figure 13.10 Drain characteristics for the same combinations of parameters given in Fig. 13.9.

As noted earlier in this subsection, the effects of charge-sharing and DIBL can be captured by defining an **effective threshold voltage**

$$V_{T,\text{eff}} = V_T + \Delta V_T(y_j, L, V_{DS}). \tag{13.9}$$

The electrostatic interaction between the drain and the channel via the depletion region in the body can be viewed as a capacitive phenomenon, as illustrated in Fig. 13.11. Therefore, the most obvious way to reduce $|\Delta V_T|$ is to reduce the size of the drain-source capacitor. As the sidewall of the np-junction defines one of the lengths of this 'plate' of the capacitor, then it would be helpful to reduce y_j. The simulation results of Fig. 13.9 and Fig. 13.10 confirm this. In fact, y_j has been shrinking over the years, going from about 40 nm at the 130-nm node, to about 15 nm at the 45-nm node. One drawback to reducing y_j is that the lateral access resistance to the channel from the source and drain contacts is increased. For this reason, only the part of the source and drain closest to the channel is thinned. These shallow **source- and drain-extensions** are evident in Fig. 13.1.

Another way of reducing the influence of the drain on $\psi_s(0)$ would be to shield the source from the field issuing from the drain. This could be done by raising N_A throughout the body. However, a high doping density at the semiconductor surface would lead to an unacceptably high V_T. The present industry solution is to use a non-uniform doping profile $N_A(x, y)$. The doping is increased in the y-direction: the desired high doping is achieved deeper into the substrate using fast-diffusing dopants such as

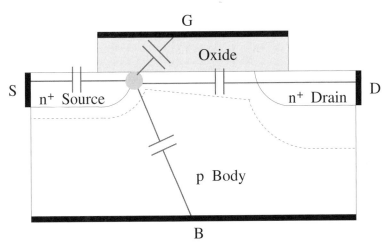

Figure 13.11 Illustrating how the drain-source capacitance abets the gate-source capacitance in determining the potential ψ_s at the source, thereby giving rise to the short-channel effect. Adapted from a figure kindly supplied by Alvin Loke, AMD.

B, As and P, whereas doping of the sensitive surface region is accomplished by more lightly doping using slower diffusants such as In and Sb. The resulting doping profile is called **retrograde**. In the x-direction, N_A is increased only close to the source and drain junctions. The feature is shown in Fig. 13.1 and is called **halo doping**. Another name for these highly doped regions is **pocket implants**.

The three features of y_j reduction, retrograde doping and pocket implants, have proved sufficient to control the short-channel effect, at least down to the 45-nm node. Future schemes may involve multiple gates, of which a 'wrap-around' gate would be the ultimate solution (see Chapter 18).

13.1.8   Threshold voltage: a quantum-mechanical effect

Recall that one of the trends in scaling has been to increase the doping density in the body to prevent V_T from getting too small due to reductions in t_{ox}. For high substrate doping densities, the potential profile in the y-direction of the body becomes steep enough for the electrons to be restricted to a very shallow region at the oxide/semiconductor surface. Of course, we recognized this before when we invoked the channel-sheet approximation, however, the added feature here is that this restriction creates a 'potential well'. If the sides of this potential well are high enough, the conduction band splits up into sub-bands, as described in detail in Section 11.3.1. The situation is illustrated in Fig. 13.12. The sub-bands are at higher energies than E_C, and, as there is no longer a continuum of energy levels near the quasi-Fermi level, it takes more energy to populate the quasi-bound states,[9] so V_T is raised slightly.

---

[9] The states are properly termed 'quasi-bound', rather than just 'bound', because the electrons in them are not completely confined. They must escape if there is to be a drain current.

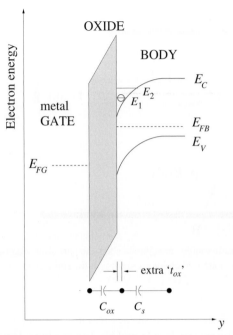

Figure 13.12 Illustration of how the fact that the centroid of charge in the first conduction sub-band E_1 is distal from the interface leads to an increase in the effective oxide thickness.

An appreciation of the effect can be gained by considering the charge distribution in the first sub-band. As shown in Fig. 11.8, the centroid of charge is near the middle of the well, i.e., distal from the actual semiconductor surface. This increase in separation between the channel charge and the gate electrode can be viewed as an increase in oxide thickness. Thus, C_{ox} is reduced and V_T is raised. The magnitude of the effect is examined in Exercise 13.2.

13.1.9   Silicon-on Insulator FET

The Si MOSFET that we have discussed so far is a **bulk-CMOS** FET; it is by far the dominant transistor in high-performance CMOS circuitry. Here, we briefly mention the **silicon-on-insulator** (SOI) FET, which has some interesting attributes.[10] In the cross-section shown in the top part of Fig. 13.13, a layer of silicon oxide is implanted into the silcon wafer, and the CMOS FETs are then defined in the overlying surface layer of silicon. Understandably, the oxygen implant disturbs the crystallinity of the Si surface layer, and it is costly to recover the perfection of this critical region. Presently, a more widely used way of forming an SOI structure is to rely on van der Waals forces to bond one Si wafer to the oxide-coated surface of another Si wafer (wafer B), and then to reduce the thickness of wafer B to end up with a structure similar to that shown in Fig. 13.13. This process is known as 'Smart Cut' [7, p. 146], and is also relatively expensive.

---

[10] Both AMD and IBM use SOI in some of their high-performance offerings.

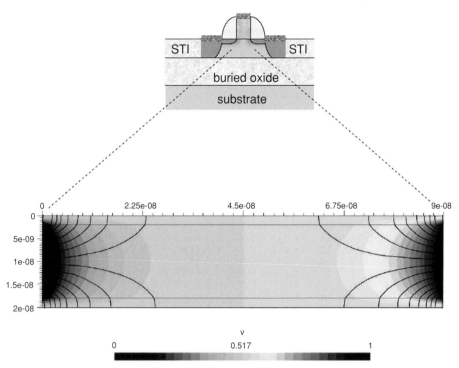

Figure 13.13 Silicon-on-insulator (SOI) CMOS. The top part is a cross-section of a thin-body device. The bottom part shows the field lines from a 2-D solution to Laplace's Equation for an arrangement similar to that of the N-FET. Specifically, the simulation is for a 90-nm channel in a 16-nm Si body, with 2-nm oxides and gates on the top and on the bottom. The top figure is courtesy of Alvin Loke, AMD, and the bottom figure is courtesy of Daryl Van Vorst, UBC.

The source and drain regions reach through to the buried oxide, so there is no 'floor' component of parasitic capacitance at the source/body and drain/body np-junctions. An additional speed improvement is possible from the **dynamic threshold-voltage effect**, which is a consequence of the body of the transistor being isolated from the substrate. When the FET is turned on by a positively increasing V_{GS}, the potential V_B of the floating body also rises, momentarily augmenting the forward bias across the source-channel junction, and increasing the current.

Another useful property can result if both the upper silicon layer and the buried oxide are thin, and if the underlying substrate is heavily doped. Under these circumstances, the field issuing from the drain preferentially penetrates the two thin oxides and terminates on the gate and the substrate, as shown in the bottom part of Fig. 13.13. Thus, the potential $\psi_s(0)$ at the source end of the channel is screened from the applied drain potential V_D, and the short-channel effect is reduced. The thin oxide/substrate combination acts like a 'bottom' gate, making the SOI FET a precursor of the anticipated **multiple gate** FET, one example of which is the FINFET [8].

A truly 'three-terminal' FET, i.e., one with no body-effects to be concerned with, would result if the silicon body layer could be made sufficiently thin for it to be fully depleted. However, obtaining such a thin semiconducting layer with reproducible and

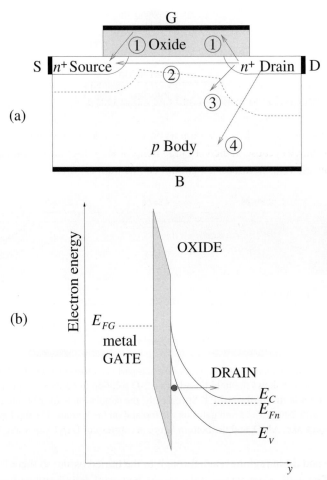

Figure 13.14 Leakage and off currents: (a) 1. oxide tunnelling; 2. sub-threshold conduction; 3. gate-induced drain leakage; 4. drift of minority carriers across the reverse-biased drain/body junction. (b) Detail of gate-induced drain leakage GIDL.

useful properties is proving difficult. In the meantime, the insulating layer of partially depleted SOI FETs isolates devices from each other, which is good for RF circuitry. It can also help de-couple noisy logic blocks from sensitive analogue circuitry, which is good in mixed-signal applications.

13.1.10   Power dissipation

The various DC leakage and sub-threshold currents in an N-FET are sketched in Fig. 13.14. When the device is ON, a high drain current is required and, ideally, this will be the current supplied by the source, i.e., the current drawn from the power supply. If the latter is more than I_D, it is indicative of the presence of current leakage paths in the transistor. Leakage to the gate via tunnelling is one such path (see Section 13.1.6).

One way to mitigate this effect is to use thin oxides only where they are absolutely necessary, e.g., for the transistors in the core of a microprocessor. For other transistors, such as those in the input/output parts of the chip, a thicker oxide is used. Other possibilities for leakage are via the substrate. One such component is the usual current in a reverse-biased np-junction, while another is referred to as **gate-induced drain leakage** (GIDL). The latter is due to band-to-band tunnelling in the depletion region where the gate overlaps the drain (see Fig. 13.14b). The mechanism is that of **Zener breakdown**, as occurs in np diodes with heavy doping on either side of the junction. GIDL has assumed importance because of the very high fields arising from the heavy doping of the pocket implant. As V_{GB} increases, the bands become more bent, and the leakage current increases.

The above leakage currents exist when the transistor is ON, i.e., when $V_{GS} > V_T$. Current can also be drawn from the power supply when the transistor is OFF, i.e., when $V_{GS} < V_T$, but $V_{DS} > 0$. Such currents are the junction leakage current, and the sub-threshold current due to injection from the source. The latter is non-zero because of the influence of the field from the drain on $\psi_s(0)$, as discussed above.

As we have mentioned before, it is necessary in highly integrated circuitry to make the OFF/ON ratio as small as possible. Ideally, switching between OFF and ON would be marked by a vertical transition in the $I_D - V_{GS}$ plane. In fact, the transition has a finite slope (see Fig. 13.9). The usual metric for this transition is the **inverse sub-threshold slope**: it is the change in V_{GS} needed to reduce I_D by a factor of 10. From (10.55)

$$S = \left(\frac{\partial \log_{10} I_D}{\partial V_{GS}} \right)^{-1} \equiv 2.303 m\, V_{\text{th}} . \tag{13.10}$$

Evidently, in this type of transistor, the minimum value of S is about 60 mV/decade. In practical MOSFETs, values are slightly higher than this (closer to 100 mV/decade) because m is greater than its ideal value of unity. Recall that $m(0) = 1 + C_b(0)/C_{ox}$, so any increase in the ionic space charge in the substrate under the source-end of the channel will raise $C_b(0)$ and, consequently, increase m. This is why Fig. 13.9 shows a significant degradation (increase) in S when L is reduced. It is further evidence of the need to control the short-channel effect.

Dynamic power dissipation

A basic CMOS switch is shown in Fig. 13.15a: it comprises a power supply, an inverter, and two capacitors, C^+ and C^-, which, together, constitute the total capacitance C_X at the output node. C_X represents the intrinsic and extrinsic capacitances of the transistors in both the logic gate and in the element that it drives, and, also, the interconnect capacitance. C_X is charged through the P-FET and discharged through the N-FET.[11] The switching events over several clock periods are shown in Fig. 13.15b. During discharge

---

[11] For these processes to occur over similar time periods it helps if the two transistors have similar saturation currents, e.g., using LEVEL 1 equations ((10.34)), $|V_T|$ and the ratio $Z\mu/L$ should be similar for each transistor.

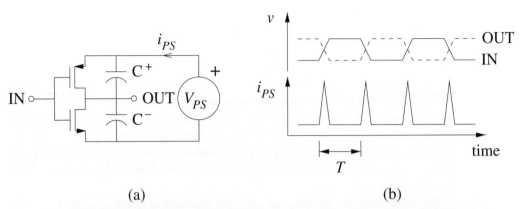

Figure 13.15 Switching in a simple CMOS logic gate. (a) Equivalent circuit. (b) Voltage and current waveforms.

of the capacitor C^-, v_{IN} is HI and the P-FET is OFF. The output goes LO and C^+ charges up. The current drawn from the power supply during this event is

$$i_{PS} = C^+ \frac{\partial}{\partial t}(V_{DD} - v_{OUT}). \tag{13.11}$$

The average power dissipated during this half-period $T/2$ is

$$P_{av} = \frac{1}{T/2} \int_0^{T/2} (V_{DD} - v_{OUT}) i_{PS} \, dt$$

$$= 2f \int_0^{V_{DD}} (V_{DD} - v_{OUT}) C^+ \, d(V_{DD} - v_{OUT})$$

$$= f C^+ V_{DD}^2, \tag{13.12}$$

where f is the clock frequency. Over the next half-period, C^+ is discharged and C^- is charged, thus, over one clock period, the average power dissipated per gate is

$$P_{av} = f V_{DD}^2 (C^+ + C^-) = f V_{DD}^2 C_X. \tag{13.13}$$

This equation highlights the importance of reducing the power-supply voltage as clock frequencies increase.

Static power dissipation

Recall that when a MOSFET is supposedly OFF ($V_{GS} = 0$), there can be a sub-threshold current if $V_{DS} > 0$. From the description of sub-threshold current in Section 10.5, it follows that the static power dissipation per inverter is

$$P(\text{static}) \approx V_{DD} I_{D,\text{sub-threshold}} . \tag{13.14}$$

In a CMOS technology capable of placing 10^8 transistors per cm^2, it is easy to appreciate how minuscule static power dissipation in a single transistor can become a major concern

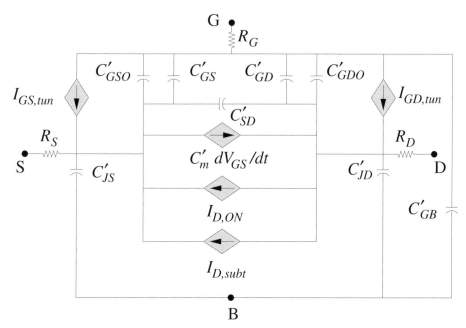

Figure 13.16 Large-signal equivalent circuit for the MOSFET.

at the chip level. The continued control of it via suppression of the short-channel effect, principally DIBL, is vital for the progression to future technology nodes.

13.1.11 Large-signal equivalent-circuit model

To capture the effect of switching, the parasitic resistances and the intrinsic and extrinsic capacitances of the MOSFET must be added to the DC equivalent circuit of Fig. 10.16.

The source and drain resistances arise mainly from the lateral path between the contact metallization and the channel. Obviously, reducing the junction depth y_j to combat the short-channel effect is not helpful in reducing R_S and R_D. In practice, the source and drain regions are thickened after the obligatory narrow region near to the channel (see Fig. 13.1). The gate resistance shown in the equivalent circuit of Fig. 13.16 is that of the gate metallization. The full suite of intrinsic and extrinsic capacitances described in Section 12.2 are represented by appropriate capacitors. Also shown is the gate/body intrinsic capacitance, which serves to remind us that we neglected body-related capacitances in Chapter 12.[12]

Finally, the single current source of the DC circuit in Fig. 10.16 has been expanded into sources representing the ON current, the sub-threshold current, and the tunnelling currents to the gate.

---

[12] Note that because of the inexorable scaling of the transistor, its capacitances can now be smaller than the capacitances of the wiring that connects devices together. To reduce this latter capacitance, low-k dielectric materials are used between the wiring levels of modern, high-performance ICs.

13.2   Emitter-coupled logic

For reasons given elsewhere in this chapter, CMOS is unrivalled when it comes to ultra-dense digital circuitry. However, if the issue is one of raw speed only, then **emitter-coupled logic** (ECL), fashioned from high-performance HBTs, can outperform CMOS. The price paid for this speed advantage is a higher transistor count per gate (see Fig. 13.2b), and a higher static power drain.

Whereas the ECL NOR gate of Fig. 13.2b has the same number (4) of switching transistors as the comparable CMOS gate shown in Fig. 13.2a, it requires additional circuitry to provide the current bias, and to shift the level of the output signal so that it is appropriate for driving other gates. The current-bias circuitry provides a constant current that is switched between the main, emitter-coupled transistors of the gate, depending on which of these transistors is ON (according to the logic levels of the input signals). Because in HBTs I_C increases exponentially with input bias, it doesn't take much change in V_{BE} to switch a large current.[13] Further, because charge flow in an HBT is not constricted by a narrow channel, the current can be large. Thus, a given charge can be sourced or sunk in a very short time. As the current is always present in one branch or another of the circuit, the static power drain is large. The speed advantage of ECL can only be realized if the transistors are biased in the active region when they are ON, and not in the saturation regime.

To illustrate the drawback of switching from the saturation regime in HBT digital circuits, consider the example of a simple inverter using resistor-transistor logic (RTL), as shown in Fig. 13.17. It can be seen that I_C at the HI bias point does not increase with I_B, but is limited by the voltage drop across the load resistor R_L. Thus, 'saturation' in an HBT is a circuit phenomenon, and does not refer to the constant-current portion of the actual transistor characteristic, which is the active regime. In saturation, both junctions are forward biased, so there is a large concentration of excess electrons in the base.[14] Turning the logic gate OFF is initiated by switching the input voltage V_{BE} to zero (see Fig. 13.18a). The excess carriers now start to flow out of the base, and there is some net recombination now that the supply of minority carriers from the emitter and collector has been turned off. The minority carrier concentrations at the edges of the quasi-neutral base depend on the actual applied-voltage drop across the junctions (see (6.29)). Put another way, V_{aj} depends logarithmically on the carrier concentrations. Thus, as these carriers exit the base, the actual junction biases change only slowly. Therefore, the collector current maintains its value. The base current is also approximately constant, but at a negative value determined by $V_{aj,BE}/R_B$, where R_B is the resistance in the base lead (see Fig. 13.19). Only when the excess electron concentration at the collector edge of the base quasi-neutral region has been reduced to zero (at time t_3 in Fig. 13.18b) does i_C start to fall towards its cut-off value, which characterizes the OFF, or logic-LO,

---

[13] As discussed in Section 10.5, $\Delta V = 0.24$ V will cause a 10 000-fold change in current.
[14] Recall from Chapter 6 that the base remains quasi-neutral, so there are also many excess holes present in the base in the saturation mode.

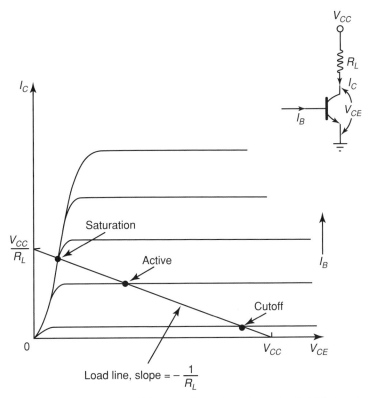

Figure 13.17 Determination of the operating point, using a simple RTL inverter as an example (see inset), by the superposition of the transistor's collector-current characteristic and the resistor's load line. From Pulfrey and Tarr [9].

state.[15] At this time, i_B finally inclines towards the zero value that is to be expected when $V_{BE} = 0$.

In the sequence of excess-electron profiles shown in Fig. 13.18b, the negative slope near the emitter edge is indicative of the diffusive outflow to the emitter lead. Also shown is the initial excess electron profile for a logic HI state in the active regime. As the excess electron concentration at W_B is already zero, because of the reverse-biased base/collector junction, i_C starts to decay immediately after V_{BE} is switched to zero. Thus, the delay when switching from the saturation state can be viewed as the time taken to remove the excess charge shown by the shaded region in Fig. 13.18b.

13.2.1   Large-signal equivalent-circuit model

The large-signal equivalent-circuit model for an HBT is shown in Fig. 13.19. Capacitors representing all the various capacitances discussed in Section 12.3 have been added to the dc equivalent circuit of Fig. 9.9. Diodes representing the back-injection and

---

[15] The symbol i_C is used to indicate a time-dependent collector current. I_C is a dc current.

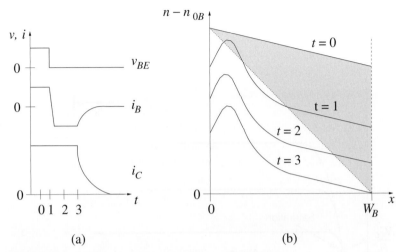

Figure 13.18 Switching out of the saturation mode. (a) Voltage and current waveforms. (b) Evolution of the excess electron profile in the base during switching. The profile for biasing in the active mode of operation is also shown (stippled line).

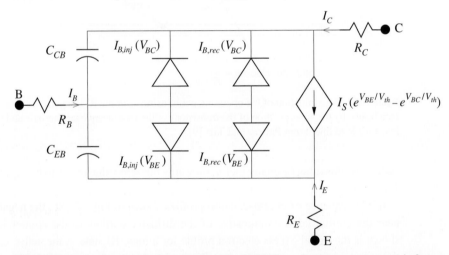

Figure 13.19 Large-signal equivalent circuit of the HBT. The collector current source is from Fig. 9.9.

base-recombination components of the base current are also shown. By providing a set of diodes for each junction, the circuit can be used for all modes of operation of the HBT.

Exercises

13.1 Tensile strain can improve the electron mobility in Si N-FETs by lowering the energy of the Δ_2 valleys with respect to the Δ_4 valleys (see Fig. 13.5).

If a particular stress results in 80% of the electrons residing in the Δ_2 valleys, estimate the percentage improvement in electron mobility with respect to the unstrained case.

13.2 For a CMOS65 N-FET the band bending in the semiconductor is steep enough that the potential-energy profile near the silicon/oxide interface can be approximated as a rectangular potential well of infinite height.

All the electrons can be taken as residing in the first energy sub-band (see Fig. 13.12), which is located 0.377 eV above the conduction-band edge. The electron effective mass can be taken as m_0.

(a) Estimate the percentage change in effective oxide capacitance due to consideration of this electron confinement.

(b) Estimate the change in threshold voltage due to this quantum-mechanical effect.

13.3 In a particular CMOS technology, C_{ox} can be taken as $2 \times 10^9 \epsilon_0$ F m^{-2}, and the electron affinity of the semiconductor Si can be taken as 4 eV. Two dielectrics, Oxide P and Oxide Q, are being considered as replacement gate oxides to double C_{ox}. The electron affinity and relative permittivity for Oxide P are 1 eV and 8, respectively, whereas the corresponding values for Oxide Q are 4 eV and 16.

Which material would make the more *practical* dielectric, and why?

13.4 For a symmetrical, rectangular potential barrier of thickness d and height U_2 the tunnelling transmission probability can be approximated by (5.56)

$$T \approx \exp\left[-\frac{2d}{\hbar}\sqrt{2m^*(U_2 - E)}\right], \tag{13.15}$$

where the electron effective mass m^* is assumed to be the same in all regions of the structure. Assume that this situation applies to tunnelling through the oxide in a silicon-gate Si MOSFET.

In Si CMOS technology the gate oxide has been silica ($\epsilon_r = 3.9$ and electron affinity $\chi = 0.9$ eV) but the change is being made to a hafnium-silicon oxynitride, for which we guess the relative permittivity to be $\epsilon_r \approx 4 \times 3.9$, and the electron affinity to be $\chi = 2.9$ eV.

What thickness of hafnia is needed for $T(E = 0.2 \text{ eV})$ to be equal to that for silica of thickness 2 nm?

13.5 Compute the gate leakage current due to tunnelling from a continuum of states for the two insulators considered in the previous question at $V_{GB} = 1$ V.

13.6 A microprocessor made from CMOS65 FETs operates at a clock frequency of 3 GHz. What percentage of transistors on the chip would need to be switching simultaneously if the static power dissipation were to equal the dynamic power dissipation?

13.7 An *Npn* GaAs-based HBT has a quasi-neutral basewidth that is much shorter than the minority-carrier diffusion length, but that is much greater than any changes in depletion-region widths that may occur on switching. The transistor is biased in the saturation regime, with V_{BE} slightly greater than V_{BC}.

(a) Sketch the electron concentration profile in the quasi-neutral base.

(b) On the same sketch, add a line showing the electron concentration profile at the instant the collector current would begin to decrease after V_{BE} had been reduced to zero.

(c) Let the base doping density be $N_A = 10^{19}\,\text{cm}^{-3}$, the quasi-neutral basewidth be 50 nm, and the cross-sectional area be $2\,\mu\text{m} \times 10\,\mu\text{m}$. Consider the transistor to be in the common-emitter configuration with $V_{BE} = 1.3$ V. The conditions are such that the transistor is in the saturation regime with $V_{BC} = 1.2$ V.

On switching-OFF the transistor, assume that electrons are removed from the quasi-neutral base only by recombination.

How long does it take before the collector current begins to change?

References

[1] International Technology Roadmap for Semiconductors (ITRS). Online [http://public.itrs.net/].

[2] G.E. Moore, Cramming More Components onto Integrated Circuits, *Electronics Magazine*, vol. 38, 114–117, 1965.

[3] C. Kittel, *Introduction to Solid State Physics*, 3rd Edn., Chap. 4, John Wiley & Sons Inc., 1968.

[4] S.E. Thompson, G. Sun, Y.-S. Choi and T. Nishida, Uniaxial-process-induced Strained-Si: Extending the CMOS Roadmap, *IEEE Trans. Electron Dev.*, vol. 53, 1010–1018, 2006.

[5] S.E. Thompson, M. Armstrong, C. Auth, S. Cea, R. Chau, G. Glass, T. Hoffman, J. Klaus, Z. Ma, B. Mcintyre, A. Murthy, B. Obradovic, L. Shifren, S. Sivakumar, S. Tyagi, T. Ghani, K. Mistry, M. Bohr and Y. El-Mansy, A Logic Technology Featuring Strained-silicon, *IEEE Electron Dev. Lett.*, vol. 25, 191–193, 2004.

[6] S.-E. Ungersböck, *Advanced Modeling of Strained CMOS Technology*, Ph.D. dissertation, Technical University of Vienna, 2007.
Online [http://www.iue.tuwien.ac.at/phd/ungersboeck/].

[7] J.D. Plummer, M.D. Deal and P.B. Griffin, *Silicon VLSI Technology: Fundamentals, Practice and Modeling*, Prentice-Hall, 2000.

[8] J. Kretz, L. Dreeskornfeld, R. Schröter, E. Landgraf, F. Hofmann and W. Rösner, Realization and Characterization of Nano-scale FinFET Devices, *Microelectronic Engineering*, vol. 73–74, 803–808, 2004.

[9] D.L Pulfrey and N.G. Tarr, *Introduction to Microelectronic Devices*, Fig. 11E.7, Prentice-Hall, 1989.

14 Transistors for high frequencies

The large-signal models of Fig. 13.16 and Fig. 13.19 are to be used when considering DC and switching operations. High-frequency operation is governed by the small-signal performance of transistors. It is insightful, and computationally efficient, to develop a small-signal model of the transistor by linearizing the large-signal model. Our approach is to linearize via a Taylor series expansion, and then manipulate the resulting equivalent circuit so that it can be used to define and evaluate two important high-frequency metrics: f_T and f_{max}.

Small-signal operation is illustrated in Fig. 14.1, which shows a simple amplifier. The base-emitter DC bias is supplied from the power-supply V_{BB} via a resistor, and the small AC signal is supplied from v_{in} via a capacitor. The transistor is of the bipolar variety, in recognition of the fact that bipolar transistors (BTs) have traditionally been dominant in high-frequency applications. This dominance has been due principally to the BT's transconductance being inherently superior to that of FETs, as we will show later in the chapter.

Presently, the world record for f_T is 710 GHz [1], and is held by an HBT based on the sleek-looking device shown in Fig. 14.2. However, nowadays, FETs are encroaching into the high-frequency domain: HEMTs using high-mobility semiconductors based on GaAs and InP can yield devices with very high transconductance; and MOSFETs employing silicon's matchless technology can yield very small devices, for which the capacitance is very low. Because of this involvement of both transistor types (bipolar and field-effect) in high-frequency circuits, we start the chapter with the development of a generic small-signal equivalent circuit, before going on to discuss the particular merits of each transistor type.

14.1 Quasi-static analysis

In this chapter the analysis we perform is of the quasi-static variety. The meaning of this was presented in Section 12.1 during the defining of capacitance. Recall that the implication of this type of analysis is that information on the behaviour of the device cannot be obtained for time periods approaching that of the transit time of charge carriers within the device. As we indicated in Section 12.1, this is not a particularly severe restriction. However, if one does need to know the current in a device over periods of 10s–100s of femtoseconds, for example, then it is necessary to perform a

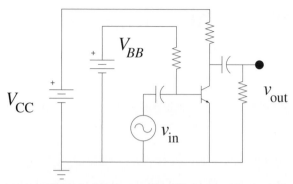

Figure 14.1 Simple amplifier circuit showing a small-signal AC input v_{in}, and the DC biasing of the input via V_{BB}.

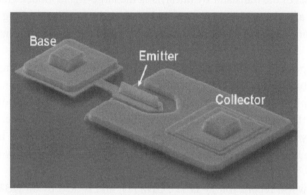

Figure 14.2 The record-breaking HBT from the University of Illinois. Reused with permission from Walid Hafez and Milton Feng, Applied Physics Letters, 86, 152101 (2005) [2]. Copyright 2005, American Institute of Physics.

non-quasi-static analysis. This involves the solution of the full, time-dependent version of our master equation (5.24). A sub-set example of this is the Hydrodynamic Equations (5.22). The presence of time derivatives and time dependencies in these equations makes their exact solution difficult to obtain. Even using a simple device structure, and after making many assumptions, obtaining an analytical solution is a lengthy procedure [3]. Generally, numerical methods are required to carry out a full, non-quasi-static analysis.

In the following, we rely on the quasi-static approach; the implications being that the charge at any point r in the device, at any time t', is determined by the instantaneous value of the applied voltages, irrespective of the previous history of the biasing. It's worth emphasizing this by expressing it in equation form:

$$q(r, t') = f(V_{\text{terminals}}, t')$$

$$\neq f(V_{\text{terminals}}, t < t'). \tag{14.1}$$

14.2   The generic small-signal model

Recall the generic transistor of Fig. 12.1, and let us use the notation

$$i_J(t) = I_J + i_j(t) \qquad J, j = 1, 2, 3, \tag{14.2}$$

where i_J is the total current at one of the three terminals J of a transistor, and comprises a DC component I_J, and a small-signal AC component i_j.

Let us take $J, j = 2$ as an example, and take it to represent either the collector current in an HBT or the drain current in a FET. Further, let us use terminal 1 (the emitter or source) as the reference for potential. Thus, using the quasi-static approximation, the total current i_2 depends on the instantaneous values of the terminal voltages $V_{21} + v_{21}$ and $V_{31} + v_{31}$.

The next step is to linearize the expression for the current by taking the Taylor series expansion for the total current i_2 to first order,

$$
\begin{aligned}
i_2 &= I_2(V_{21} + v_{21}, \ V_{31} + v_{31}) \\
&= I_2(V_{21}, V_{31}) + \frac{\partial I_2}{\partial V_{21}} v_{21} + \frac{\partial I_2}{\partial V_{31}} v_{31} \\
&\equiv I_2 + g_{22} v_{21} + g_{23} v_{31} ,
\end{aligned}
\tag{14.3}
$$

where the g's are conductances, names for which will be assigned later.

By restricting the Taylor series expansion to produce a linear relationship, it is implicit that the second- and higher-order terms can be neglected. For FETs, this is not much of an approximation because $I_2 \equiv I_D$, for example, is already linear in $V_{31} \equiv V_{GS}$ and $V_{21} \equiv V_{DS}$ in the resistive regime, and the exponent of any relations in the saturation regime never exceeds 2. However, for bipolar devices, $I_2 \equiv I_C$ depends exponentially on V_{31}, so the linearization places a severe restriction on the allowable magnitude of the small signal:

$$v_{31} \equiv v_{be} \ll 2kT/q \approx 50\,\text{mV at 300 K.} \qquad \text{Applicable to BTs.} \tag{14.4}$$

Performing a similar Taylor series expansion for i_3, the gate or base small-signal current, yields

$$
\begin{aligned}
i_3 &= \frac{\partial I_3}{\partial V_{31}} v_{31} + \frac{\partial I_3}{\partial V_{21}} v_{21} \\
&= g_{33} v_{31} + g_{32} v_{21} ,
\end{aligned}
\tag{14.5}
$$

where the new g's are to be named later.

The equivalent circuit representing the small-signal components is shown in Fig. 14.3a. It is often convenient to turn this two-generator equivalent circuit into a one-generator circuit. One version of the latter is shown in Fig. 14.3b.[1] Don't be alarmed by the

---

[1] It is instructive to convince yourself that the two circuits of Fig. 14.3 are, indeed, equivalent. Do this by showing that the terminal currents are the same in each case.

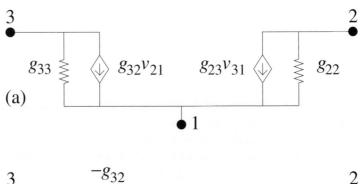

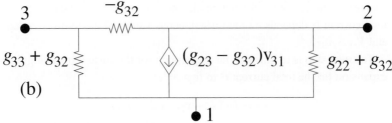

Figure 14.3 Small-signal AC equivalent circuits for all transistors. (a) Two-generator model. (b) One-generator model with the same terminal characteristics as the circuit in (a).

appearance of a negative sign in front of g_{32}, as, in fact, g_{32} itself is negative.[2] Thus, the conductor in the top branch of the circuit is real. g_{32} is often called the **reverse feedback conductance**. Generally speaking, $|g_{32}|$ is much smaller than the other conductances, so it is common practice to name the other conductances as follows:

$$g_{33} + g_{32} \approx g_{33} \qquad \text{input conductance}$$
$$g_{23} - g_{32} \approx g_{23} \qquad \text{transconductance}$$
$$g_{22} + g_{32} \approx g_{22} \qquad \text{output conductance.}$$

The naming of the **transconductance** deserves some comment. Recall the term 'transcapacitance' in Chapter 12: it referred to a circuit element that allowed for the non-reciprocity of capacitances, i.e., the fact that the effect of terminal 2 on terminal 3 as regards charging currents may be different from the effect of terminal 3 on terminal 2. The transconductance performs exactly the same function, but for the transport currents, i.e., those currents involving only conductive components. Transconductance is denoted by the symbol g_m and it follows that

$$g_m \equiv g_{23} - g_{32} \approx \frac{\partial I_2}{\partial V_{31}}. \tag{14.6}$$

Interestingly, the 'approximate' sign in (14.6) is usually replaced by an equals sign, implying a definition of transconductance that we have just shown to be strictly incorrect. However, practically, g_m can be evaluated this way because $g_{23} \gg |g_{32}|$.

---

[2] Taking the bipolar case as an example, recall $g_{32} = \partial I_B / \partial V_{CE}$. As V_{CE} is increased (in the active mode), the quasi-neutral basewidth shrinks and there is less recombination in the base, so I_B decreases.

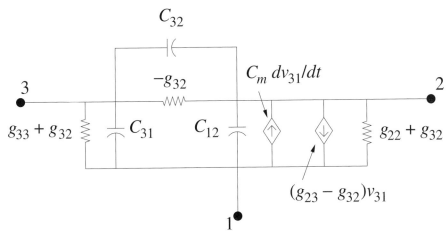

Figure 14.4 Generic, linearized, hybrid-π, small-signal, equivalent circuit.

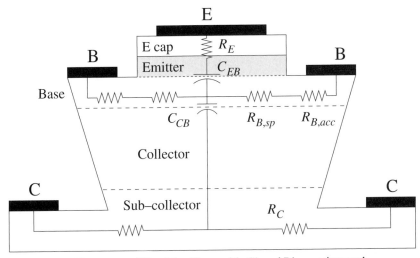

Figure 14.5 HBT structure of Fig. 9.1 with parasitic C's and R's superimposed.

With some imagination, the circuit of Fig. 14.3b can be viewed as having the shape of the Greek letter π. The next step in the development of the small-signal equivalent circuit is to add capacitors to represent the various charge-voltage relationships derived in Chapter 12. The resulting circuit is shown in Fig. 14.4. It is a hybrid circuit because it brings together two sets of components that were derived using different models. Because of this, and its topology, the circuit is known as the **hybrid-π** equivalent circuit.

14.3   Hybrid-π small-signal model for HBTs

To make the generic circuit of Fig. 14.4 specific to HBTs, first consider Fig. 14.5, which shows a superposition of the capacitors and parasitic resistors on an actual HBT structure.

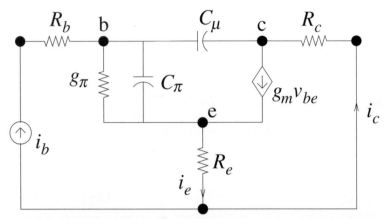

Figure 14.6 The circuit of Fig. 14.4 with some omissions (see text), and with new labelling for the remaining components and for the terminals. Note the AC short between the collector and emitter terminals, as applies to the measurement of f_T.

The resistors represent: the resistance of the quasi-neutral emitter; the resistance of the quasi-neutral collector and the path followed by the collector current to the actual collector contact; and the resistance of the base. The latter comprises an access resistance ($R_{B,acc}$) due to the path from the base contact to the edge of the base quasi-neutral region, which is easy to envisage, and a less obvious component ($R_{B,sp}$) that must account for the spreading nature of the resistance in the quasi-neutral base. $R_{B,sp}$ is considered in more detail in Section 14.6.1.

Transferring the resistors and capacitors to the hybrid-π circuit yields Fig. 14.6. The circuit is for the case of an HBT biased in the active mode of operation, so it is permissible to ignore g_{32}. Also, the output conductance has been omitted because it has a near-infinite value in the active mode (see Fig. 9.6). As we saw in Section 12.3, transcapacitance is not an issue in HBTs, at least as regards junction and storage capacitances, so we omit it here. C'_{EC} is also omitted as there is negligible effect at the emitter of any change in V_{CB} when the transistor is in the active mode. Following tradition, the input conductance is labelled as g_π, the base-emitter capacitance is labelled as C_π, and the base-collector junction capacitance is labelled as C_μ. The AC short-circuit at the output is the effective result of holding constant the DC bias V_{CE}. This is the condition under which the frequency metric f_T, is determined, as we now describe.

14.4 f_T: the extrapolated unity-current-gain frequency

Fig. 14.7 shows experimental data for the frequency dependence of the square of the magnitude of the current gain $|i_c/i_b|^2$ of an HBT. Note how the measured data ends at a frequency that is imposed by the capabilities of the measuring equipment. However, for a decade or so before this frequency limit, the gain appears to be rolling-off at -20 dB/decade, which is indicative of an RC-circuit with a dominant single pole. The

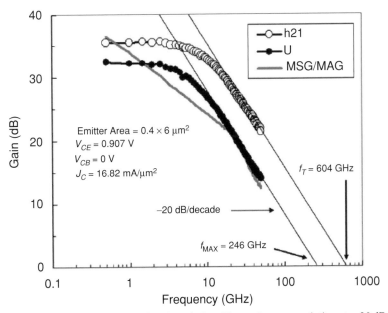

Figure 14.7 Experimental data for f_T and $f_{\max}$, illustrating extrapolation at -20 dB/decade from lower frequencies. The label for $|i_c/i_b|^2$ should be $|h_{21}|^2$, rather than h_{21}. Reused with permission from Walid Hafez and Milton Feng, Applied Physics Letters, 86, 152101 (2005) [2]. Copyright 2005, American Institute of Physics.

gain can be extrapolated at this slope to higher frequencies, and the frequency at which the gain becomes unity (0 dB), is called **the extrapolated** f_T, or, simply, f_T. The object of the following subsection is to derive an expression for f_T, starting from the equivalent circuit that we have developed. This expression will identify the components of the transistor that need to be considered if high-frequency performance is to be attained. Because we are seeking a frequency range at which there is a single-pole roll-off, we will systematically simplify the expressions in our derivation to arrive at a gain that is proportional to $1/\omega^2$.

You may be wondering why anyone would wish to know a frequency that cannot be measured, and that is associated with an impractically low value of gain? The answer is two-fold: it is a definite figure that can be used to compare transistors and, by extrapolating back from f_T at $+20$ dB/decade, the current gain at useful operating frequencies can be readily ascertained.

14.4.1 An expression for f_T

Measurements of the current gain for evaluation purposes are performed with the collector and emitter terminals held at constant potentials. Thus, as far as the AC signal is concerned, the emitter is shorted to the collector. This is the case shown in Fig. 14.6.

Using the notation of this figure, it follows that

$$i_b = [g_\pi + j\omega C_\pi]\, v_{be} + j\omega C_\mu v_{bc}$$

$$i_c = g_m v_{be} - j\omega C_\mu v_{bc}\,. \tag{14.7}$$

To eliminate v_{bc} from the above expressions, note

$$\begin{aligned} v_{bc} = v_{be} + v_{ec} &= v_{be} + i_e R_e + i_c R_c \\ &= v_{be} + (i_b + i_c) R_e + i_c R_c \\ &\approx v_{be} + i_c (R_e + R_c)\,, \end{aligned} \tag{14.8}$$

where, to simplify the following algebra, the assumption has been made that

$$i_b R_e \ll i_c (R_e + R_c), \quad \text{i.e.,} \quad i_c/i_b \gg R_e/(R_e + R_c)\,. \tag{14.9}$$

As we expect the current gain to be much greater than unity, this assumption is justifiable. Substituting for v_{bc} in the expression for i_c yields

$$i_c = \frac{g_m - j\omega C_\mu}{1 + j\omega C_\mu R_{ec}}\, v_{be}\,, \tag{14.10}$$

where $R_{ec} = R_e + R_c$.

The square of the magnitude of the collector current is

$$|i_c|^2 = \frac{g_m^2 + \omega^2 C_\mu^2}{1 + \omega^2 C_\mu^2 R_{ec}^2}\, v_{be}^2\,. \tag{14.11}$$

To reduce this to an expression having the desired frequency independence, the following two conditions must be met:

$$\omega^2 \ll \frac{g_m^2}{C_\mu^2} \qquad \text{Condition 1,}$$

$$\omega^2 \ll \frac{1}{C_\mu^2 R_{ec}^2} \qquad \text{Condition 2.}$$

These conditions set an upper limit to the frequency at which the extrapolation can start. Turning now to the base current; substituting for v_{bc} in the expression for i_b yields

$$i_b = \left(g_\pi + j\omega C_\pi + j\omega C_\mu (1 + R_{ec} g_m)\right) v_{be}\,. \tag{14.12}$$

The square of the magnitude of the base current is

$$|i_b|^2 = \left[g_\pi^2 + \omega^2 (C_\pi + C_\mu (1 + R_{ec} g_m))^2\right] v_{be}^2\,. \tag{14.13}$$

To reduce this to an expression having the desired dependence on frequency of ω^2, the following condition must be met:

$$\omega^2 \gg \frac{g_\pi^2}{\left(C_\pi + C_\mu (1 + R_{ec} g_m)\right)^2} \qquad \text{Condition 3}\,. \tag{14.14}$$

This condition sets a lower limit to the frequency at which the extrapolation can start.

Finally, assuming that the three conditions above are satisfied, the expression for the square of the magnitude of the current gain is

$$\left|\frac{i_c}{i_b}\right|^2 = \frac{g_m^2}{\omega^2(C_\pi + C_\mu(1 + g_m R_{ec}))^2} \, . \tag{14.15}$$

This expression is in the desired form as the square of the magnitude of the current gain has a -20 dB/decade roll-off with increasing frequency. When extrapolated to a current gain of unity, the resulting frequency, in this common-emitter, short-circuit condition, is called **the extrapolated** f_T, or simply f_T. From (14.15)

$$2\pi f_T = \frac{g_m}{C_\pi + C_\mu(1 + g_m R_{ec})} \, . \tag{14.16}$$

14.5   Designing an HBT for high f_T

For design purposes it is helpful to break-up (14.16) into components that can be identified with various regions of the device, which can then be targeted for improvement if necessary. To do this, the reciprocal of f_T is considered, so the resulting components have the dimensions of time, and are known as **signal delay times**. Each time is in the form of an RC-time constant, and can be thought of as the time taken for the charge in a particular region to adjust to the new value demanded by the small-signal at the input of the transistor.

From (14.16) an overall signal delay time τ_{EC} can be expressed as

$$\begin{aligned}
\tau_{EC} &= \frac{1}{2\pi f_T} = \frac{C_\pi}{g_m} + \frac{C_\mu(1 + g_m R_{ec})}{g_m} \\
&= \frac{C'_{EB,j}}{g_m} + \frac{C'_{EB,b}}{g_m} + \frac{C'_{BE,t}}{g_m} + C'_{CB}\left(\frac{1}{g_m} + R_{ec}\right) \\
&\equiv \tau_E + \tau_B + \tau_C + \tau_{CC} \, , \tag{14.17}
\end{aligned}$$

where the components of C'_{EB} and C'_{BE} discussed in Section 12.3 have been substituted for C_π, and the collector-base junction capacitance has been substituted for C_μ. From (12.23), the emitter signal delay is the time taken to charge the emitter/base junction capacitance via the dynamic resistance $(1/g_m)$ of the transistor:

$$\tau_E = \frac{\epsilon_s A_E}{g_m W} \, . \tag{14.18}$$

The signal delay associated with the quasi-neutral base follows from (12.27) and (9.11):

$$\tau_B = \frac{W_B}{2}\left(\frac{W_B}{D_n} + \frac{1}{v_R}\right) \, . \tag{14.19}$$

The signal delay associated with the change in field in the base/collector depletion region due to the passage of the electrons carrying the signal current follows from (12.36):

$$\tau_C = \frac{w}{2v_{\text{sat}}} \, . \tag{14.20}$$

Finally, the charging of the base/collector junction capacitance via the dynamic and parasitic resistances follows from (12.37):

$$\tau_{CC} = \frac{\epsilon_s A_C}{w} \left(\frac{1}{g_m} + \frac{W_E}{A_E \sigma_E} + \frac{W_C}{A_C \sigma_C} \right), \tag{14.21}$$

where A_E and A_C are the areas of the two junctions, W_E and W_C are the widths of the quasi-neutral-emitter and collector, respectively, and the σ's are the relevant conductivities (see Section 5.4.2).

As in a MOSFET, the key to better performance is scaling. Lateral scaling reduces A_E and A_C. Vertical scaling reduces W_B and the base/collector depletion-region width w. In high-performance InGaP/GaAs HBTs, for example, W_B is so short ($\ll 50$ nm) that the major delay in the device is no longer the base delay. That dubious honour is shared by the two delays associated with the collector. Note from (14.20) and (14.21) that a compromise must be reached regarding w as it affects τ_C and τ_{CC} differently. To get an idea of how short the delays must become, realize that for $f_T = 800$ GHz, the overall signal delay τ_{EC} would have to be ≈ 200 fs.

14.5.1   SiGe HBT

Fig. 8.3 indicates that the choice of semiconductors for combining with silicon to make a near-lattice-matched HBT is limited. Germanium has many similarities to silicon, but its lattice constant is too different for it to be used directly as a low-bandgap base with a Si emitter. However, by varying the Ge mole fraction x in a compound of $Si_{1-x}Ge_x$, a graded-composition base can be formed. An example is shown in Fig. 14.8, where the Ge content is varied from 0 to about 15% over a region spanning the p-type base. The bandgap decreases as the Ge mole fraction increases, yet the valence-band edge remains essentially flat in the base because of the very high p-type doping therein. Thus, the change in bandgap is manifest as a change in conduction-band edge, leading to the creation of an electric field that adds a component of drift to the usual diffusion current of electrons. This increase in current reduces the time needed to charge the capacitance $C'_{EB,b}$, i.e., τ_B is reduced.

The derivation of an expression for τ_B in a graded-base HBT is outlined in Exercise 14.8. The result is

$$\tau_B = \frac{Q_{nB}}{J_e} = \frac{W_B^2}{b D_n} \left[1 - \frac{1 - e^{-b}}{b} \right] + \frac{W_B}{v_{sat}} \frac{1 - e^{-b}}{b}, \tag{14.22}$$

where $b = \Delta E_g / k_B T$, with ΔE_g being the bandgap change across the base due to a linear compositional grading, and the electrons are assumed to exit the base at their saturation velocity v_{sat}.

The improvement in τ_B due to base-grading is illustrated in Fig. 14.9. The stored charge is not much affected by base-grading, so the improvement in τ_B is due almost entirely to the increase in current, which stems from the higher, field-assisted, electron velocity.

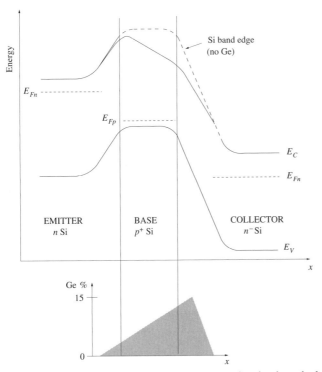

Figure 14.8 SiGe HBT. Top: energy-band diagram showing how the bandgap reduction in the base, as the Ge content increases, leads to a field that aids the passage of electrons. Bottom: a typical Ge profile.

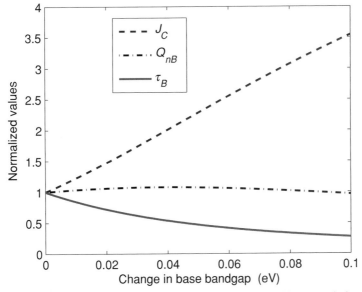

Figure 14.9 Dependence of electron current, base charge, and base transit time on the bandgap difference between the ends of a linearly graded base of quasi-neutral basewidth $W_B = 100\,\text{nm}$, and for an electron diffusivity of $30\,\text{cm}^2\,\text{s}^{-1}$. These parameters are typical of SiGe HBTs. J_e, Q_{nB} and τ_B are all normalized to their values in an HBT with a uniform base-composition.

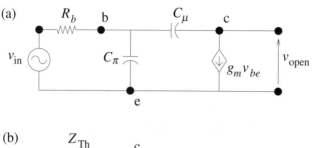

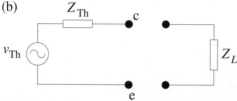

Figure 14.10 Equivalent circuits for estimating f_{max} for bipolar transistors. (a) Simplified form of the circuit in Fig. 14.6. (b) The Thévenin equivalent.

The ability to provide high currents and good high-frequency performance are bipolar-transistor attributes. On the other hand, FETs, in the form of Si CMOS, are unrivalled for dense, digital circuitry. The SiGe HBT provides silicon technology with the capability of integrating bipolar- and field-effect-transistors on the same chip. The result is **BiCMOS**, a powerful technology for mixed-signal applications.

14.6 f_{max}: the extrapolated unity-power-gain frequency

One important device parameter missing from the expression for f_T is R_b, the base resistance. This is because f_T relates to the current gain, and any desired current can be forced through any resistance, provided enough voltage can be applied. In practice, there are limits to the available input voltage, so some metric that takes this into account is desirable: f_{max}, an extrapolated frequency related to the power gain, is such a metric. Following the general procedure used in Section 14.4, we seek to develop an expression for the power gain that rolls-off with frequency at -20 dB/decade. Extrapolation at this gradient to a power gain of 0 dB yields f_{max}.

To develop an expression for f_{max} consider the hybrid-π circuit of Fig. 14.6 and, for simplicity, ignore R_e and R_c.[3] The resulting circuit is shown in Fig. 14.10a. Note also that the input conductance is missing from the circuit: this is because we have supposed that the frequency from which we are going to make the extrapolation is high enough such that

$$\omega^2 \gg \frac{g_\pi^2}{C_\pi^2} \qquad \text{Condition 4}, \qquad (14.23)$$

where the numbering of the conditions follows-on from Section 14.4.

---

[3] Including R_e and R_c leads to a very cumbersome expression for f_{max} [4].

The Thévenin equivalent of this circuit is shown in Fig. 14.10b. Expressions for the Thévenin voltage source v_{Th} and impedance Z_{Th} can be obtained using standard procedures from Network Analysis. The results are

$$v_{\text{Th}} = v_{\text{in}} \frac{j\omega C_\mu - g_m}{j\omega C_\mu (1 + g_m R_b) - \omega^2 C_\mu C_\pi R_b}$$

$$Z_{\text{Th}} = \frac{R_b^{-1} + j\omega C_T}{j\omega C_\mu (g_m + R_b^{-1}) - \omega^2 C_\mu C_\pi} , \tag{14.24}$$

where $C_T = C_\pi + C_\mu$.

For maximum power transfer to the load, Circuit Theory tells us that the load impedance Z_L should be the complex conjugate of Z_{Th}. Under this circumstance, the total impedance of the circuit in Fig. 14.10b is

$$Z_{\text{circ}} = Z_{\text{Th}} + Z_L = 2\Re(Z_{\text{Th}}) , \tag{14.25}$$

where the real part of Z_{Th} is

$$\Re(Z_{\text{Th}}) = \frac{-\omega^2 C_\mu C_\pi R_b^{-1} + \omega^2 C_\mu C_T (g_m + R_b^{-1})}{\omega^2 C_\mu^2 (\omega^2 C_\pi^2 + (g_m + R_b^{-1})^2)} \approx \frac{C_T}{C_\mu g_m} , \tag{14.26}$$

where the considerable simplification is achieved by imposing the following assumptions:

$$C_\pi \approx C_T \qquad \text{in the numerator}$$

$$g_m \gg \frac{1}{R_b} \qquad \text{in the denominator,}$$

and by invoking the following condition:

$$\omega^2 \ll \frac{(g_m + R_b^{-1})^2}{C_\pi^2} \qquad \text{Condition 5}.$$

It is understood that the magnitudes of the small-signal currents and voltages we are using are RMS quantities. Thus, we need $|v_{\text{Th}}|^2$, which is given from (14.24) by

$$|v_{\text{Th}}|^2 = |v_{\text{in}}|^2 \frac{\omega^2 C_\mu^2 + g_m^2}{\omega^2 C_\mu^2 (1 + g_m R_b)^2 - \omega^4 C_\mu^2 C_\pi^2 R_b^2} \approx \frac{|v_{\text{in}}|^2}{R_b^2} \frac{1}{\omega^2 C_\mu^2} , \tag{14.27}$$

where the simplification is achieved by invoking the following conditions:

$$\omega^2 \ll \frac{g_m^2}{C_\mu^2} \qquad \text{Condition 6}$$

$$\omega^2 \ll \frac{g_m^2}{C_\pi^2} \qquad \text{Condition 7} , \tag{14.28}$$

and by assuming, once again, that $g_m \gg 1/R_b$. Thus, the output power under these conjugately matched conditions is

$$P_{\text{out, max}} = \frac{|v_{\text{Th}}|^2}{4\Re(Z_L)} \approx \frac{|v_{\text{in}}|^2}{4R_b^2} \frac{g_m}{\omega^2 C_\mu C_T} . \tag{14.29}$$

Turning now to the input power, we make use of an earlier assumption ($C_\pi \approx C_T$), which is equivalent to $C_\mu \ll C_\pi$. This allows the impedance looking into the input of the transistor to be written as

$$Z_{\text{in}} \approx R_b - \frac{j}{\omega C_\pi}. \tag{14.30}$$

Replacing v_{in} in Fig. 14.10a with a power supply of voltage v_S, and with a series impedance of Z_S that is conjugately matched to Z_{in}, the condition for maximum power transfer to the transistor is met, and the input power is

$$P_{\text{in,max}} = \frac{|v_S|^2}{4\Re(Z_{\text{in}})} = \frac{|v_{\text{in}}|^2}{R_b}. \tag{14.31}$$

The **maximum available gain** is

$$\text{MAG} = \frac{P_{\text{out,max}}}{P_{\text{in,max}}} = \frac{g_m}{4\omega^2 C_\mu C_T R_b}. \tag{14.32}$$

The frequency at which MAG $= 1$ is the **extrapolated** f_{max}, or, simply f_{max}. It is given by

$$f_{\text{max}} = \sqrt{\frac{f_{T,i}}{8\pi C_\mu R_b}}, \tag{14.33}$$

where $f_{T,i} = g_m/2\pi C_T$ is the intrinsic f_T, i.e., f_T from (14.16) without considering parasitic resistances.

Equation (14.33) emphasizes the need to have not only a transistor with good intrinsic high-frequency performance, but also a low base resistance to reduce absorption of the input power, and a high impedance $1/j\omega C_\mu$ to reduce feedback of power from the output. HBTs are particularly suited to decreasing R_b because the heterojunction at the emitter/base interface can be specifically designed to impede back-injection of holes into the emitter. This allows the base doping density to be increased, thereby reducing R_b without compromising the current gain. Reduction of C_μ is largely a question of making the active part of the base/collector junction as small as possible.

14.6.1   Base-spreading resistance

A significant contribution to R_b comes from the base-spreading resistance ($R_{b,sp}$ in Fig. 14.5). To estimate this resistance, consider the HBT with two base contacts shown in Fig. 14.11. The total base current I_B is split between the two contacts, and the current $I_B/2$ at the left-hand contact is represented by arrows penetrating laterally under the emitter. Let us assume that this component of base current falls off linearly in the lateral direction y due to recombination with electrons injected from the emitter. Thus, the current associated with the left contact is given by

$$I_B(y) = \frac{I_B}{2}[1 - y/h], \tag{14.34}$$

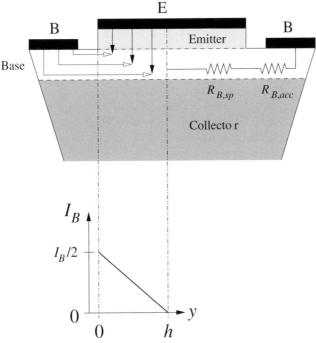

Figure 14.11 Base spreading resistance. Top panel: illustrating the recombination in the quasi-neutral base of holes, laterally injected from the base lead, with electrons injected from the emitter. Bottom panel: showing the assumption of a base current varying linearly with y.

where h is one half of the lateral extent of the base (see Fig. 14.11). We now relate this current to a power loss in the left-hand side of the device via

$$P_{\text{left}}(y) = I_B(y)^2 R$$

$$P_{\text{left}} = \int_0^h \frac{I_B^2}{4} \left(1 - \frac{y}{h}\right)^2 \frac{\rho \, dy}{A} , \tag{14.35}$$

where ρ is the resistivity of the base material and $A = Z W_B$ is the cross-sectional area, with Z being the length into the page of the base and W_B being the quasi-neutral basewidth. Performing the integration gives

$$P_{\text{left}} = \frac{1}{24} I_B^2 \frac{\rho 2h}{A} , \tag{14.36}$$

where

$$\frac{\rho 2h}{A} \equiv R_{B,QNB} , \tag{14.37}$$

where $R_{B,QNB}$ is the actual resistance of the full quasi-neutral base region. Adding-in the power loss due to Joule heating in the right-hand side of the base, the total power dissipation is

$$P = \frac{I_B^2}{12} R_{B,QNB} \equiv I_B^2 R_{B,sp} , \tag{14.38}$$

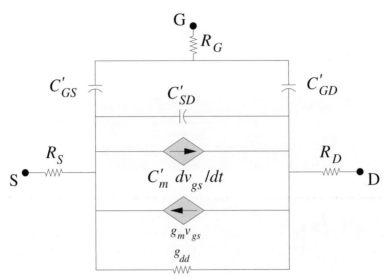

Figure 14.12 The high-frequency, small-signal, linearized, equivalent circuit for a FET, with no leakage currents to the gate, and with no substrate effects.

which defines the **base spreading resistance** $R_{B,sp}$. Thus, the provision of two base contacts achieves the desired goal of reducing the effective base resistance. A value for $R_{B,sp}$ less than $R_{B,QNB}$ is to be expected as not all of the base current penetrates the full lateral width of the base. The total base resistance in Fig. 14.4 is

$$R_B = R_{B,acc} + R_{B,sp}.$$ (14.39)

14.7 f_T and f_{max} for FETs

The small-signal, high-frequency, equivalent circuit for FETs follows directly from the general, hybrid-π circuit of Fig. 14.4. All one needs to do is: change the subscripts (1,2,3 become s,d,g, respectively); add resistors to represent the parasitic resistances of the source, drain, and gate; and make any appropriate simplifications. One can also maintain the topology of the large-signal equivalent circuit of Fig. 13.16, as we have done in Fig. 14.12. We have omitted the input conductance g_{33}, which implies that there is no tunnelling or leakage through the oxide in the case of a MOSFET, nor any transport current in the Schottky diode of an HJFET. Additionally, the reverse-bias feedback conductance g_{32} has been neglected, as was the case for the HBT. However, the output conductance ($g_{22} \equiv g_{dd}$) has been retained because of the resistive nature of the channel that connects the source to the drain. Note that we have not included any components to represent the substrate: this is warranted for HJFETs on semi-insulating substrates, Si MOSFETs in the SOI technology, and coaxial nano-FETs (see Chapter 18), but it is optimistic for planar Si MOSFETs on substrates of appreciable conductivity.

14.7.1 f_T

By following the procedure used for HBTs in Section 14.4.1, an equation very similar to (14.16) can be derived for the extrapolated, common-source, unity-current-gain frequency of FETs:

$$2\pi f_{\mathrm{T}} = \frac{g_m}{C_{gs}(1 + g_{dd}R_{sd}) + C_{gd}(1 + (g_m + g_{dd})R_{sd})}, \tag{14.40}$$

where $R_{sd} = R_s + R_d$. Observe that the equivalence of (14.16) and (14.40) becomes exact if g_{dd} is less than both $1/R_{sd}$ and g_m. This correspondence emphasizes the basic similarity of transistor types.

Obviously, a high transconductance is desirable for obtaining a high f_T. Having a high mobility helps in this regard, and it is this attribute that makes InP and GaAs HEMTs and MESFETs well-suited to high-frequency applications. This is particularly the case for HEMTs, where the high-mobility of the starting semiconductor material is preserved by the device features of: undoped channel material; confinement of the channel charge, to a large extent, away from the interface; undoped spacer layer of barrier material for electrons that do spread into the barrier; finally, a point not mentioned in Section 11.3, less scattering than in the 3-D case because of there being one less dimension of freedom for electron movement.

Low values for the gate capacitances C_{gs} and C_{gd} are also desirable. In HJFETs these capacitances are primarily reverse-biased junction capacitances, for which the dielectric is thicker than the oxide in MOSFETs. This is another reason for the dominance of HJFETs in high-frequency applications where field-effect transistors are employed. However, the superlative technology of Si CMOS means that very small structures can be realized, so low values of capacitance are also possible in Si MOSFETs. This is clear from measurements on a prototype Si MOSFET from IBM, for which the effective gate length was 27 nm and the extrapolated f_T was 220 GHz [5]. For comparison, an f_T of 562 GHz has been reported for a similar length (25 nm) HEMT [6].

In this era of ubiquitous wireless communication there is considerable interest in improving the high-frequency capability of Si CMOS FETs, as this would enable smaller and more powerful mixed-signal integrated circuits. The main drawback for MOSFETs is the inherently low transconductance. Recall that $g_m \approx dI_2/dV_{31}$, and that for a FET in the saturation mode $I_2 = I_{Dsat} \propto (V_{GS} - V_T)^n$, where the exponent n is somewhere between 1 and 2. Contrast this with an HBT, for which $I_C \propto \exp(V_{BE}/V_{th})$. The corresponding transconductances per unit current are

$$
\begin{aligned}
\frac{g_m}{I} &= \frac{1}{V_{\mathrm{th}}} && \text{HBT} \\[2mm]
&= \frac{n}{V_{GS} - V_T} && \text{MOSFET}.
\end{aligned} \tag{14.41}
$$

For equal ratios in the two transistor types, it is required that the overdrive voltage be

$$V_{GS} - V_T = nV_{\mathrm{th}}. \tag{14.42}$$

At 300 K the operative voltage for the MOSFET would be $\approx 50\,\text{mV}$, which would mean a very low bias current. If an even lower current could be tolerated, then it would be advantageous to operate in the sub-threshold regime, in which an exponential $I_D - V_{GS}$ relationship holds. In this case a g_m/I ratio rivalling that of HBTs is possible naturally, i.e., by virtue of the transport mechanism (thermionic emission over a barrier), as opposed to having to control $(V_{GS} - V_T)$ to be $\approx V_{\text{th}}$.

14.7.2 f_{max}

For HBTs, the derivation of an expression for f_{max} proceeded from the equivalent circuit of Fig. 14.10b. The conditions that must be met for this circuit to be a reasonable approximation to the full, small-signal, equivalent circuit of Fig. 14.4 were stated earlier in this chapter. If comparable conditions are met for the operation of a MOSFET, then it follows that an approximate expression for f_{max} for a MOSFET will have exactly the same form as (14.33) for the HBT. Specifically:

$$f_{\text{max}} = \sqrt{\frac{f_{T,i}}{8\pi C_{dg} R_g}}, \tag{14.43}$$

where the intrinsic value of f_T is $2\pi f_{Ti} = g_m/C_{gg}$. If the conditions required to derive this equation cannot be met, then it is still possible to arrive at an expression for the power gain that rolls off at $-20\,\text{dB/decade}$, but the equation is much lengthier [7]. For our present purpose, (14.43) serves to draw attention to the need to minimize the gate capacitance and resistance. This is achieved in modern MESFETs and HEMTs by using the 'mushroom' structure for the gate illustrated in Fig. 11.2. The small contact region between the gate metal and the underlying semiconductor allows a short gate length to be achieved, and the wider top region keeps the access resistance low.

14.8   Power gain, oscillation and stability

Our presentation of the high-frequency metrics f_T and f_{max} has focused on their relation to the physical properties (capacitance and resistance) of transistors. In Circuit Analysis, it is usual to treat the transistor as a two-port network, and to employ small-signal parameters that refer to either admittance (y-parameters), impedance (z-parameters), or power 'scattering' (s-parameters) [8]. Experimentally, it is possible to determine the y- and z-parameters by measurements under specific short- or open-circuit conditions. For very high frequency measurements, which are demanded by modern high-performance transistors, terminating the network ports via finite impedances is generally done. Terminating the output by the characteristic impedance of the transistor, for example, results in no power reflection from the load, and this allows direct determination of the **reverse feedback** parameter, which is s_{32} in our notation.

Earlier, we defined f_{max} as the frequency at which the power gain is unity. It marks the boundary between when the device is active (power gain > 1) and when it is passive (power gain < 1). At such a frequency, if the output were fed back through an external circuit to the input, the gain of the system could become infinite, i.e., an output could be sustained with no input. Such a circuit is an oscillator. At frequencies higher than f_{max} no amount of external feedback via passive components can create this condition, so f_{max} is also known as the **maximum frequency of oscillation**.

If we want MAG > 1, and no oscillation, as we do in an amplifier, then passivity has to be maintained at frequencies lower than f_{max}. It is not sufficient to merely disconnect the external feedback circuitry, because there is always feedback internally through C_{bc} or C_{gd}. In the absence of external feedback, but allowing for passive terminations that maintain the conjugate matching but do not cause oscillation, the transistor is said to be **inherently stable**. For even lower frequencies, the transistor is potentially unstable, i.e., it could oscillate, so it has to be stabilized by additional terminating components, or by appropriate external feedback. Thus, the gain is no longer the maximum available gain, but is, instead, the **maximum stable gain** MSG.

Another frequently used power gain in transistor measurements is **Mason's Unilateral Gain U**. This metric refers to the situation when feedback is employed to ensure that, *at some particular frequency*, there is no contribution to the current or voltage at the input from any current or voltage appearing at the output. This condition is achieved at the specified frequency by addition of suitable components to the circuit [9]. In terms of the small-signal parameters, this means that, for the entire network (transistor and additional components), $s_{32} = z_{32} = y_{32} = 0$. Under these circumstances the circuit is said to be **unilateralized**. Evidently, both U and MAG extrapolate to the same value of f_{max} (see Fig. 14.7). The z-parameter version of U is

$$U = \frac{|z_{23} - z_{32}|^2}{4[\Re(z_{33})\Re(z_{22}) - \Re(z_{32})\Re(z_{23})]} . \tag{14.44}$$

This expression is used in Exercise 18.3.

Exercises

14.1 Consider an *Npn* $In_{0.49}Ga_{0.51}P/GaAs/GaAs$ HBT operating in the active mode with $V_{BE} = 1.25$ V and $V_{BC} = -3.0$ V. The emitter doping density is 10^{18} cm^{-3} and the width of the emitter quasi-neutral region is 100 nm. The corresponding values for the base are 10^{19} cm^{-3} and 25 nm. The cross-sectional area of the HBT is (1×1) mm^2. The minority carrier properties of InGaP can be taken to be the same as for correspondingly doped GaAs.

Estimate the transconductance and the input conductance of the HBT under the stated operating conditions.

14.2 For the HBT used to generate Fig. 14.7 it appears that the current gain ($h_{21} \equiv$ our y_{23}/y_{32}) flattens out at 'low' frequencies, i.e., below about 3 GHz in this case.

Calculate the low-frequency gain for the device of the previous question.

14.3 For the HBT shown in Fig. 14.2, Hafez *et al.* [2] quote a figure of 72 fs for the base/collector signal delay time τ_C. The semiconductor is InGaAs, for which the drift velocity is shown in Fig. 11.1.

If the area of the collector/base junction is $0.4 \times 6\,\mu m^2$, estimate the collector/base junction capacitance C'_{CB}.

14.4 The width of the InGaAs base in the high-performance transistor of the previous question can be taken to be 20 nm. The base signal delay time τ_B is quoted as being 65 fs.

Estimate the electron mobility in the base.

14.5 An HBT of similar structure to that shown in Fig. 14.2 has been reported with $f_T = 710\,\text{GHz}$ and $f_{max} = 340\,\text{GHz}$ [1]. The base appears to be contacted on either side of the emitter stripe.

What might be the changes in f_{max} and in f_T if there was only a base contact on one side of the emitter?

14.6 The collector of the HBT of Question 14.1 has a doping density of $10^{17}\,\text{cm}^{-3}$ and is quite wide (vertical dimension). Ignore all resistances.

(a) Evaluate the combined signal delay time $(\tau_C + \tau_{CC})$ associated with the base/collector space-charge region of this HBT.

(b) Determine the width of depletion region at the base/collector junction that would minimize $(\tau_C + \tau_{CC})$.

(c) Evaluate the reverse bias that should be applied to the base/collector junction to realize this minimum delay.

14.7 For the HBT of the previous question, what value of base resistance R_b would be needed to make $f_{max} = f_{Ti}$?

14.8 This question concerns the base signal delay in a graded-base HBT.

(a) Derive (14.22) for τ_B in a graded-base HBT.

Start with the expression for the electron current in a graded base [10]

$$J_e = q D_e \left(\frac{dn}{dx} - \frac{bn}{W_B} \right), \tag{14.45}$$

where $b = \Delta E_g / k_B T$.

Assume there is no recombination in the base $(dJ_e/dx = 0)$, and solve for $n(x)$. Hence find J_e from (14.45), and the base charge Q_B by integration, both in terms of $n(0)$ and $n(W_B)$.

Use the boundary condition (9.7) for $n(0)$, and for $n(W_B)$ use $J(W_B) = -qn(W_B)v_{sat}$.

The expression for τ_B in (14.22) follows from Q_B/J_e.

(b) Use L'Hôpital's rule to show that (14.22) reduces to (14.19) in the case of no base-grading and $v_{sat} = 2v_R$.

14.9 Consider a typical CMOS90 Si N-MOSFET with $V_{DS} = 1\,\text{V}$. Estimate the ratio g_m/I_D at two gate-source biases: (i) $V_{GS} = (V_T - 0.5)\,\text{V}$, and (ii) $V_{GS} = (V_T + 0.5)\,\text{V}$.

14.10 One of the two values computed in the previous question is equal to that for an HBT. Does this fact alone mean that MOSFETs can challenge HBTs in practical high-frequency applications?

14.11 Fig. 14.3a can be turned into a y-parameter 2-port by simply replacing all the conductances by admittances. Taking terminal 1 as reference, the circuit is then described by

$$\begin{pmatrix} i_2 \\ i_3 \end{pmatrix} = \begin{pmatrix} y_{22} & y_{23} \\ y_{32} & y_{33} \end{pmatrix} \begin{pmatrix} v_{21} \\ v_{31} \end{pmatrix}. \tag{14.46}$$

This description applies to any 2-port, such as Fig. 14.12 for a MOSFET.

Here, ignore the parasitic resistances and then, by comparing the two circuits, show that the so-called *intrinsic y*-parameters are:

$$y_{22} = g_{dd} + j\omega(C_{sd} + C_{gd})$$

$$y_{23} = g_m - j\omega(C_m + C_{gd})$$

$$y_{32} = -j\omega C_{gd}$$

$$y_{33} = j\omega(C_{gs} + C_{gd}), \tag{14.47}$$

where we have used lower-case subscripts in place of the upper-case subscripts and the primes used in Fig. 14.12, i.e., the capacitances in the above equation are total capacitances, not capacitances per unit area.

14.12 Equation (14.46) reveals that, when $v_{21} = 0$, the current gain is simply y_{23}/y_{33}.

Use this fact to show that the expression for the intrinsic f_T for a MOSFET is of exactly the same form as for an intrinsic HBT ((14.16) with $R_{ec} = 0$).

References

[1] W. Hafez, W. Snodgrass and M. Feng, 12.5 nm Base Pseudomorphic Heterojunction Bipolar Transistors Achieving $f_T = 710$ GHz and $f_{max} = 340$ GHz, *Appl. Phys. Lett.*, vol. 87, 252109, 2005.

[2] W. Hafez and M. Feng, Experimental Demonstration of Pseudomorphic Heterojunction Bipolar Transistors with Cutoff Frequencies above 600 GHz, *Appl. Phys. Lett.*, vol. 86, 152101, 2005.

[3] W. Liu, *Fundamentals of III-V Devices: HBTs, MESFETs, and HFETs/HEMTs*, pp. 226–231, John Wiley & Sons Inc., 1999.

[4] M. Vaidyanathan and D.L. Pulfrey, Extrapolated f_{max} for Heterojunction Bipolar Transistors, *IEEE Trans. Electron Dev.*, vol. 46, 301–309, 1999.

[5] Sungjae Lee, Lawrence Wagner, Basanth Jagannathan, Sebastian Csutak, John Pekarik, Noah Zamdmer, Matthew Breitwisch, Ravikumar Ramachandran and Greg Freeman, Record RF Performance of Sub-46 nm L_{gate} NFETs in Microprocessor SOI CMOS Technologies, *IEEE IEDM Tech. Digest*, 241–244, 2005.

[6] Y. Yamashita, A. Endoh, K. Shinohara, K. Hikosaka, T. Matsui, S. Hiyamizu and T. Nimura, $In_{0.52}Al_{0.48}As/In_{0.7}Ga_{0.3}As$ HEMTs with an Ultrahigh f_T of 562 GHz, *IEEE Electron Dev. Lett.*, vol. 23, 573–575, 2002.

[7] L.C. Castro and D.L. Pulfrey, Extrapolated f_{max} for CNFETs, *Nanotechnology*, vol. 17, 300–304, 2006.

[8] G. Gonzalez, *Microwave Transistor Amplifiers: Analysis and Design*, 2nd Edn., Chap. 1, Prentice-Hall, 1984.

[9] S.J. Mason, Power Gain in Feedback Amplifier, *IRE Trans. Circuit Theory*, vol. 1, 20–25, 1954.

[10] S.C.M. Ho and D.L. Pulfrey, The Effect of Base Grading on the Gain and High-frequency Performance of AlGaAs/GaAs Heterojunction Bipolar Transistors, *IEEE Trans. Electron Dev.*, vol. 36, 2173–2182, 1989.

15 Transistors for memories

Semiconductor memory is one of the main drivers of the semiconductor industry. Competition is fierce to increase density, and to reduce power dissipation and access times. In this chapter we describe two very important and much-used types of memory cell, each based on the Si MOSFET: flash memory, and dynamic random access memory (DRAM). Both memories, and, indeed, all semiconductor memories, have the basic organizational structure shown in Fig. 15.1a. The feature of flash memory and DRAM that gives them their capability of exceptionally high density is the use of a single transistor for the memory element (see Fig. 15.1b).

Flash memory is used in computer BIOS, and has also enabled many popular products, e.g., memory sticks, digital cameras, and personal digital assistants. There is some interesting physics in this non-volatile memory element: data is stored as charge on a floating gate within the insulator of a MOSFET. Charging and discharging is achieved 'in a flash' via field-assisted tunnelling.

DRAM is the main memory in PCs and workstations. Data is stored in a capacitor attached to the source of a pass transistor. As we show later, the action of reading a ZERO changes the stored data, so the memory has to be refreshed periodically, hence the appellation 'dynamic'. Leakage of charge from the storage capacitor also dictates that there be frequent refreshing of the memory's contents.

15.1 Flash memory

The MOSFET in a flash-memory cell has two polysilicon gates, one of which is completely surrounded by insulating oxide, and is, therefore, floating electrically. The structure is shown in Fig. 15.2.

Programming of the cell is achieved by applying an appropriate voltage to the top gate (also called the control gate) via the word line. The intent is to either place electrons on the floating gate (writing), or to remove electrons from the floating gate (erasing). The state of charge on the floating gate determines the threshold voltage of the transistor and, therefore, the magnitude of the drain current that will pass into the bit line when the cell is interrogated. A current sensor in the bit-line circuitry detects this current, and interprets from it the state of the memory cell.

The effect of floating-gate charge density Q_F on the charge density in the channel Q_n is illustrated in Fig. 15.3. Applying Gauss's Law, or by invoking directly the conservation

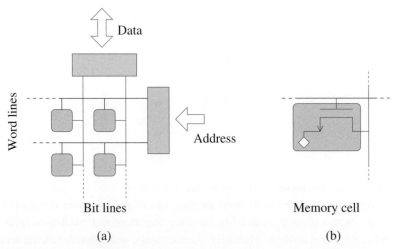

Figure 15.1 Basic organization of a semiconductor memory. (a) Each memory element is *xy*-addressable, and can be written to, and read from, via the bit lines. (b) Single-MOSFET memory element. The gate connects to the word line, the drain connects to the bit line, and the connection to the source differs according to the type of memory, as discussed in the text.

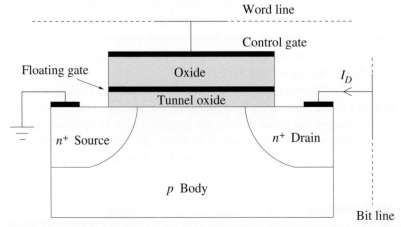

Figure 15.2 Individual flash-memory cell, showing the two gates, one of which is floating, and the connections to the word line and bit line of a memory array.

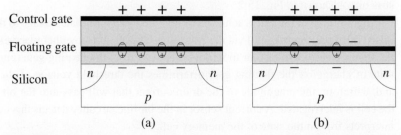

Figure 15.3 Illustrating how electronic charge on the floating gate affects the charge in the MOSFET channel. (a) No net charge on the floating gate, only polarization charges induced by the charge on the top gate. (b) Reduction in electron charge in the channel due to the presence of electron charge on the floating gate.

of charge:

$$Q_n + Q_F + Q_G = 0\,, \tag{15.1}$$

where Q_G is the charge density on the control gate. Thus, to maintain a given Q_n in the presence of a change in floating-gate charge ΔQ_F, the change in control-gate charge density is simply $-\Delta Q_F$. The associated change in control-gate voltage follows from Gauss's Law, and is the change in threshold voltage:

$$\Delta V_T = \frac{\Delta Q_G}{C_{ox,\text{top}}} = -\Delta Q_F \frac{t_{ox,\text{top}}}{\epsilon_{ox}}\,, \tag{15.2}$$

where $t_{ox,\text{top}}$ is the thickness of the upper oxide.

The word-line voltage used in addressing the cell is not sufficient to alter Q_F, so the act of reading the contents of the cell is **non-destructive**. If the read voltage is less than ΔV_T, then the presence of charge on the floating gate puts the transistor in the OFF state, i.e., the storage of electrons on the floating gate is interpreted as the storage of a ZERO in the memory cell. The read voltage is such that it is greater than the threshold voltage when $Q_F = 0$. Thus, in this case the FET is ON, drain current is received at the bit line, and the storage of a ONE is recognized.

In the procedure just described, each cell carries one bit of information. This is **single-level-cell** operation. Note that, if Q_F could be precisely controlled, then a variable ΔV_T would be obtained. In this case, different values of read voltage would be required to turn-on the FET, depending on the amount of Q_F present. This leads to **multi-level-cell** operation and the option of increasing the number of stored bits per cell. For example, if the maximum drain current, which occurs at $Q_F = 0$, is I_{max}, and the minimum current change that can be detected by the sensing circuitry is ΔI, then the number of possible bits is

$$n = \log_2 \left(\frac{I_{\text{max}}}{\Delta I} + 1 \right)\,. \tag{15.3}$$

Increasing the number of bits per cell is obviously desirable from a memory-density point of view, but it places greater demands on the placement of charge and on its discernment. Presently (early 2009), 64 Gbit chips using 4bits/cell have been reported by SanDisk and Toshiba [1].

The actual programming of a flash-memory cell is illustrated in Fig. 15.4. The top-left band diagram shows the equilibrium condition when there is no charge stored on the floating gate. Direct tunnelling of electrons to the floating gate from either the control gate or the body (substrate) is not possible because the oxides on either side of the gate are too thick. Writing occurs when a large bias V_{GS}^W is applied to the control gate. This voltage has to be large enough to: (a) invert the semiconductor surface to create a source of electrons; (b) produce an electric field in the thinner oxide that is sufficient to enable field-assisted tunnelling of the channel electrons to the floating gate (see Fig. 15.4b). The 'field-assistance' comes from a reduction of the thickness of the tunnelling barrier at higher energies; the phenomenon is also known as Fowler-Nordheim tunnelling, after the two men who first identified it in 1928. The polysilicon floating gate has to be thick

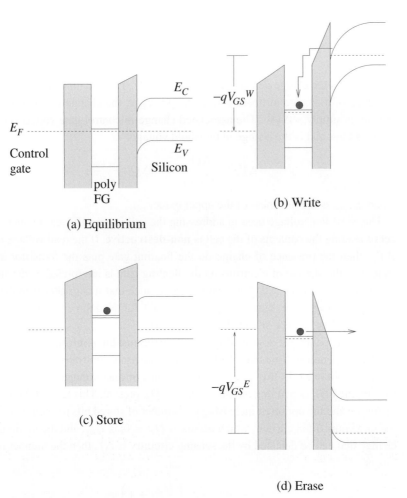

E_C

$-qV_{GS}{}^W$

E_F

Control
gate

E_V

Silicon

poly
FG

(b) Write

(a) Equilibrium

(c) Store

$-qV_{GS}{}^E$

(d) Erase

Figure 15.4 Energy-band diagrams for a flash-memory call. The control gate is assumed to be metal, the floating gate is n^+-polysilicon, and the body of the cell is p-silicon. (a) Equilibrium. (b) Writing a ZERO by tunnelling from the channel to the floating gate. (c) Storing a ZERO via charge on the floating gate. (d) Erasing the ZERO by tunnelling from the floating gate to the channel.

enough for the injected electrons to lose energy by collisions and 'fall into' the potential well. On removing the write voltage, the charge is trapped (see Fig. 15.4c).

To erase the cell, a large, negative voltage V_{GS}^E is applied to facilitate Fowler-Nordheim tunnelling of the stored electrons to the body (see Fig. 15.4d). This erasure process is fast, hence the name 'flash'.[1] On the other hand, storage is long-lasting because the electrons are trapped in a deep potential well, from which direct tunnelling is not probable. As the data can be retained with no applied voltages to the cell (see Fig. 15.4c), this memory is **non-volatile**.

---

[1] 'Flash' is attributed to Shoji Ariizumi, a colleague of Fujio Masuoka, who invented Flash Memory at Toshiba in 1984.

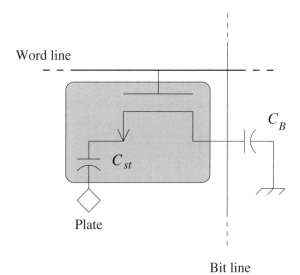

Bit line

Figure 15.5 Schematic of an individual DRAM cell, showing the relevant capacitors (storage and bit-line), and the MOSFET acting as a pass-transistor.

15.2 Dynamic Random Access Memory

Another type of semiconductor memory in which a single-transistor is employed as the memory cell is dynamic random access memory (DRAM). In this case, the MOSFET acts as a switch between the bit line and a storage capacitor. The charge stored on the capacitor determines the state (ONE or ZERO) of the cell. Schematically, the circuit is as shown in Fig. 15.5, where we label the storage capacitor as C_{st}, rather than C_S, to emphasize that in modern DRAMs the storage capacitance is not that of the source, but is that of a parallel-plate capacitor, which is connected to the source and has a much larger capacitance than that of the source/body np-junction. The stored charge connects to the bit line via the inversion layer in the channel when the FET is turned ON. The bit line is floating during this part of the operation, so its voltage may change in response to the new charged state of the bit-line capacitance C_B. Any voltage change is detected by a sensitive amplifier attached to the bit-line circuitry. In the DRAM we now describe, no voltage change is indicative of a stored ONE, whereas a slight decrease in bit-line voltage is associated with a stored ZERO.

The terminal marked $\diamond$ on Fig. 15.5 is called the **plate** electrode of the capacitor, and it is held at $V_{DD}/2$, where V_{DD} is the usual CMOS supply voltage. Labelling the actual source potential as V_S, the basic charge-sharing equation is

$$C_{st} V_{st} + C_B V_B = (C_{st} + C_B) V_B' , \qquad (15.4)$$

where $V_{st} = (V_S - V_{DD}/2)$ is the voltage across the storage capacitor, V_B is the voltage on the pre-charged bit line, and V_B' is the bit-line voltage after accessing the storage capacitor.

When writing a ONE, the bit-line voltage is raised by internal, charge-pumping circuitry to $(V_{DD} + V_T)$, where V_T is the threshold voltage of the FET. This means that when the word line is enabled, turning on the N-channel pass transistor, the source node rises to V_{DD}. Thus, $V_{st} \rightarrow +V_{DD}/2$.

To read this ONE, the bit-line is pre-charged to $V_{DD}/2$ and left floating as the transistor is turned on. Charge sharing occurs, and the new bit-line voltage V_B' follows from (15.4):

$$V_B' = \frac{V_{DD}}{2} \left[\frac{C_{st} + C_B}{C_{st} + C_B} \right] = \frac{V_{DD}}{2}. \tag{15.5}$$

Thus, the bit-line voltage is unchanged, and this is interpreted as a ONE.

To write a ZERO to the cell, V_B is set at 0, the word line is enabled, turning on the pass transistor, and the source node falls to 0, which means that $V_{st} \rightarrow -V_{DD}/2$.

To read this ZERO, again the bit-line is pre-charged to $V_{DD}/2$ and left floating as the transistor is turned on. Charge sharing results in the bit-line voltage changing to V_B', which from (15.4) is now

$$V_B' = \frac{V_{DD}}{2} \left[\frac{C_B - C_{st}}{C_{st} + C_B} \right] \rightarrow \; < \frac{V_{DD}}{2}. \tag{15.6}$$

The smaller V_B' can be made the easier it is to distinguish between a ONE and a ZERO.

Note that, on reading a ZERO, the source potential is changed, so the memory cell needs to be refreshed before the next read operation. This is the reason for this type of memory being called 'dynamic'. Besides having to refresh after reading a ZERO, it is also necessary to periodically refresh the entire memory because of the inevitable leakage of charge from the storage capacitor. Leakage via the source node principally involves: sub-threshold conduction, gate-induced-drain-lowering and the reverse-bias current of the source/body diode. These mechanisms are illustrated in Fig. 13.14.

In view of the need to lower V_B' below $V_{DD}/2$ to detect a ZERO, it is not suprising that the DRAM cell has evolved over the years with the aim of keeping C_{st} high while simultaneously shrinking the cell size to increase memory density (see Fig. 15.6). The challenge has been great because increasing the density means that the number of cells attached to a bit line increases, thereby increasing C_B.

The first DRAMs used an MOS capacitor as the storage element: a positive potential applied to the plate electrode (see Fig. 15.6a) pulsed the underlying silicon into **deep depletion**. In this condition, the surface potential ψ_s exceeded the usual limit in MOSFETs of $\approx 2\phi_B$ because there were no contiguous n^+ regions to supply electrons and create an inversion layer at the silicon surface. The large ψ_s was interpreted as the storage of a ONE. Eventually, ψ_s would reduce to $\approx 2\phi_B$ due to the thermal generation of electron-hole pairs within the depletion region, and their separation by the field therein. Thus, the cell had to be refreshed every few milliseconds or so in order to restore the stored ONE. The problem of electron-hole pair generation was avoided in the next generation of DRAM cells by replacing the MOS storage capacitor with a conductor/oxide/conductor parallel-plate capacitor in which the bottom electrode was n^+ silicon (see Fig. 15.6b).

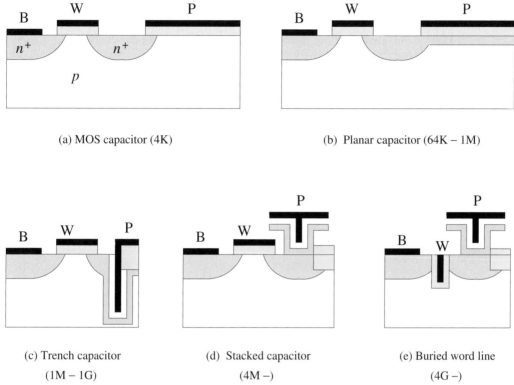

(a) MOS capacitor (4K) (b) Planar capacitor (64K – 1M)

(c) Trench capacitor (1M – 1G) (d) Stacked capacitor (4M –) (e) Buried word line (4G –)

Figure 15.6 Illustrative example of the evolution of the storage capacitor within a DRAM cell. 'P' is the plate electrode, 'W' the word line, and 'B' the bit line. The numbers indicate typical maximum values or ranges of values for the number of bits stored in each embodiment. (a) Plate-oxide-p-silicon MOS capacitor. (b) Poly-oxide-n^+-silicon capacitor in a planar arrangement. (c) Poly-oxide-poly capacitor formed in a trench within the silicon wafer. (d) Poly-oxide-poly capacitor stacked on top of the cell. (e) Buried word-line cell.

Both of the above cells were 'planar', and this placed a limitation on how small the cell could be made. Therefore, to increase cell density, and yet maintain a high storage capacitance, the next generation of DRAM cells employed 3-D structures, with the storage capacitor being either buried in the substrate (Fig. 15.6c), or stacked above the transistor (Fig. 15.6d). Cells with these structures are still used today.

The latest development in DRAM-cell structuring seeks to reduce the bit-line capacitance C_B, rather than to increase C_{st}. In the embodiment shown in Fig. 15.6d, this is achieved by burying the pass transistor below the surface of the silicon, thereby physically distancing it from the bit line and, consequently, reducing the bit-line-to-word-line capacitance, which has been one of the major contributions to C_B. This structural arrangement is called 'recess-channel array transistor' (RCAT) by Samsung, and 'buried word-line' (BWL) by Qimonda [2]. The example shown in Fig. 15.6d is based on Qimonda's design; note how the channel forms a U-shape around the word line, thereby reducing short-channel effects, which would otherwise be significant, given that the word line width is that of the technology node, e.g., presently (2009), 65 or 45 nm.

Exercises

15.1 A floating-gate flash memory cell of the construction shown in Fig. 15.2 uses a word-line voltage of 1.5 V during the READ operation. The upper insulator is silicon dioxide and has a thickness of 20 nm. The threshold voltage when there is no charge stored on the floating gate is 1.0 V. The area of the floating gate is (100×100) nm^2.

How many electrons need to be stored on the floating gate to represent a ZERO?

15.2 The maximum current that can be delivered to the bit line by one cell of a floating-gate flash memory is 1.5 mA μm^{-1}. The cell width is 100 nm.

If each cell is required to store 4 bits, what level of current discrimination must the bit-line sensing circuitry have?

15.3 Consider a stacked-capacitor DRAM with its storage capacitor of 1 pF charged to 1 V. This represents a stored ONE.

The FET in the DRAM is a CMOS65 N-FET of width $Z = 100$ nm.

If charge leakage from the storage capacitor is due to the sub-threshold current of the FET, how long will it take for the storage capacitor to lose 50% of its charge?

15.4 The bit-line sensing circuitry in the DRAM of the previous question can detect a voltage change of 10 mV, and this is used to distinguish between a stored ONE and a stored ZERO in a single cell.

Compute the magnitude of the bit-line capacitance.

References

[1] C. Trinh *et al.*, A 5.6MB/s 64Gb 4b/cell NAND Flash Memory in 43 nm CMOS, *ISSCC Digest Tech. Papers*, 246–248, 2009.

[2] T. Schloesser *et al.*, A 6F^2 Buried Wordline DRAM Cell for 40 nm and Beyond, *IEEE IEDM Tech. Digest*, paper 33.4, 2008.

16 Transistors for high power

Power amplifiers and switch-mode power supplies are two instances where the constituent transistors are required to deliver higher currents, and to withstand higher voltages, than are encountered in the digital, high-frequency and memory transistors discussed previously. In this chapter we briefly describe the structural details of, and discuss the principal properties of, several types of high-power transistor: the GaAs HBT and GaN HJFET for power amplification, and the Si MOSFET and hybrid transistor for power supplies.

High currents mean high carrier densities, which can lead to a modification of the space-charge region at the output junction (collector/base or drain/body), with consequences for the frequency response and/or the breakdown voltage. We begin this chapter with a description of the breakdown process (avalanche breakdown), and of the high-current, space-charge-modifying effect (the Kirk Effect).

16.1 Avalanche breakdown

Operating transistors at high V_{CE} or high V_{DS} can lead to electrical breakdown of the collector/base or drain/body junction, respectively. Breakdown is characterized by the sudden onset of a large current which, if it is not interrupted, can lead to thermal destruction of the transistor. The breakdown process in a reverse-biased *pn*-junction is illustrated in Fig. 16.1.

Electrons entering the high field of the junction rapidly gain kinetic energy. If this energy is allowed to exceed the bandgap energy E_g, then, when the electron finally collides with a lattice atom, an energy $E \geq E_g$ can be transferred to another electron, thereby exciting it into the conduction band. Thus, one electron creates another electron (and a hole). This process is the generation-equivalent of Auger recombination. The newly generated electron and hole are, in turn, accelerated by the junction field, leading to the possibility of creation of more electron-hole pairs, and a rapidly increasing current. For obvious reasons this phenomenon is called **avalanche breakdown**.

The value of the electric field at which the avalanche is initiated is called the **breakdown field strength** $\mathcal{E}_{br}$; as is to be expected, perhaps, there is some correlation between $\mathcal{E}_{br}$ and E_g (see Table 16.1). The Table lists several semiconductor properties that are important to power transistors: κ_L is the thermal conductivity; JFOM is Johnson's figure-of-merit [2], which is intended to characterize devices for applications in which

Table 16.1 Important, power-related, properties of various semiconducting materials. κ_L is the thermal conductivity; JFOM is Johnson's figure-of-merit (see text), which is normalized to that of Si in this Table. Adapted from DiSanto [1], courtesy of David DiSanto, ex-SFU.

Material	E_g (eV)	$\mathcal{E}_{br}$ (MV cm^{-1})	v_{sat} (10^7 cm s^{-1})	κ_L W (cm.K)$^{-1}$	JFOM ($\mathcal{E}_{br} v_{sat}$)/Si
Si	1.12	0.3	1.0	1.5	1
InP	1.35	0.6	1.5	0.7	3
GaAs	1.43	0.4	1.0	0.5	2
SiC	3.26	3.5	2.5	5.0	29
GaN	3.39	2.0	2.5	1.5	17
Diamond	5.45	10.0	2.7	20	90

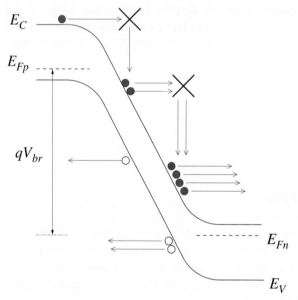

Figure 16.1 Energy-band diagram for large reverse bias across a *pn*-junction, showing avalanche multiplication of both electrons and holes, following collisions with atoms at the sites marked X.

operation at both high power and high frequency is required. The large-bandgap semi-conductor diamond comes out very well in both of the above categories, and the making of transistors from it is presently the subject of much research interest.

In a *pn*-junction, the maximum electric field occurs at the interface between the two, differently doped regions. The left panel of Fig. 16.2 shows the case of an abrupt junction with uniform doping on each side of the junction, and under low-current conditions. Note that the *n*-side is less heavily doped than the *p*-side. The area under the $\mathcal{E}$-x curve is the total voltage drop across the junction:

$$V_{bi} + V_a = \frac{1}{2}\mathcal{E}_{br}\, W\,, \tag{16.1}$$

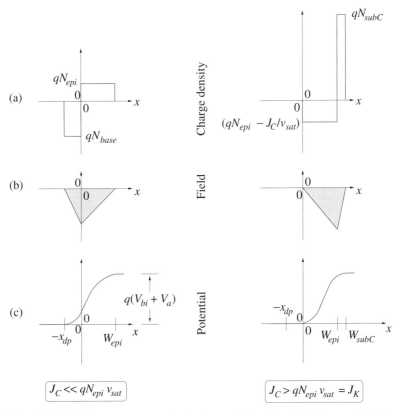

Figure 16.2 Variation across a pn^-n^+ structure with two abrupt junctions of: (a) charge density, (b) electric field, (c) electrostatic potential. The set of figures on the left is for the case of a current density that is low enough for the space-charge due to the mobile electrons to be negligible, and for the n^- layer (labelled 'epi') to be just fully depleted. The set on the right is for the case when this mobile charge is sufficient to reduce the electric field at the pn^- interface to zero. The depletion region spreads into the n^+ layer, the sub-collector. The applied bias is the same in both sets of figures.

where W is the width of the space-charge region at the junction, and V_a is the applied, reverse-bias voltage (assumed to be dropped entirely across the space-charge region), and $\mathcal{E}_{br}$ is identified with $\mathcal{E}(0)$. Invoking the Depletion Approximation, we use (6.22) to estimate W, and simplifying by assuming $|V_a| \gg V_{bi}$, a useful expression for the breakdown voltage results:

$$V_{br} = \mathcal{E}_{br}^2 \frac{\epsilon}{2q} \left(\frac{1}{N_D} + \frac{1}{N_A} \right). \qquad (16.2)$$

This equation emphasizes the importance of employing a semiconductor with a high breakdown field strength. Also, if other factors dictate the use of a semiconductor with a relatively low $\mathcal{E}_{br}$, such as Si, then the equation shows the necessity of using a low doping density for at least one side of the junction. So as not to compromise the principal transport-determining region of the transistor (the base or channel), it is the collector (drain) in which the doping is reduced.

16.2   The Kirk Effect

Consider a structure in which a highly doped n-region has been added to the right-hand side of our pn-junction (see the right panel of Fig. 16.2). This pn^-n^+ arrangement arises at the base/collector junction in Npn-HBTs, and in the body/drain region of lateral-diffused MOSFETs. The lightly doped region is usually deposited by vapour-phase epitaxy, and is referred to as the epi-layer. Generally, we are interested in the electric-field profile in and around the epi-layer, and how it changes in response to an increasing electron current. Analytically, it is easier to treat the HBT case as the situation is essentially, one-dimensional. Thus, let us consider an HBT in the forward-active mode of operation with electrons being injected from the emitter, and passing through the p-type base into the lightly n-doped epi-layer, before being collected in the heavily n-doped sub-collector.

Neglecting the holes in the epi-layer, Poisson's Equation in 1-D is

$$\frac{d\mathcal{E}}{dx} = \frac{1}{\epsilon_s}\left[qN_{\text{epi}} - \frac{J_C}{v_{\text{sat}}}\right], \tag{16.3}$$

where we assume that the electrons are moving in the epi-layer at their saturation velocity. For low collector current density J_C the field gradient is positive, as shown in the left-side of Fig. 16.2b for $x > 0$. At $J_C \equiv J_{\text{crit}} = qN_{\text{epi}}v_{\text{sat}}$ the field becomes constant. At higher J_C the field gradient in the epi-layer becomes negative, as shown in the right-side of Fig. 16.2b for $x > 0$. What is happening, of course, is that the positive space charge of the donor ions in the epi-layer is being swamped by the negative space charge of the electrons carrying the current. Eventually, a current density is reached at which the field goes to zero at the base/epi-layer boundary. This means that there is then no field to prevent holes from the base moving into the epi-layer!

The current density at which the field disappears at the p/epi junction is known as the Kirk current J_K. An expression for J_K is easily derived (see Exercise 16.3), but it should be clear that the onset of the **Kirk Effect** can be delayed by using a semiconductor for which v_{sat} is high, and by choosing a high doping density for the epi-layer. Of course, to avoid the effect altogether, the epi-layer thickness could be reduced to zero, but then the breakdown issue comes to the fore. The implications of operating at or above J_K for HBTs and lateral diffused MOSFETs are discussed in Section 16.3.1 and Section 16.4.1, respectively.

16.3   Transistors for power amplifiers

The modern craving for wireless electronics has led to the development of transistors suited to the provision of high power at high frequencies. Here, we consider briefly two types of transistor that meet these specifications: InGaP/GaAs HBTs, and AlGaN/GaN HJFETs. The former, operating at modest voltages and power, are used in the transmitter stage of cell phones, and are also candidates for the final stages of power amplifiers in radio base-stations, where operation is at tens of volts and hundreds of watts. GaN-based

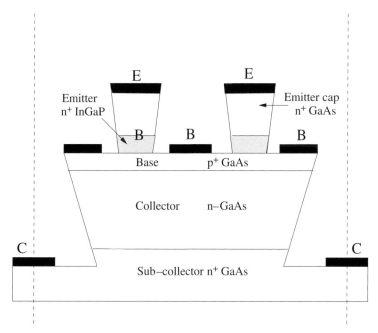

Figure 16.3 A power-HBT cell. The dashed lines delineate one cell, which can be replicated to place more transistors in parallel, thereby increasing the current-handling capability.

HJFETs are also being considered for the latter application, as well as for implementation in situations where the high bandgap of the material gives this transistor an advantage for operation in harsh environments, e.g., in automobiles and war zones.

16.3.1 GaAs HBTs

The essential elements of a high-power, high-frequency HBT are illustrated in Fig. 16.3. They are: semiconducting materials chosen for their high mobility; interdigitated emitter and base contacts to enable a large current via the large total emitter area whilst simultaneously reducing the base-access- and base-spreading-resistances; a thick, lightly doped collector to ensure a high breakdown voltage.

In the embodiment shown, the splitting of the base current between three contacts reduces the power dissipation in the base spreading resistance to 1/12 of its value in the single-base case (see Exercise 16.4). The tall emitter helps to separate the emitter and base metallizations, and the thickening of the emitter towards its top helps reduce the parasitic emitter resistance. Fig. 16.3 forms a 'power cell', many of which can be connected in parallel to increase the output current.

By opting to make the collector thick and lightly doped, the design gives precedence to attainment of a high V_{br} over a high J_K. Recall that at $J_C = J_K$ there is no field at the base/collector junction, i.e., there is no potential barrier to constrain the holes to the base. Therefore, on further increasing J_C, holes flood into the epi-layer, effectively extending the width of the quasi-neutral base region (the p-region in Fig. 16.2). This increases the

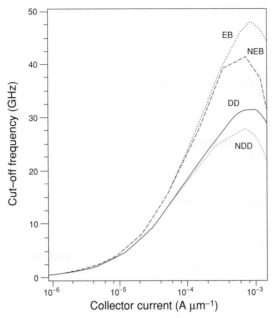

Figure 16.4 Predictions from ATLAS of f_T for an AlGaAs/GaAs HBT using the four transport models described in the text. From Apanovich *et al.* [3], © 1995 IEEE, reproduced with permission.

base transit time τ_B. The associated reduction in the width of the base/collector space-charge region may improve the collector signal delay time τ_C, but this is countered by the increase in the collector charging time τ_{CC}. Thus, the overall effect is to increase τ_{EC} (see (14.17)), and this leads to a decrease in f_T at high currents, as illustrated in Fig. 16.4. The data here, and in the next three figures, are from numerical simulations using Silvaco's ATLAS. The device is an AlGaAs/GaAs HBT with emitter, base, and collector thicknesses of 150, 100 and 500 nm, respectively, and associated doping densities of 5×10^{17}, 10^{19} and 10^{17} cm^{-3}.

Prior to the onset of the Kirk Effect, f_T increases with current due to the improvement in transconductance. This can be appreciated from (9.12) and (14.6),

$$g_m \approx \frac{\partial I_C}{\partial V_{BE}} = \frac{I_C}{\gamma(I_C)V_{th}}, \tag{16.4}$$

where γ is the **junction ideality factor**. At emitter current densities that are high enough for recombination in the emitter/base space-charge region to be neglected, as we considered in Section 9.2, $\gamma = 1$, at least while low-level injection conditions apply. In power HBTs, this condition can be violated.

As an example of high-level injection conditions, let us assume that the injected minority carrier concentration at the edge of the depletion region in the base of an *Npn* transistor is such that $n(x_{dp}) = p_{p0}$. Under this condition, (6.29) becomes

$$n(x_{dp})p_{p0} = n_i^2 \exp^{V_a/V_{th}} \rightarrow n(x_{dp}) = n_i \exp^{V_a/2V_{th}}. \tag{16.5}$$

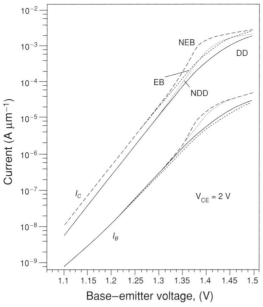

Figure 16.5 Predictions from ATLAS of the collector and base currents for the same HBT and models as used in Fig. 16.4. From Apanovich *et al.* [3], © 1995 IEEE, reproduced with permission.

The increase in minority carrier concentration is matched by a corresponding increase in majority carrier concentration in order to maintain charge neutrality. The altering of the conductivity by changing the carrier concentrations is known as **conductivity modulation**. At the collector-end of the base the minority carriers are extracted, so this means that there will be large gradients in the concentrations of both carriers. Thus, both electrons and holes will diffuse; the condition is known as **ambipolar diffusion**. As holes and electrons diffuse at different rates, an electric field is set up that retards the diffusion of the faster carrier. The situation is clearly quite complicated, but it transpires that diffusion dominates, with an effectively higher diffusivity. This is known as the **Webster Effect**. The importance of all this for our discussion is that a diffusive electron current with the boundary condition of (16.5) leads to an ideality factor of $\gamma = 2$ in (16.4), and another contribution to the reduction in f_T at high currents.

An increase in γ brings about a decrease in current for a given bias; the effect is clearly seen in Fig. 16.5. This figure also shows the effect of different transport models on the predictions for the collector and base currents. The four models referred to in this figure, and in Fig. 16.4, Fig. 16.6, and Fig. 16.7 are:

- DD - Drift Diffusion. This model uses Poisson's equation, and the first four equations from (5.24) with $T_e = T_h = T_L$, and $\nabla T_L = 0$.
- NDD - Non-isothermal Drift Diffusion. This is the same as DD but with the lattice temperature allowed to vary with position. Heat flows out of the device via one of the contacts, and heat balance is accounted for by (5.25).

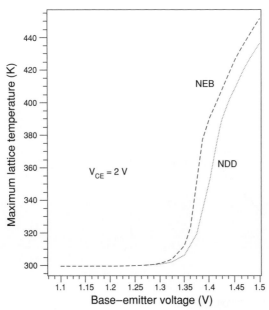

Figure 16.6 Predictions from ATLAS of the maximum lattice temperature for the same HBT and models as used in Fig. 16.4. From Apanovich *et al.* [3], © 1995 IEEE, reproduced with permission.

- EB - Energy Balance. This model uses Poisson's equation and the set of equations (5.22). The carrier temperatures are allowed to be functions of position, so there is a contribution to the current from any gradient in kinetic energy. However, the lattice temperature is held constant.
- NEB - Non-isothermal Energy Balance. This is the same as EB, but the lattice temperature is allowed to vary with position.

The inclusion of energy balance leads to prediction of an improved f_T, as Fig. 16.4 clearly shows. This is because electrons injected into the high field region of the reverse-biased base/collector space-charge region are accelerated to velocities above v_{sat} before thermalizing collisions occur. This phenomenon of **velocity overshoot** reduces the signal delay time in the collector space-charge region, thereby improving f_T. The increased velocity also increases the current, as evinced by comparison of the DD and EB models in Fig. 16.5. The non-isothermal models predict a significant rise in lattice temperature (see Fig. 16.6). This increases the minority-carrier concentrations (see (4.19) and (4.20)), leading to a further increase in current at high bias.

The prediction of a higher current by the non-isothermal models may seem like a good thing. However, as Fig. 16.7 shows, this is not necessarily so. The 'current droop' that is evident is a consequence of the device heating up. Note that the parameter held constant in this figure is the base current. To achieve this as the temperature rises, the base/emitter voltage has to be reduced, and it is this that leads to the undesirable reduction in collector current.

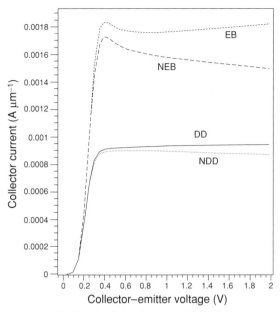

Figure 16.7 Predictions from ATLAS of the collector-current characteristic for the same HBT and models as used in Fig. 16.4. From Apanovich *et al.* [3], © 1995 IEEE, reproduced with permission.

The results displayed in the previous four figures show the importance of including energy balance and lattice heating in the modelling of high-power bipolar transistors. The former allows for **non-local** transport, i.e., it predicts currents due to kinetic energy changes which cannot be accounted for in the usual drift and diffusion currents, as these are determined by the local electric field via a field-dependent mobility $\mu(\mathcal{E})$ and the Einstein Relation. Lattice heating is particularly important in bipolar transistors because of the exponential dependence of minority-carrier concentrations on temperature.

16.3.2 GaN HJFETs

High-power HJFETs based on GaN are relative newcomers to the power-transistor scene. With reference to Table 16.1, two properties of the material that are superior to GaAs, the material on which rival high-power HBTs are based, are the breakdown field strength and the thermal conductivity. The high electron saturation velocity is also advantageous for high-frequency applications, at least in devices where the field is high enough for this velocity to be attained.

A typical GaN HJFET is shown in Fig. 16.8. The gate length and the separations between gate and source/drain electrodes are about 100–500 nm, but the operating voltages can be high (28 V for base-station applications, for example) so velocity saturation is likely. As we explain below, an AlGaN surface is electrically active, hence the use of a passivation layer such as silicon nitride. The gate electrode extends over the passivating layer towards the drain, forming a **field plate**. This serves to reduce the field in the

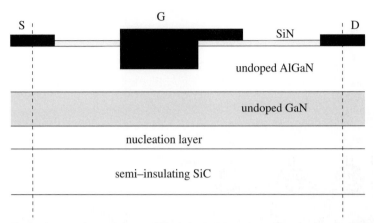

Figure 16.8 AlGaN/GaN power HJFET. The dashed lines delineate one cell, which can be replicated to place more transistors in parallel, thereby increasing the current-handling capability.

channel at the edge of the gate, thereby improving the breakdown voltage. However, the field plate adds capacitance to the device, which is not desirable for high-frequency performance. This drawback can be mitigated somewhat by recessing the gate, as shown in Fig. 16.8. The benefit comes from increasing g_m through the closer coupling of the gate to the channel. The latter takes the form of a 2-D electron gas at the AlGaN/GaN interface.

Presently, single-crystal GaN wafers, on which high-quality GaN and AlGaN layers might be grown, are not available. Instead, one choice of substrate is SiC, which has excellent thermal conductivity (see Table 16.1). A buffer layer of GaN is deposited before commencement of epitaxial growth of the actual layers of the device, which crystallize in the wurtzite structure. Typically, growth is from the {0001} basal plane in the [0001] direction, which is known as the c-axis (see Fig. 16.9). The atoms are arranged in two, repeating, closely packed bilayers, each layer of which is an hexagonal arrangement of either Ga ions or N ions. In the example shown, the top layer in each bilayer is gallium, so this structure is known as 'Ga-face' GaN. The Ga-N bond is strongly ionic, and because the electronegativity of N is higher than that of Ga, the electron probability density is greater in the vicinity of the N atom. Thus, the material is spontaneously polarized, with the polarization vector $\vec{P}_{Sp}$ pointing from N to Ga, i.e., towards the substrate in this example.

For the ternary material $Al_xGa_{1-x}N$, the lattice constant is given by

$$a = (0.3189 - 0.0077x) \text{ nm}. \tag{16.6}$$

Therefore, AlGaN grown pseudomorphically on GaN is under tensile strain. This adds a piezoelectric polarization $\vec{P}_{Pz}$ to the spontaneous polarization that is inherent to the material. The situation for AlGaN on Ga-face GaN is illustrated in Fig. 16.10. The surface polarization charge densities are negative on the top of the AlGaN (region 2) and positive on the bottom of the GaN (region 1). At the interface between the two materials the surface polarization density is

$$\sigma_{int} = (\vec{P}_{Sp,2} + \vec{P}_{Pz,2}) \cdot \hat{n}_2 + \vec{P}_{Sp,1} \cdot \hat{n}_1, \tag{16.7}$$

and it is positive for the case under consideration.

Ga-face

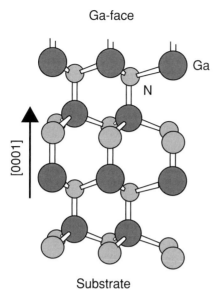

Substrate

Figure 16.9 Wurtzite crystal structure showing the bilayers of N and Ga arising from growth on a Ga-face substrate. Reused with permission from O. Ambacher, J. Smart, J. R. Shealy, N. G. Weimann, K. Chu, M. Murphy, W. J. Schaff, L. F. Eastman, R. Dimitrov, L. Wittmer, M. Stutzmann, W. Rieger and J. Hilsenbeck, Journal of Applied Physics, 85, 3222 (1999) [4]. Copyright 1999, American Institute of Physics.

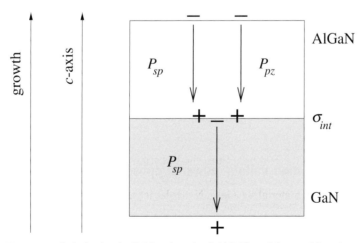

Figure 16.10 Polarization in GaN and strained AlGaN, and the resulting charge density at the interfacial layer, for films grown on Ga-face GaN.

The sheet polarization charges are bound charges, i.e., they are fixed in space, rather like donor or acceptor ions. In the case under consideration, electrons are drawn to the interface, perhaps during the period of cooling after the growth of the layers, or from the ohmic contacts in the subsequently fabricated HJFET. Thus, it is possible to create a sheet of electrons at the interface without doping either of the layers! This sheet of charge

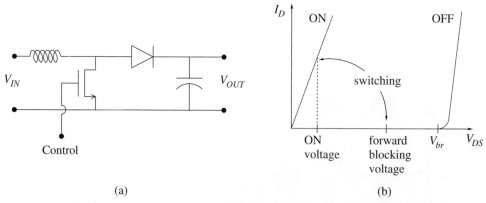

(a)   (b)

Figure 16.11 Switched power supply. (a) Basic circuit for the transformation of a DC input voltage to a higher DC output voltage. (b) The switching cycle.

is actually a two-dimensional gas because the electrons are confined in a potential notch, one side of which is due to the electon-affinity mismatch between the two materials. The relation, in terms of the mole fraction x for Al, is

$$\chi(x) = 4.1 - 1.87x \ \text{eV}.\tag{16.8}$$

A typical value is $x = 0.15$, which is small enough for the lattice-constant mismatch to be acceptable; it yields a barrier of about 0.28 eV.

The spontaneous and piezoelectric polarizations in AlGaN are so large that the surface concentration of electrons n_s at the interface can be around $10^{13} \ \text{cm}^{-2}$. This very large value gives AlGaN/GaN HJFETs a high current-carrying capability. Additionally, the large bandgap and breakdown voltage allow operation at typical base-station voltage levels of 28 V. This means that additional voltage-conversion circuitry is unneccesary. These factors, allied with the previously discussed attributes of high electron velocity and high thermal conductivity, combine to produce a transistor with considerable merit for power amplifiers operating at high frequencies.

16.4   Transistors for high-voltage power supplies

Fig. 16.11a shows a very rudimentary example of the use of a switch in a circuit that can transform DC voltages. Basically, when the switch is closed energy is stored in the inductor, and when the switch is opened, this energy is transferred to the capacitor, which charges up to a voltage that depends on the duty cycle, i.e., the fractional time that the switch is closed during one cycle of the control signal.

Naturally, transistors are used for the switch. When the switch is closed, the transistor needs to pass a large current (recall that the energy stored in an inductor is $\propto L I^2$). As this current is derived from the input voltage source, then it follows that the **ON-resistance** of the transistor must be low. This resistance should also be low to avoid excessive power dissipation within the transistor. When the switch is open, the transistor

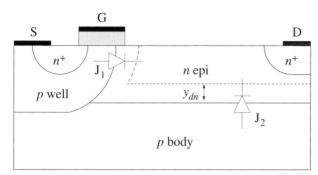

Figure 16.12 High-power laterally diffused (LD) Si MOSFET. The dashed line marks the edge of the space-charge layer at the p-body/n-epi junction; for RESURF operation its thickness y_{dn} equals the thickness of the epi-layer.

must be able to withstand a voltage at least equal to that of the output voltage. The rating that specifies this property is the **forward blocking voltage**. The two states of the transistor are illustrated in Fig. 16.11b. Being able to switch quickly between these two states is obviously advantageous.

Originally, in the 1960s, the Si BJT was the most developed transistor and it was employed for the switching transistor in power supplies. However, bipolar transistors (BTs) are current-controlled devices, and this leads to complexity in the pulse-width-modulated control circuitry used to adjust the duty cycle. Also, to get the required low ON-resistance, BTs must be operated in the saturation mode; as we saw in Section 13.2, this is not conducive to fast switching. MOSFETs are voltage-controlled devices, which simplifies the control circuitry, and are unipolar, so switching times are not determined by slow recombination processes. Therefore, with the advent of high-current-handling Si MOSFETs in the 1970s, this transistor gradually became the dominant transistor in switch-mode power supplies. Today, in 2009, in the form of a lateral-diffused device, it is used in most high-speed, medium-power applications. For very high power situations the insulated-gate BT is used. Both these transistors are described below.

16.4.1 Si L-DMOSFETs

The basic lateral-diffused MOSFET is illustrated in Fig. 16.12. The source is embedded in a p-well, and these two regions are formed in a sequential diffusion process, hence the word 'diffused' in the transistor title. 'Lateral' comes from the lateral layout of the device, which allows the three terminal contacts to be placed on the top of the structure. This is the major difference from the earlier vertically oriented device, the V-DMOSFET. The lateral arrangement greatly facilitates the integration of the power transistor with standard CMOS FETs, which can be used for the control circuitry. The drain consists of the usual n^+-contact region, and a lightly doped drain extension in the form of an n^--epi layer grown on the p-body.

The junction between the p-well and the epi-layer forms a pn-diode (J1 on the figure). The junction between the epi-layer and the p-body forms a second diode (J2 in the figure).

The breakdown voltage of J1 is given by (16.2), with the relevant doping densities being N_{epi} and $N_{\text{p-well}}$. The latter is the greater because of the need to counter-dope the epitaxial layer, so the breakdown voltage of J1 can be approximated as

$$V_{br,1} = \mathcal{E}_{br}^2 \frac{\epsilon}{2q} \frac{1}{N_{\text{epi}}}. \tag{16.9}$$

The desirability of a low value for N_{epi} is mitigated by the need to reduce the ON-resistance, which is determined to a large extent in this device by the lateral resistance of the epi-layer. One solution, you might imagine, would be to use a thick epi-layer. However, it turns out that a thin epi-layer is advantageous because of an interesting 2-D effect, the reduced surface field (**RESURF**) effect [5].

At the diode junction J2, the space charge region extends into the n-epi layer by a distance y_{dn}, which can be estimated from applying the Depletion Approximation. With reference to Fig. 16.12,

$$y_{dn} = \sqrt{\frac{2\epsilon_s}{q} V \frac{N_{\text{body}}}{N_{\text{epi}}(N_{\text{epi}} + N_{\text{body}})}}, \tag{16.10}$$

where $V = V_{bi,J2} + V_{DS}$. The basic idea is to choose the thickness of the epi-layer to be less than y_{dn} at the desired forward blocking voltage. In this way, the depletion region from J2 reaches through to the surface of the FET and augments the depletion region surrounding J1, effectively extending it in the x-direction. Thus, the voltage drop across J1 is spread over a longer region, and this has the effect of reducing the electric field $\mathcal{E}_x$ below $\mathcal{E}_{br}$. This then shifts the region of likely breakdown to junction J2. By making $N_{\text{body}} < N_{\text{epi}}$, then $V_{br,2}$ can exceed $V_{br,1}$. This desirable state of affairs increases the forward blocking voltage. However, note that there is a parasitic n^+pn^- BJT formed by the source/body/epi regions, and if the current is high enough, the Kirk Effect will come into play and affect the voltage distribution in the structure (see Fig. 16.2). In the L-DMOSFET the point of highest field moves to the n^+n^- junction between the drain and epi-layer regions, and this becomes the determinator of the breakdown voltage.

It can be appreciated that the L-DMOSFET poses an interesting design problem involving the choice of epi-layer thickness and doping, and the interplay between blocking voltage and operating current. If very high operating currents and voltages are required, then the device of choice is the insulated-gate bipolar transistor, a lateral version of which is described in the following section.

16.4.2   Lateral insulated-gate bipolar transistor

As its name suggests, the insulated-gate bipolar transistor (IGBT) is a hybrid device. It combines the high output-current capability of a BJT with the high input-impedance- and voltage-control-attributes of a FET. A lateral version of the insulated-gate bipolar transistor (LIGBT) is shown in Fig. 16.13. The structure differs from that of the L-DMOSFET in Fig. 16.12 at the drain-end of the device: the n^+ region is replaced by a p^+ region within an n-type buffer zone. The additional pn-junction is labelled J3.

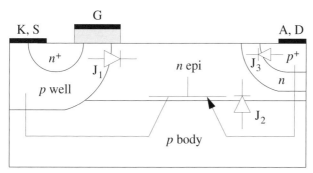

Figure 16.13 Lateral IGBT, showing the biolar transistor with its base connected to the cathode K only when the MOSFET is ON.

There are name changes too: the drain contact becomes the anode (A), and the source contact becomes the cathode (K). The bipolar part of the device is traced on Fig. 16.13; it is a p^+np_{well} BJT, with the base being the n-epi layer. The cathode doubles as the contact to the collector and to the source of the N-MOSFET. The gate-source voltage V_{GK} determines the state (ON or OFF) of the device.

When $V_{GK} < V_T$, the threshold voltage of the MOSFET, there is no inversion layer connecting the n^+-source to the n-epi region, which is now the drain of the MOSFET. A forward bias on the IGBT ($V_{AK} > 0$) results in the reverse biasing of junction J1, which provides the LIGBT with its forward blocking voltage. The depletion region in the epi-layer extends towards the anode, but is prevented from reaching the p^+ region under the anode by the n-buffer layer. In other words, **punch-through** of the epi-layer, now in its role as the base, is averted. In punch-through there is no quasi-neutral base region to limit the current, so the blocking condition would be lost. The p^+n_{buffer} junction J3 provides the IGBT with its reverse blocking voltage, i.e., when $V_{AK} < 0$ in the OFF state.

The IGBT turns on when $V_{GK} > V_T$; an inversion layer forms under the gate, and this establishes a terminal contact (K) to the base epi-layer, into which electrons are injected. If, additionally, $V_{AK} > 0$, then the p^+n_{buffer} junction J3 becomes forward biased, and holes are injected into the base. This hole current is collected by the cathode. Thus, large hole conduction is modulated by a much smaller electron conduction. This is another example of conductivity modulation, and can lead to a situation where the injected hole density exceeds the background doping density of the n-epi layer. This leads to a larger current, and to a smaller ON-resistance, than is possible in the L-DMOSFET, where the current is due to the drift of majority carriers.

Note that, before conductivity modulation can occur, the turn-on voltage of junction J3 has to be reached. Thereafter, the familiar BT collector characteristic unfolds, albeit with the MOS-derived voltage V_{GK} being the control voltage, rather than the usual pn-junction-derived voltage V_{BE}. The general characteristics are shown in Fig. 16.14. Specific characteristics are shown in Fig. 16.15, where yet another name for the source/cathode has been used, namely, the emitter.

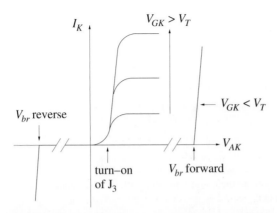

Figure 16.14 General collector characteristic for an IGBT.

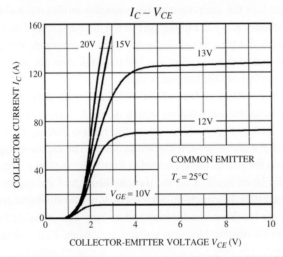

Figure 16.15 Collector characteristic for Toshiba MG75J1ZS50 IGBT.

Si IGBTs are capable of supplying 100's of amps and watts in the ON state, and of withstanding 100's of volts in the OFF state. They are not fast-switching devices, though, on account of the substantial minority carrier storage that occurs in the long n-epi base region.

Exercises

16.1 A silicon n^+p^-p power diode has an emitter with a donor doping density of 10^{20} cm^{-3}. The weakly doped region of the two-part p-type base is 2 μm long and has a doping density of 10^{16} cm^{-3}. The remainder of the base has a doping density of 10^{17} cm^{-3}. This arrangement gives a high breakdown voltage and a low series resistance.

A reverse bias of 40 V is applied to the diode, and the junction capacitance is measured to be $3.48 \times 10^{-5}\,\mathrm{F\,m^{-2}}$.

(a) Show that the lightly doped part of the base is completely depleted.

(b) Calculate the maximum electric field in the device.

16.2 The breakdown voltage of a particular GaAs *pn* junction is 3 V.

If the doping densities remained the same but the semiconductor were changed to GaN, what would the breakdown voltage be?

16.3 Integrate (16.3) to show that the current density at which the field disappears at the *p*/epi interface in a pn^-n^+ junction is

$$J_K = q \left[\frac{2\epsilon_s}{q W_{\mathrm{epi}}^2}(V_a + V_{bi}) + N_{\mathrm{epi}} \right] v_{\mathrm{sat}}. \qquad (16.11)$$

16.4 Fig. 16.3 shows a power HBT in which the emitter is divided into two fingers, each of lateral width $h/2$. Also, the base current is split between three base contacts that are interdigitated with the emitter fingers.

Show that this arrangement leads to a 12-fold reduction in the power dissipation associated with the base-spreading resistance, as compared to an HBT with a single emitter of width h and a single base contact.

16.5 Consider an $Al_{0.3}Ga_{0.7}N/GaN$ HJFET. Data on the polarization properties of this material system are given in Appendix B.

Using this data, show that if 10% of the polarization charges at the interface between the two materials attract electrons, then the electron sheet density would be $\approx 1.7 \times 10^{12}\,\mathrm{cm^{-2}}$.

16.6 A Si L-DMOSFET has a body doping density of $5 \times 10^{16}\,\mathrm{cm^{-3}}$. The body/epi junction is required to withstand a voltage of 100 V, under which conditions y_{dn} in Fig. 16.12 can be assumed to have penetrated through the entire epi-layer thickness. The sheet resistance of the epi-layer is to be $2 \times 10^3\,\Omega$/square, and can be taken to exist at the normal ON voltage, at which y_{dn} can be taken to be zero.

Using these specifications, calculate the thickness and doping density of the epi-layer.

Note: an analytical solution can be easily obtained if the assumption of $N_{\mathrm{body}} < N_{\mathrm{epi}}$ is made in (16.10).

References

[1] D. DiSanto, Aluminum Gallium Nitride/Gallium Nitride High Electron Mobility Transistor Fabrication and Characterization, Table 1-1, Ph.D. Thesis, Simon Fraser University, 2005.

[2] E.O. Johnson, Physical Limitations on Frequency and Power Parameters of Transistors, *RCA Review*, vol. 26, 163–177, 1965

[3] Y. Apanovich, P. Blakey, R. Cottle, E. Lyumkis, B. Polsky, A. Shur and A. Tcherniaev, Numerical Simulation of submicrometer Devices Including Coupled Nonlocal Transport and Nonisothermal Effects, *IEEE Trans. Electron Dev.*, vol. 42, 890–898, 1995.

[4] O. Ambacher, J. Smart, J.R. Shealy, N.G. Weimann, K. Chu, M. Murphy, W.J. Schaff, L.F. Eastman, R. Dimitrov, L. Wittmer, M. Stutzmann, W. Rieger and J. Hilsenbeck, Two-dimensional
Electron Gases Induced by Spontaneous and Piezoelectric Polarization Charges in N- and
Ga-face AlGaN/GaN Heterostructures, *J. Appl. Phys.*, vol. 85, 3222–3233, 1999.
[5] A. Ludikhuize, A Review of RESURF Technology, *Proc. 12th IEEE Int. Symp. Power Semiconductor Devices and ICs*, pp. 11–18, 2000.

17 Transistors for low noise

In this modern age of increasing wireless communications there are many applications requiring transistors that have low inherent noise, and can operate at very high frequencies, e.g., low-noise amplifiers for satellite communications systems, mixers and multipliers for point-to-point and point-to-multi-point radio, multi-port phase/frequency discriminators for automobile collision-avoidance radar.

This chapter gives a very brief overview of the main sources of noise in transistors: thermal noise, shot noise, and flicker noise. How each of these noise sources is manifest in the various types of transistor discussed in this book is mentioned. It becomes apparent that the HJFET has superior low-noise performance at high frequencies, so this transistor is then used to show how a transistor's noise sources are incorporated into its small-signal equivalent circuit. A metric for noise performance, the noise figure NF, is then introduced.

17.1   Noise: general properties

We are interested in how noise can originate in the inherent properties of a transistor, rather than arise due to interference from other sources.

Inherent noise is random in nature, often with a time-averaged value $\langle v_n(t) \rangle \approx 0$; so noise is usually described in terms of a root-mean-square (rms) value

$$\sqrt{\overline{v_n^2}} = \sqrt{\lim_{T \to \infty} \left(\frac{1}{T} \int_0^T v_n(t)^2 \, dt \right)}, \tag{17.1}$$

where T is the time period of observation. Recall that the rms formalism is used in relation to power:

$$P_{\text{average}} = \frac{1}{T} \int_0^T \frac{v(t)^2}{R} \, dt \equiv \frac{v_{\text{rms}}^2}{R}, \tag{17.2}$$

i.e., an rms AC voltage and a DC voltage would lead to the same power dissipation in a resistor of resistance R, provided $v_{\text{rms}} = V_{DC}$. Thus, in noise considerations, one often talks of a noise power proportional to $\overline{v_n^2}$ or $\overline{i_n^2}$. Further, one is usually interested in the signal-to-noise ratio, which is customarily expressed in decibels

$$SNR = 10 \log_{10} \left[\frac{\text{signal power}}{\text{noise power}} \right]. \tag{17.3}$$

Often, a power level is referred to 1 mW, e.g., a signal power of $1\,\mu W$ is -30 dBm.

Whereas a domestic power supply delivers power at 50 or 60 Hz, random noise signals have, or can have, power over a wide frequency range. If one measures the noise power over a bandwidth Δf, the **spectral power density** is

$$S = \frac{\overline{v_n^2}}{\Delta f}. \tag{17.4}$$

$\sqrt{S}$ is the root spectral density, and is expressed in volts per root Hertz.

17.2 Noise inherent to transistors

17.2.1 Thermal noise

Electrons in a semiconductor are thermally agitated, and, under equilibrium conditions, move randomly with Brownian motion and a mean thermal speed given by (4.29). Thus, in equilibrium, the necessary condition of no net current is satisfied by the cancellation of many, tiny, randomly directed currents. These currents are manifest as noise in any resistive element, e.g., in the contact resistances of any transistor, in the base resistance of an HBT, and in the channel resistance of a FET.

In 1928, Johnson determined experimentally that a voltage appeared across an open-circuited resistor. The mean-square value of this voltage was found to be proportional to the resistance and to the temperature. A theoretical basis for this result from thermo-dynamical considerations was provided by Nyquist in the same year [1, Chapter 5]. The spectral power density of **thermal or Johnson noise** is

$$S_{th} = 4k_B T R, \tag{17.5}$$

and the corresponding mean-square noise voltage is

$$\overline{v_{th}^2} = 4k_B T R \Delta f. \tag{17.6}$$

Therefore, to represent a noisy resistor in an equivalent circuit, one can replace the real resistor R with the Thévenin equivalent of an ideal, noiseless resistor R in series with a voltage source of magnitude $\sqrt{4k_B T R \Delta f}$. Alternatively, a parallel combination of noise-current source and noiseless resistor could be used, with the current source being specified by

$$\overline{i_{th}^2} = \frac{4k_B T \Delta f}{R}. \tag{17.7}$$

Note that thermal noise has a flat frequency response; hence its popular name of 'white noise'. It is heard as 'audio hiss', and seen as 'TV snow'. As a yardstick, note that a $50\,\Omega$ resistor generates a noise of ≈ 1 nV over a bandwidth of 1 Hz.

Generally, the resistance of a block-shaped object of length L and cross-sectional area A is

$$R = \frac{L}{\sigma A}, \tag{17.8}$$

where, using an n-type semiconductor as an example, the conductivity σ due to the majority carrier electrons is (5.32)

$$\sigma_n = q n \mu_e. \tag{17.9}$$

Immediately, one can appreciate the importance of employing high-mobility material for transistors intended for low-noise operation. HBTs, MESFETs, and HJFETs that utilize III-V compound semiconductors (see Fig. 11.1) take advantage of this fact. The design features of the HEMT discussed in Section 11.3 (no ionized-impurity scattering, carrier confinement away from the heterointerface, carrier motion restricted to two dimensions) give this transistor the capability of attaining the highest mobility of all, which contributes greatly to its exceptional noise performance. Competition comes from the HBT, which capitalizes on its geometrical features of large area and short base width to produce a current path that is not constricted by a narrow channel.

17.2.2   Shot noise

Shot noise arises when there is a direct current in a material. It is due to the fact that the charges, whose flow determines the current, are discrete and independent, i.e., the flow is not continuous on a microscopic scale.

    Let us consider electrons as the charge carriers. If one could observe the net number of electrons crossing a particular plane in a conductor in a particular direction during a specified time interval, then that number would vary slightly about some mean value. This variation about the mean value would be the **shot noise**. It is not possible to detect shot noise in a uniform conductor because it is obscured by the random thermal motion of the electrons. For shot noise to be evident, the conductor must be part of a structure that allows essentially unidirectional flow. An np-junction is such a structure. Under forward bias, for example, the injection of each electron from the hemi-Maxwellian distribution on the n-side of the junction into the p-side is an independent and random event. Each injection event causes a random current pulse that can be detected in the external circuit. If you could hear it, it would sound like buck**shot** (pellets) striking a hard floor.

    For an electronic current made up of random, independent pulses of average value I_{DC}, Fourier Analysis reveals that the electron shot noise current is given by [1, Chapter 2]

$$\overline{i_{\text{shot}}^2} = 2q I_{DC} \Delta f. \tag{17.10}$$

Evidently, this is also a white noise source. For a DC current of 1 mA, a noise of $\approx 18\,\text{pA}$ is generated over a bandwidth of 1 Hz.

Forward biased *np*-junctions occur in HBTs, solar cells, LEDs and in MOSFETs when operating in the sub-threshold regime. Thus, shot noise arises in all these cases, e.g., it originates in the base and emitter currents of HBTs, and in the source current of MOSFETs. For MOSFETs in the above-threshold regime, the source/body *np*-junction is shorted by the inversion layer, in which case thermal noise in the resistive channel dominates. Tunnelling is another transport mechanism in which the charge flow is due to random movements of independent carriers. Therefore, gate leakage currents in MOSFETs are another source of shot noise. HJFETs are less prone to this because of the relatively large thickness of the barrier layer. However, if the gate/barrier Schottky diode is forward biased, as in enhancement-mode HJFETs, there will be shot noise in the gate current.

17.2.3   Flicker noise

If there are imperfections or defects in the crystalline semiconductor through which the charge carriers flow, then the moving electrons (say) can be impeded, giving rise to a fluctuation in the current that is called **flicker noise**. The deviations from crystalline perfection can be regarded as producing local variations in, for example: the recombination-generation rate in the bulk; the recombination velocity at a surface; the carrier mobility, the carrier density or the bandgap in the bulk or at a surface; the tunnelling rate between different regions in a device. Irrespective of the particular mechanism that is operative in any one device, the characteristic signature of flicker noise is an inverse dependence on frequency

$$S_{\text{flick}} \propto \frac{1}{f^\alpha}, \tag{17.11}$$

where, often, $\alpha \approx 1$, hence the alternate name of **1/f noise**. Sometimes this noise is referred to as 'pink noise', in view of its existence at long wavelengths (low frequencies).

Let us consider, for example, the situation at the semiconductor/insulator interface of a N-MOSFET. This is obviously an inhomogeneous region, within which local variations in transport-related properties are to be expected. We will consider the situation depicted in Fig. 17.1, where local variations in the electrostatic potential give rise to electron traps of varying depth. Electrons can be captured in these potential wells, before being released (by thermal emission) at some variable time τ later. It is reasonable to assume that τ is longer for deeper wells; a relationship that has been used is [2]

$$\tau = \tau_0 e^{E_a/k_B T}, \tag{17.12}$$

where E_a is the activation energy (depth of a potential well (trap)). For a normalized distribution of activation energies $g(E_a)$ within the range bounded by the deepest trap E_{a2} and the lowest trap E_{a1}, and for a uniform distribution of activation energies within this range,

$$\int_{E_{a1}}^{E_{a2}} g(E_a)\, dE_a = 1 \quad \text{and} \quad g(E_a) = \frac{1}{E_{a2} - E_{a1}}. \tag{17.13}$$

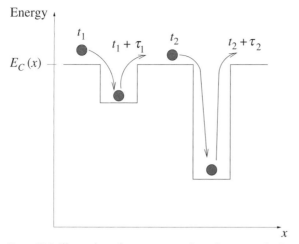

Figure 17.1 Illustration of capture, at various times t, and release, after various times τ, of electrons from traps of varying depths caused by crystalline imperfections at a semiconductor surface.

It can then be shown [2] [1, p. 126], that the spectral number density is

$$S_N(f) = \overline{\Delta N^2} \frac{k_B T}{(E_{a2} - E_{a1})} \frac{1}{f}, \tag{17.14}$$

where $\overline{\Delta N^2}$ is the variance of the number N of untrapped carriers contributing to conduction in the channel. The key point here is that the noise spectrum has a $1/f$ dependence.

Traps and imperfections are often associated with interfaces, particularly between poorly lattice-matched materials. Hence, $1/f$ noise is much in evidence in MOSFETs. In MESFETs, one side of the channel is defined by the 'smooth' edge of a depletion region, rather than by a 'rough' semiconductor/oxide heterojunction, so flicker noise is less evident in this transistor. In HJFETs, the use of lattice-matched materials, and a 2-DEG that keeps the centroid of channel charge away from the interface (see Fig. 11.8b), lead to the possibility of very low $1/f$ noise. In all FETs, the interface that defines the channel has a length that is about as long as the gate length, so another requirement of low-noise FETs is that L be small.

17.2.4 Induced gate noise

Noise in the channel of FETs, whether it be due to thermal noise from the resistive nature of the channel, or to 1/f noise due to the 'charging' and 'discharging' of localized traps, creates a time-varying local charge that couples capacitively to the gate electrode via C_{gs} and C_{gd}. The impedance of these capacitive components decreases with frequency, so the **induced gate noise** becomes prominent at high frequencies.

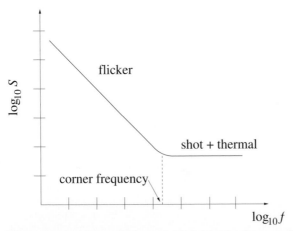

Figure 17.2 General form of the spectral density for the case of uncorrelated noise sources.

17.2.5   Adding-up the noise

Imagine that two noise sources in a transistor are effectively in series. Because of the mean-square representation of the noise, the total noise is

$$\overline{v_n^2} = \overline{v_{n1}^2} + \overline{v_{n2}^2} + 2\overline{v_{n1}v_{n2}}\,. \tag{17.15}$$

If the noise sources arise from separate physical mechanisms they are independent of each other, and are said to be uncorrelated. This is the case for all the noise mechanisms discussed in this chapter, with the exception of gate-induced noise, which is **correlated** with the noise in the drain current. In the uncorrelated case, the total noise is simply the algebraic sum of each individual noise component. The overall spectral density will then have the general form shown in Fig. 17.2. The frequency at which the extrapolations of the white and pink noise terms intersect is called the **1/f corner frequency**. In HJFETs specifically designed to have low noise, this frequency may approach 1 GHz for the drain-current noise. This is primarily because in HJFETs designed for high mobility (HEMTs), the correspondingly high transconductance makes for a low channel resistance, thereby reducing the principal intrinsic source of thermal noise. Additionally, HJFETs are well-suited to low-noise applications because they have less shot noise than HBTs and less flicker noise than MOSFETs. For these reasons we will concentrate on the HJFET in the remainder of this chapter.

17.3   Representation of noise in an equivalent circuit

Noise is, essentially, a small signal, so it is natural to account for it in a transistor by adding noise-voltage sources and noise-current sources, where appropriate, to the small-signal equivalent circuit of the transistor. We consider the case of an HJFET intended for a low-noise amplifier application at high-enough frequencies for flicker noise not to be an issue. Additionally, for simplicity, we neglect any shot noise, such as might arise

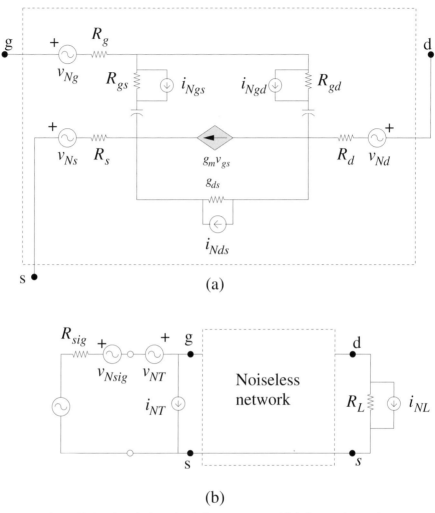

Figure 17.3 (a) Small-signal equivalent circuit for an HJFET at high frequencies. Noise sources have been added to account for thermal noise in all the resistive components of the circuit. (b) Equivalent representation of (a): the noise sources within the dashed box of (a) have been replaced by two noise sources in the external input circuit. A signal generator at the input and a resistive load at the output are also included.

from leakage current to the gate. Thus, only thermal noise sources are included in the equivalent circuit shown in Fig. 17.3a. Each noise source is designated by the inclusion of 'N' in its subscripted notation.

The basis for this circuit is the equivalent circuit for a FET shown in Fig. 14.12. The differences are: the upper-case subscripts have been changed to the more conventional lower-case; the prime has been dropped from the capacitor labels, but these elements still represent total capacitances; the transcapacitance element has been omitted, which implies that non-reciprocity is not important and justifies writing C_{sd} as the more usual C_{ds}; g_{dd} has been written as the more traditional g_{ds}; 'intrinsic charging resistors'

R_{gd} and R_{gs} have been included in series with the respective components of the gate capacitance. These resistors are not easy to quantify a priori, and their values are usually determined from measurements on an actual device. They represent the resistive paths within the actual device connecting the gate to either the source or the drain. They are relevant here because they are sources of thermal noise. R_{gs} is particularly important because it is on the input side of the device, so its associated noise is amplified by the transistor.

The noisy network in Fig. 17.3a can be replaced by an equivalent arrangement in which the noise sources are separated from the physical components of their origin, and are represented by two noise generators at the input to the now noiseless transistor network (see Fig. 17.3b). For the two arrangements to be equivalent, the terminal currents and voltages (noise and signal) must be the same. For this to happen, the sources representing the total noise, i_{NT} and v_{NT} are likely to be correlated, even though the individual noise sources in Fig. 17.3a are not.

Also shown in Fig. 17.3b are a signal source, which adds some noise to the system via its series resistance, and a noisy resistor for the load.

17.4   Noise figure

Here, we employ the **noise figure** NF as a figure-of-merit by which a transistor's noisiness can be characterized. It does this by comparing the noise that the transistor itself generates at the output of the network, to the noise at the output that comes from the amplification of any noise in the input signal. In other words, the noise figure quantifies how the signal-to-noise ratio is degraded as the signal and the input noise pass through the transistor. The noise figure can be derived from the ratio of input and output signal-to-noise-ratios:

$$NF = 10 \log_{10} \left(\frac{S_i/N_i}{S_o/N_o} \right)$$

$$\equiv 10 \log_{10} \left(\frac{1}{G_A} \frac{N_o}{N_i} \right)$$

$$= 10 \log_{10} \left(\frac{\text{total output noise}}{\text{amplified input noise}} \right)$$

$$= 10 \log_{10} \left(1 + \frac{\text{inherent noise}}{\text{amplified input noise}} \right)$$

$$\equiv 10 \log_{10}(F), \tag{17.16}$$

where G_A is the small-signal power gain that is available on account of the particular matching at both the input and the output, and F is defined as the **noise factor**. Measurements of NF and F are usually reported at a standard temperature of $T_0 = 290$ K.

We showed in Section 14.6 how the maximum available gain MAG of a transistor depended on correctly matching the impedance of the signal source to that of the transistor's input, and also on correctly matching the load to the output impedance. Similarly, minimum noise is added to the output of a transistor when there is appropriate matching at the input and the output [3]. When NF is minimized in this way we have the **minimum noise figure** $NF_{\min}$.

To derive an expression for $NF_{\min}$ is an exercise in Circuit Analysis: the noise sources in Fig. 17.3a have to be transferred to the input, the signal-source properties have to be adjusted to give minimum noise, and the noise and signal at the output have to be determined. To obtain a tractable, analytical solution various simplified models have been proposed [4, Chapter 17]. Here, we start with the already approximate circuit of Fig. 17.3a, and further simplify by ignoring the drain-related components, C_{gd}, R_d, and R_{gd}. The remaining resistive/conductive components, R_g, R_{gs}, R_s, and g_{ds}, are then treated as contributing thermal noise, each at a temperature that may not necessarily be T_0. In essence, this is the approach followed by Roblin and Rohdin [4, p. 592], and their resulting equation is

$$F_{\min} = 1 + 2 \left(\frac{f}{f_T'} \right) \sqrt{ g_{ds} \frac{T_d}{T_0} \left[R_{gs} \frac{T_g}{T_0} + R_g \frac{T_a}{T_0} + R_s \frac{T_a}{T_0} \right] }, \tag{17.17}$$

where f_T' is the intrinsic f_T when C_{gd} is ignored (see (14.40)), T_a is the ambient temperature, and T_d and T_g are adjustable model parameters for characterizing thermal noise in the channel, at the drain-end and under the gate at the source-end, respectively. Usually, $T_g \approx T_a$, because R_{gs} is associated with the source end of the channel where the field $\mathcal{E}_x$ is relatively weak. Towards the drain end, where the field is higher, velocity saturation can occur. As we saw in Section 5.4, this phenomenon is associated with significant transfer of energy to the lattice; this is modelled here by attributing a large value to T_d.

Fig. 17.4 shows plots of $NF_{\min}$ from (17.17) for an HJFET with the parameters listed in the figure caption. The upper and lower curves of the simulated set show the effect of decreasing g_m alone, and of decreasing both g_m and T_d, respectively. The intention is to indicate that $NF_{\min}$ is dependent on the applied biases: decreasing V_{GS} while remaining in the saturation mode would reduce the drain current and g_m, whereas reducing V_{DS} would reduce $\mathcal{E}_x$ and, presumably, T_d. Thus, there will be some set of bias conditions at which $NF_{\min}$ has a minimum value. The minimum noise figure for an actual low-noise HEMT is also shown in Fig. 17.4.

17.4.1   Associated gain

Unfortunately, the source impedance that minimizes the noise to allow attainment of $NF_{\min}$ is not the same as the source impedance that maximizes the power gain. Thus, the available gain under minimum-noise conditions is less than the maximum possible gain MAG, which was introduced in Section 14.6. The available gain associated with operating under minimum-noise conditions is called the **associated gain**. Fig. 17.5 illustrates the difference between the two gains for a high-performance, low-noise HJFET from Triquint.

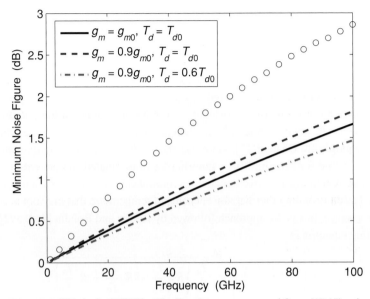

Figure 17.4 NFmin for HJFETs. The line data are computed from (17.17) using the following parameters kindly supplied by Hans Rohdin of Avago Technologies: $g_{m0} = 78.7\,\mathrm{mS}$, $g_{ds} = 3.25\,\mathrm{mS}$, $C_{gs} = 42.65\,\mathrm{fF}$, $R_s = 7\,\Omega$, $R_{gs} = 0.1\,\Omega$, $R_g = 6.42\,\Omega$, $f_T' = 264\,\mathrm{GHz}$, $T_g = T_a = T_0 = 290\,\mathrm{K}$, $T_{d0} = 3100\,\mathrm{K}$. The symbol data is from a low-noise, 150-nm gate-length, AlGaAs/GaAs HEMT from Triquint, courtesy of Tony St. Denis. The gate is 200 µm wide, $V_{DS} = 1\,\mathrm{V}$, $I_D = 10\,\mathrm{mA}$.

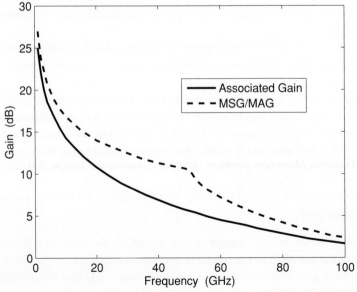

Figure 17.5 Associated gain, maximum stable gain, and maximum available gain of a low-noise, 150-nm gate-length, AlGaAs/GaAs HEMT from Triquint, courtesy of Tony St. Denis. The gate is 200 µm wide, $V_{DS} = 1\,\mathrm{V}$, $I_D = 10\,\mathrm{mA}$ for the associated gain measurement and 15 mA for the MSG/MAG measurement.

Exercises

17.1   Fig. 17.4 compares minimum-noise-figure calculations from (17.17) with measured data from a low-noise HEMT from Triquint. This measured data is given in the table below. The measured f_T for this transistor is 135 GHz, and can be taken as the intrinsic value.

Use this value of f_T in (17.17), along with values for the other parameters from the caption to Fig. 17.4, and plot both the measured and calculated data on a graph similar to Fig. 17.4.

Perhaps you will be as surprised as I was that the two curves are in such good agreement.

Improve the fit by slightly varying the adjustable parameters T_g and T_d.

f (GHz)	NF_{min} (dB)	f (GHz)	NF_{min} (dB)	f (GHz)	NF_{min} (dB)
1	0.04	36	1.32	72	2.28
4	0.16	40	1.46	76	2.38
8	0.32	44	1.58	80	2.48
12	0.48	48	1.68	84	2.56
16	0.64	52	1.8	88	2.64
20	0.78	56	1.9	92	2.7
24	0.92	60	2	96	2.78
28	1.06	64	2.1	100	2.86
32	1.2	68	2.2		

References

[1] A. van der Ziel, *Noise in Solid State Devices and Circuits*, John Wiley & Sons, Inc., 1986.

[2] M. Chertouk, A. Chovet and A. Clei, 1/f Trapping Noise Theory with Uniform Distribution of Energy-activated Traps and Experiments in GaAs/InP MESFETs Biased from Ohmic to Saturation Region, *AIP Conf. Proc.*, vol. 285, 427–432, 1993.

[3] G. Gonzalez, *Microwave Transistor Amplifiers: Analysis and Design*, 2nd Edn., Appendix L, Prentice-Hall, 1984.

[4] P. Roblin and H. Rohdin, *High-speed Heterostructure Devices*, Cambridge University Press, 2002.

18 Transistors for the future

Predicting the future is the realm of prophets and gamblers. Generally speaking, engineers belong to neither of these groups. However, it is evident that the trend in high-performance transistors is towards smaller and smaller devices, so we might reasonably expect this to continue into the near future. This miniaturizing trend has led, in the early part of the 21st century, to much talk about nanoelectronics. In my opinion, nanoelectronics does not mean, contrary to popular belief, merely the scaling down of transistors to devices with feature sizes of the order of tens of nanometres. If it did, then present-day devices, such as InGaP/GaAs HBTs with basewidths of around 30 nm, Si MOSFETs with channel lengths of $\approx$40 nm, and InAlAs/InP HJFETs with channel thicknesses of a few nanometres, would already qualify as nanotransistors.

What nanoelectronics implies, to me at least, is the fabrication of transistors from the 'bottom up', rather than from the 'top down'. The latter process is the one that applies to all the HBTs, MOSFETs, and HJFETs discussed in earlier chapters: it involves the making of something small from something large using photolithographic masking techniques. Contrarily, a true nanotransistor is built-up from something even smaller, such as a molecule or a nanoparticle. This opens up opportunities for new methods of self-assembly (perhaps using biological recognition), and for new 3-D circuitry (perhaps using a molecular backbone), that cannot be matched using conventional processing techniques [1].

Smaller transistors, as explained throughout this book, are desired because they enable portable, higher performance electronic products, for which much of the world's population appears to have a craving. One of the difficulties that arises with densely packed transistors is heat dissipation, as discussed in Section 13.1.10. One way in which future transistors might avoid this problem is by using very few electrons. An extreme example of this is the single-electron field-effect transistor (SET), in which the drain current is due to tunnelling of electrons that are accumulated singly in a capacitor 'island' between the closely spaced source and drain [2]. Another example is the molecular switch, involving a voltage-controlled oxidation/reduction reaction of a complex molecule, each end of which is connected to an electrode [3]. Another way of reducing power would be to not rely on electron transport, but rather on the change of spin of electrons [4]. Clearly, such a device would also be very fast. Although 'spin-FETs', SETs, and molecular switches have been under investigation for many years, as the dates of the cited references attest, the prospect of commercial devices is not imminent.

Still within the realm of nanotechnology, but returning now to the traditional, gate-controlled, multi-electron-transfer type of device, two candidate nanotransistors are FETs utilizing either semiconducting nanowires (Si, Ge, GaN, InAs are being tried), or nanotubes (carbon). The cylindrical geometry of these devices begs an all-around gate, which would be the ultimate electrostatic solution to the short-channel-effect problem. The use of materials such as InAs and C, in which an extremely high mobility can be obtained, would be advantageous as regards many features: ON-current, transconductance, f_T, f_{max}, noise figure, and parasitic resistance.

The small diameters of these embryonic transistors (≈ 1–5 nm), and the small channel lengths (≈ 10 nm), which can be anticipated if these devices are to be significantly smaller than planar Si MOSFETs, mean that the active volume of these transistors is going to be very small. This raises an interesting, and very practical, question: what is the achievable repeatability of dopant incorporation in such small structures? For example, a wire with a length of 10 nm, a radius of 2 nm, and a high-ish doping density of 8×10^{18} cm^{-3}, would contain approximately 1 dopant atom!

One way to circumvent the issue of fluctuations in doping density would be to avoid using extrinsic semiconductors. For example, an intrinsic semiconductor could be used for the channel, and metals could be employed for the source and drain regions. We highlight several properties of such a **Schottky-barrier nanoFET** later in this chapter. Such a transistor, if made from a nanocylinder, would be essentially a 1-D structure, and if its length were short, then there would be the possibility of ballistic transport between the source and drain. Therefore, in the next two sections, we describe some basic properties of 1-D semiconductors, and develop an expression for the drain current due to ballistic transport in a 1-D FET.

To appreciate the length scale at which ballistic transport might become operative, consult (5.45), which is an expression for the equilibrium mean-free-path length $\bar{l}_0$ between collisions in terms of the mobility and the effective mass. Taking near-intrinsic values for mobility (0.12 and 0.8 m^2 V^{-1} s^{-1} for Si and GaAs, respectively), and corresponding effective masses of $0.19\,m_0$ and $0.066\,m_0$, gives $\bar{l}_0 \approx 25$ nm for Si, and $\bar{l}_0 \approx 100$ nm for GaAs.

18.1   1-D carrier basics

In this section expressions are derived for the following one-dimensional properties: the density of states, the electron concentration, the effective density of states and the mean, unidirectional thermal velocity. These differ from the corresponding 3-D expressions given in Chapter 3 and Chapter 4 mainly by the order of the Fermi-Dirac integrals involved.

18.1.1   Density of states

We assume that the wire or tube is long enough so that periodic boundary conditions can be applied, in which case the spacing between states in k-space is $2\pi/L$ (from

Section 2.4), where L is the length of the semiconductor, and k is the wavevector in the longitudinal direction. Allowing for spin, there can be two electrons per momentum state; therefore, the number of states per unit real-space length in a length δk of k-space is simply

$$g(k)\delta k = \frac{\delta k}{\pi}. \tag{18.1}$$

To find the number of states in an equivalent element δE in energy, note that whatever variable we use, k or $(E - E_C)$ in this case, the actual spatial density of states must be unchanged. Thus,

$$\int_{-\infty}^{\infty} g(k)\,dk = \int_{0}^{\infty} g(E - E_C)\,d(E - E_C). \tag{18.2}$$

From this it is clear that

$$g(E - E_C)\delta(E - E_C) = 2g(k)\delta k = \frac{2}{\pi}\delta k. \tag{18.3}$$

Making use of the general relationship for the band-determined carrier velocity (2.23), we obtain the desired expression for the number of states per unit length and unit energy, i.e., the **1-D density of states**:

$$g(E) = \frac{4}{hv(E)} \qquad E \geq E_C. \tag{18.4}$$

Assuming a parabolic band, (2.31) can be used to obtain an explicit relation between g and E:

$$g(E) = \frac{2}{h}\sqrt{\frac{2m^*}{(E - E_C)}} \qquad E \geq E_C. \tag{18.5}$$

There is no ambiguity about the value to take for m^* as the system is one-dimensional. Note that g in the 1-D case goes to infinity at the conduction-band edge (see Fig. 18.1).

In fact, because we are considering wires or tubes of finite radius, and because this radius is very small, there will be quantum confinement of the wavevector in the radial or circumferential directions, respectively. For a tube of circumference c, for example, the allowed circumferential wavevectors are $k_c = n\pi/c$, where $n \geq 1$ is an integer. For each allowed value of k_c there is an E-k band called a sub-band. Several of these are shown in Fig. 18.1. The associated $g(E)$ profile has a sawtooth appearance.

18.1.2   Carrier density

The equilibrium carrier density for one sub-band follows from (4.9). Specifically, for electrons,

$$n_0 = \int_{E_C}^{\infty} g(E)f_0(E)\,dE \equiv \int_{0}^{\infty} g(a)f_0(a)k_BT\,da, \tag{18.6}$$

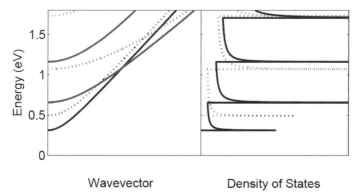

Figure 18.1 E_C-k relation for several sub-bands and positive k, and density of states g for two carbon nanotubes of different diameter: $d = 1.25\,\text{nm}$ (solid line), $d = 1.02\,\text{nm}$ (dotted line). Courtesy of Leonardo Castro, ex-UBC.

where, as used in Section 4.3, $a = (E - E_C)/k_B T$. Taking f_0 as the Fermi-Dirac distribution function gives:

$$n_0 = 2 \left(\frac{2\pi m^* k_B T}{h^2} \right)^{1/2} \mathcal{F}_{-1/2}(a_F) \equiv N_C \mathcal{F}_{-1/2}(a_F), \tag{18.7}$$

where $a_F = (E_F - E_C)/k_B T$. The second form of this equation is the 1-D equivalent of (4.10); it serves also to define the 1-D effective density of states N_C corresponding to the 3-D version in (4.11). Evidently, the two N_C's differ only in the power of the exponent.

18.1.3 Mean, unidirectional velocity of a 1-D equilibrium distribution

Following the procedure described for the 3-D case in Section 4.5, it is straighforward to show that the mean thermal speed for a 1-D equilibrium distribution is

$$v_{th} = \sqrt{\frac{2k_B T}{\pi m^*}} \frac{\mathcal{F}_0(a_F)}{\mathcal{F}_{-1/2}(a_F)}. \tag{18.8}$$

This equation is the 1-D equivalent of (4.29).

Because we are working in only one dimension, the construct for obtaining v_R from v_{th} is a line, rather than a sphere, which was the case for 3-D (see Fig. 4.5). Thus, in the 1-D case, the mean, unidirectional velocity v_R of the entire distribution is simply $v_{th}/2$. The mean, unidirectional velocity of the distribution that is moving in the positive direction is $2v_R$; thus it is exactly the same as v_{th}.

The 3-D and 1-D magnitudes of $2v_R$ are compared in Fig. 18.2. At low carrier concentrations there is essentially no difference between $2v_R$ in the two cases. But as the carrier concentration increases, the electrons in the 1-D distribution move increasingly faster.

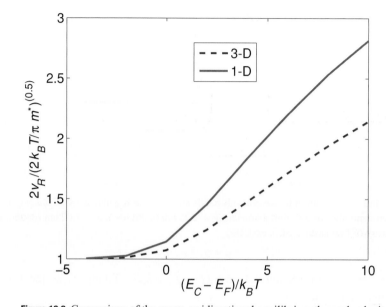

Figure 18.2 Comparison of the mean, unidirectional equilibrium thermal velocity for 3-D and 1-D semiconductors. $T = 300\,\text{K}$, $m^* = 0.1m_0$.

18.2   1-D ballistic transport

Consider the case of electrons being injected into a 1-D semiconductor from two 1-D metallic contacts: the source on the left of Fig. 18.3a, and the drain on the right. The electrons are at an energy E that is greater than the energy E_C of the edge of the first sub-band in the semiconductor. Quantum mechanical reflection at the two metal/semiconductor interfaces leads to some of the incident flux from the source J_S being reflected and some being transmitted into the drain. An analogous situation pertains to electrons in the incident flux from the drain J_D. Here, the subscript indicates the origin of the carriers (source or drain). Adding up the two fluxes in the drain, for example, gives the total spectral current of electrons:

$$J_e(E) = T(E)J_S(E) + [R(E) - 1]\,J_D(E)$$

$$= T(E)\,[J_S(E) - J_D(E)]\,, \tag{18.9}$$

where $T(E)$ and $R(E)$ are the transmission and reflection probabilities, respectively, at energy E for the entire system.

The incident electron currents at some energy E are simply the products of the charge and the velocity of electrons with $k > 0$. At the source, for example,

$$J_S(E) = -q\frac{g_S(E)}{2}\,f_S(E) \cdot v(E)\,, \tag{18.10}$$

where the factor of $1/2$ in the density of states arises because only states in the positive-going half of the distribution contribute to the injection; $f_S(E)$ is the Fermi-Dirac

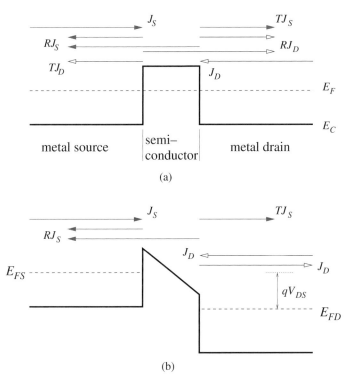

Figure 18.3 Incident, reflected, and transmitted components of electrons in currents originating in the source ($J_S(E)$) and in the drain ($J_D(E)$). $T(E)$ and $R(E)$ are transmission and reflection probabilities, respectively. (a) At equilibrium. (b) At a sufficiently large bias V_{DS} for the drain-originating electrons to be totally reflected.

distribution function for electrons in the source. $J_D(E)$ can be written analogously. Substituting into (18.9), and making use of (18.4), leads to the **Landauer** expression for the electron current:

$$J_e = -q\frac{2}{h}\int_E T(E)[f_S(E) - f_D(E)]\,dE\,. \tag{18.11}$$

Notice how g and v are no longer explicit in this equation because their product is a constant.

In Fig. 18.3a the Fermi levels in the source and drain are at the same energy, so (18.11) correctly predicts that the current is zero. Now consider applying a bias $V_{DS} > 0$. As Fig. 18.3b shows, E_{FD} drops below E_{FS}, so $|J_e|$ increases. Eventually, the barrier for electron injection from the drain is so high that the current saturates at

$$I_{Dsat} \equiv J_e \quad \text{at saturation} \quad = -q\frac{2}{h}\int_E T(E)f_S(E)\,dE\,. \tag{18.12}$$

In a FET, this current is altered by applying a gate bias: this varies $T(E)$ by changing the shape of the E_C profile.

18.2.1   Dimensions for current density

Note that in (18.12) we made no distinction between the current and the current density (except for a conventional sign). This is because, in a 1-D system they have the same units. To appreciate this refer to (3.26) for the fundamental expression for carrier concentration, multiply it by $-qv$ to get a current density, and the result is

$$J_e(\alpha) = \frac{-q}{\Omega} \sum_{\text{filled states}} v_k \Rightarrow \frac{-q}{\Omega} \frac{2}{\frac{(2\pi)^\alpha}{\Omega}} \int_{\vec{k}} f(\vec{k}) v_{\vec{k}} \, d\vec{k}, \qquad (18.13)$$

where α is the dimensionality of the volume under consideration, and f is the distribution function. Thus, the units are $A\,m^{-2}$, $A\,m^{-1}$, A, for $\alpha = 3, 2, 1$, respectively.

18.2.2   Local density of states

An informative way of following the injection of charge into the 1-D semiconductor is via the **local density of states** LDOS. This is defined as

$$\mathcal{G}_S(x, E) = \psi_S(x, E)\psi_S^*(x, E), \qquad (18.14)$$

where

$$\psi_S(x, E) = A_S(x, E)e^{ik_S(x,E)x} \qquad (18.15)$$

is the wavefunction at some point x and some energy E of an electron issuing from the source. Thus, the LDOS refers in this case to the states in the semiconductor that are available for occupancy by electrons injected from the source. An example of the LDOS for a Schottky-barrier carbon nanotube FET is shown in Fig. 18.4, where $A_S(x, E)$ has been normalized as described in the next subsection. There is a large change in k as electrons enter and leave the semiconductor, and this gives rise to resonances and quasi-bound states[1]. Five levels of quasi-bound states are readily identifiable in this example. The distortion is caused by the presence of a drain voltage: this produces a potential variation of the first conduction sub-band edge, which is superimposed on the figure. There is injection from both the source and the drain contacts, as evinced by the states within the bandgap at each end of the device.

18.2.3   Evaluating the charge

The actual charge due to source-injected electrons, for example, is

$$Q_S(x, E) = -q\mathcal{G}_S(x, E)f_S(E). \qquad (18.16)$$

To evaluate this we need to know $\psi_S(x, E)\psi_S^*(x, E)$ absolutely. This need does not arise when evaluating the current because that depends on the wavefunctions via $T(E)$, which is a ratio, not an absolute value. Thus, we need to normalize the amplitude of the wavefunction.

---

[1] The states are not fully bound, as in a deep potential well for example, because electrons in the states do escape to the contacts.

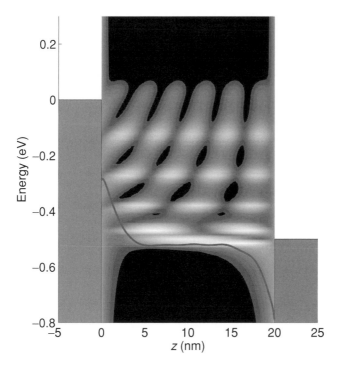

Figure 18.4 LDOS for a carbon nanotube FET with tube of length 20 nm and Schottky-barrier source and drain contacts. $V_{DS} = 0.5$ V and $V_{GS} = 0.9$ V. The profile of the edge of the first conduction sub-band is shown. Courtesy of Leonardo Castro, ex-UBC.

The usual procedure of normalizing via $\int_{\Omega} \psi \psi^* \, d\Omega = 1$ is not possible here because the electron is not confined to a known volume. We have open boundary conditions at the contacts, which are effectively regions of semi-infinite length, so we cannot assume that the entire wavefunction is confined to the vicinity of the semiconductor. Instead, we normalize via equating the Landauer current to the probability density current [5]. The latter follows from (5.52), and, for the source-originating electrons that have reached the drain, the probability density current is

$$ J_P = -q \frac{\hbar k_S(D, E)}{m^*(D)} |A_S(D, E)|^2 . \tag{18.17} $$

In the expression for the Landauer current, $T(E)$ is given in the context of our example by

$$ T(E) = \frac{k_S(D, E)}{k_S(S, E)} \frac{m^*(S)}{m^*(D)} \frac{|A_S(D, E)|^2}{|A_S(S, E)|^2} , \tag{18.18} $$

where the subscript S defines the origin of the injected electrons as the source; their wavevector is k_S and the amplitude of the incident wavefunction is $|A_S|$. Within the brackets ( ), S or D denotes the position (source or drain) at which the particular property needs to be determined. Inserting into (18.11), and equating to the current given by (5.52), we obtain a definite value for the previously arbitrary amplitude:

$$|A_S(S, E)|^2 = \frac{m^*(S)}{k_S(S, E)} \frac{1}{\pi \hbar^2} f_S(E) \equiv \frac{g_S(E)}{2} f_S(E). \tag{18.19}$$

Therefore, the desired relationship between probability density and charge for source-originating electrons is

$$Q_S(x, E) = -q \frac{g_S(E)}{2} f_S(E) \tilde{\psi}_S(x, E) \tilde{\psi}_S^*(x, E), \tag{18.20}$$

where $\tilde{\psi}_S$ is the un-normalized wavefunction.

18.3   Master set of equations for 1-D simulations

(5.24) is the master set of equations for use in 3-D systems that are large enough to be treated 'semi-classically', i.e., quantum-mechanical phenomena, such as reflection, resonance, and tunnelling, can be ignored, and microscopic phenomena, scattering for example, can be represented by some average property (mobility in this case). In our brief treatment of 1-D nanoelectronic systems we have seen that quantum phenomena are very much in evidence. In particular, ensemble-averaged quantities such as mobility and v_R are not usually employed because we try to keep track of each electron in its state k at energy E. Thus, Quantum Mechanics is very much to the fore, and this gives the new master set of equations a different appearance:

$$\frac{d^2 V(x)}{dx^2} = -\frac{Q_e(x)}{\epsilon}$$

$$Q_e(x) = -q \int_E [\mathcal{G}_S(x, E) f_S(E) + \mathcal{G}_D(x, E) f_S(E + qV_{DS})] \, dE$$

$$J_e(E) = -q \frac{2}{h} T(E) [f_S(E) - f_D(E)]$$

$$E\psi(x, E) = \left[E_C(x)\psi(x, E) - \frac{\hbar^2}{2m^*} \frac{d^2\psi(x, E)}{dx^2} \right], \tag{18.21}$$

where V is used for potential in order to avoid confusion with the wavefunction ψ. The effective-mass form of Schrödinger's equation is used in this example, and, for simplicity, the charge and the current are restricted to electrons.

A self-consistent solution to the equations of Poisson and Schrödinger is sought. The charge Q_e is the link between these two equations, and the Landauer current is used in the normalization of the wavefunction ψ, as described in the previous section. Generally, a charge profile $Q_e(x)$ is chosen, and Poisson's equation is solved for $V(x)$, subject to appropriate boundary conditions. In a truly 1-D system these would be the potentials on the end contacts. Taking the source as reference:

$$V(0) = 0 \quad \text{and} \quad V(L) = V_{DS} + \frac{\Phi_S - \Phi_D}{q}, \qquad (18.22)$$

where Φ is a work function, and the ends of the contacts ($x = 0$ on the source side, and $x = L$ on the drain side) are determined by the edges of the space-charge regions at the metal/semiconductor junctions. The work-function difference term should be recognizable from Chapter 10 as the built-in voltage of the system.

The resulting $V(x)$ is then used in the Schrödinger equation to solve for $\psi(x)$, from which a new $Q_e(x)$ is found via the LDOS. The process is iterated until convergence at the desired tolerance is obtained.

18.4   Comparison of 1-D and 2-D currents

In Fig. 18.2 we showed that there is some slight advantage as regards electron mean velocity in going from a 3-D- to a 1-D-structure. A more practical comparison would be between the currents in sytems of different dimensionality. This comparison is not easy to make for at least two reasons: (i) the potential profiles, and hence $T(E)$, are likely to be significantly different; (ii) the current densities have different dimensions. Nevertheless, we'll attempt to make a comparison between the currents in 1-D and 2-D systems. Such a comparison is relevant to FETs as the channel in MOSFETs is essentially a charge sheet, as discussed in Section 10.3.2. We circumvent the first difficulty by assuming that $T(E) = 1$ for all energies, i.e., we will be comparing the maximum theoretical currents.

Setting $T(E) = 1$ in (18.12) and expressing the integral as a Fermi-Dirac integral gives

$$I_{Dsat}(1D) = -q\frac{2}{h}k_B T \mathcal{F}_0(a_F). \qquad (18.23)$$

By setting the transmission probability to unity we are removing the quantum-mechanical features from the treatment. Therefore, we could derive (18.23) by a classical treatment of the type we employed to get the current in 3-D systems. By 'classical' we mean taking an average value for the electron velocity. The relevant expression for the current density is

$$J_{Dsat} = -q\frac{n_0}{2}2v_R. \qquad (18.24)$$

We do not have yet an expression for the current in a classical 2-D system. However, we can obtain such an expression for a 3-D system by substituting for n_0 from (4.10) and (4.11), and for v_R from (4.29):

$$J_{Dsat}(3D) = -q\frac{2^2}{h^3}(k_B T)^2 m^* \pi \mathcal{F}_1(a_F). \qquad (18.25)$$

It is possible to derive an expression for $J_{Dsat}(2D)$ directly, but, instead, let us use (18.23) and (18.25) to infer a general expression for the maximum current in terms of

Table 18.1 Values for the coefficients to be used in (18.26) for the maximum current in systems of a given dimensionality.

System	α	β	γ
1-D	1	1	0
2-D	2	$\frac{3}{2}$	$\frac{1}{2}$
3-D	3	2	1

three coefficients, α, β, γ, that are dimension-dependent:

$$J_{Dsat}(\alpha D) = -q\frac{2^\beta}{h^\alpha}(k_B T)^\beta (m^*\pi)^\gamma \mathcal{F}_\gamma(a_F).\qquad(18.26)$$

The appropriate values for the coefficients for the different dimensionalities are given in Table 18.1.

To make a comparison of currents in 1-D and 2-D, rather than current densities, we need to multiply the 2-D expression by the width Z of the channel. Here, we elect to take $Z = 2d$, where d is the diameter of the cylindrical semiconductor of the 1-D system. We use $2d$ rather than d because if we imagine assembling lots of nanowires or nanotubes in parallel, then some separation between them is needed to ensure that the gate electrode, rather than the neighbouring semiconductors, is the main determinator of the electrostatic conditions in each nanocylinder. Under these conditions, the ratio of maximum theoretical currents for 1-D and 2-D systems is

$$\frac{I_{Dsat}(1D)}{I_{Dsat}(2D)} = \frac{h}{2d}\frac{1}{\sqrt{2k_B Tm^*\pi}}\frac{\mathcal{F}_0(a_F)}{\mathcal{F}_{1/2}(a_F)}.\qquad(18.27)$$

To obtain a quantitative comparison, we consider two cases using parameters for carbon nanotubes: $d = 1$ nm, $m^* = 0.08m_0$; and $d = 3$ nm, $m^* = 0.03m_0$. Results are shown in Fig. 18.5. In this example there is some advantage in terms of maximum theoretical current to employing a 1-D semiconductor. Let's consider the case of $(E_F - E_C) = 10k_B T$, which corresponds to an electron concentration of ≈ 0.5 nm^{-1} in the 1-D case, and to $\approx 9 \times 10^{12}$ cm^{-2} in the 2-D case. The advantage in current of the 1-D case over the 2-D case, for the example of $d = 1$ nm, $m^* = 0.08m_0$, is about 3 times. The theoretical current in the nanotube in this case is ≈ 20 μA.

To put the latter number into some sort of practical perspective, realize that Si N-FETs in the 45-nm technology of 2009 can attain I_{Dsat} values of about 1.5 mA μm^{-1}. Thus, for a width equivalent to $2d = 2$ nm, the current would be 3 μA. The carbon nanotube FET looks very good in this comparison, and comes out even better when it is realized that the lowest conduction sub-band in carbon nanotubes is doubly degenerate, and that states could be populated in higher-order sub-bands. Thus, the maximum theoretical current is at least ≈ 40 μA. However, it would be prudent to keep in mind that the carbon nanotube figure is a theoretical maximum value, whereas the Si figure has been attained in practice.

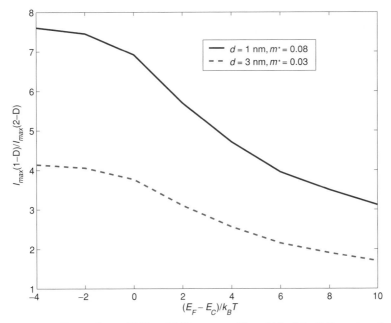

Figure 18.5 Comparison of 1-D and 2-D currents. The width of the 2-D semiconductor is $2d$, and results for two different cases are shown.

18.4.1 Energy dissipation in ballistic transistors

The picture we have presented of ballistic transport is of an electron leaving the source at energy E and arriving at the drain at the same energy. Thus, there is no energy loss in the intervening semiconductor. This means that it should be possible to operate ballistic transistors at enormous current densities without risk of the device burning out. However, with reference to Fig. 5.7, if the metallic contacts to the semiconductor are to remain in a near-equilibrium state, then the electrons injected into the drain must eventually lose their excess energy in thermalizing collisions. Thus, the contacts get hot and become the defining region for the safe, operable current density.

18.5 Novel features of carbon nanotube FETs

In the comparison of currents made in the previous section, carbon nanotube field-effect transistors (CNFETs) were pitted against Si MOSFETs. The comparison is interesting, but dangerous: it is very premature to conclude that CNFETs can take over from Si MOSFETs as the upholders of Moore's Law. Carbon nanotube development is still in the laboratory stage, whereas Si MOSFET technology is highly advanced, and has proven capable of putting hundreds of millions of transistors on a chip, and of having that chip work reliably for many years. Perhaps it would be more useful to focus on any novel features of nanotransistors that could enable applications which seem beyond both Si MOSFETs and the other modern transistors treated in this book? Two such features of CNFETs are mentioned very briefly below.

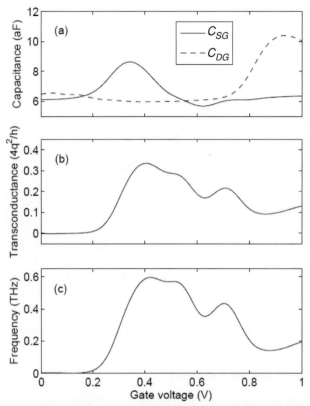

Figure 18.6 Bias-dependent small-signal properties of a Schottky-barrier CNFET using a nanotube of diameter $=1.26$ nm, length $=20$ nm; source and drain work functions $=3.9$ eV; $V_{DS} = 0.5$ V. (a) Gate-related capacitances; (b) transconductance; (c) f_T. From Castro *et al.* [6], © 2005 IEEE, reproduced with permission.

18.5.1   Quantum capacitance and transconductance

The LDOS shown in Fig. 18.4 become filled according to the positions of the contact Fermi levels E_{FS} and E_{FD}. In the instance shown, the energy of the first quasi-bound state is close to E_{FD}: this means that charge is injected into the semiconductor from the drain. The energy of the quasi-bound states is determined by the gate-source voltage V_{GS}, which is ≈ 0.9 V in this case. A change in semiconductor charge via the drain due to a change in gate voltage is represented by a capacitance C_{DG}, using the notation of Chapter 12. As V_{GS} is reduced, the edge of the first conduction sub-band is raised, and injection from the drain diminishes. Thus C_{GD} is reduced. Further reduction in V_{GS} eventually aligns the energy of the first quasi-bound states with the source Fermi level, resulting in a peak in the source/gate capacitance C_{SG}. These capacitances are called **quantum capacitances** because they arise from quantum phenomena (LDOS), rather than from the usual, merely electrostatic interactions that are responsible for metal/insulator/metal capacitance, for example. Examples of this bias dependence of the capacitances in a CNFET are shown in Fig. 18.6a.

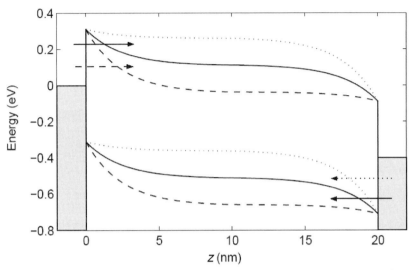

Figure 18.7 Energy band diagram for a Schottky-barrier CNFET using a nanotube of diameter $= 1.26$ nm, length $= 20$ nm; work functions of the nanotube and all metals $= 4.5$ eV; $V_{DS} = 0.4$ V. Hole injection at $V_{GS} = 0.05$ V (dotted line and arrow); hole and electron injection at $V_{GS} = 0.2$ V (solid lines and arrows); electron injection at $V_{GS} = 0.35$ V (dashed line and arrow). From Castro *et al.* [7], © 2004 SPIE, reproduced with permission.

Part (b) of the figure shows that there is bias dependence in the transconductance also. Again, as V_{GS} is increased, the band edge is 'pushed down' and g_m shows a peak when the energy of the first quasi-bound state is close to E_{FS}. Further increase in V_{GS} lowers the barrier for electron injection into the drain. Thus, the overall current of electron flow from source to drain decreases, and so does g_m.

The net effect of these changes in capacitance and transconductance is an interesting bias-dependence in f_T, as shown in Fig. 18.6c. Perhaps there is some useful application that could exploit this novel feature?

18.5.2 Ambipolarity

So far in this chapter we have neglected holes, but in CNFETs they can be important. Because the conduction-band structure and valence-band structure are symmetrical in carbon nanotubes, holes can be just as easily injected into a nanotube as electrons, providing the work functions of the two contacts are favourable.

Fig. 18.7 shows the situation for a Schottky-barrier CNFET in which the work functions of the source, drain, gate and nanotube are all the same. V_{DS} is fixed and the profiles of the first conduction and valence sub-bands for three values of V_{GS} are shown. At low V_{GS} there is hole injection from the drain via tunnelling. Thus, the device is operating like a P-FET. This current diminishes as the barrier at the drain thickens on increasing V_{GS}. However, the overall current does not keep on decreasing, as in a Si MOSFET (see Fig. 10.15 for an N-FET), because the barrier at the source becomes thinner, enabling

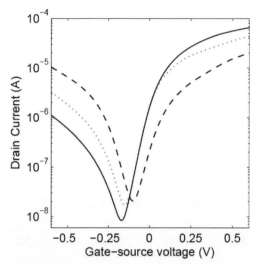

Figure 18.8 Gate characteristics at $V_{DS} = 0.4$ V for the CNFET used for Fig. 18.7, but with gate work function $= 4.2$ eV, and various source/drain work functions: 3.9 (solid line), 4.2 (dotted line), 4.5 eV (dashed line). From Castro *et al.* [7], © 2004 SPIE, reproduced with permission.

electron flow by tunnelling from the source. Therefore, the overall $\log I_D$-V_{GS} curve shows a minimum, as illustrated in Fig. 18.8.

Because of this inability to attain a very low OFF current, a CNFET with this particular set of work functions would make a poor transistor for digital-logic applications. However, the **ambipolarity** of the device, i.e., its ability to inject both electrons and holes, means that, in long devices, there is likely to be significant electron-hole recombination within the nanotube. As the bandgap is direct in carbon nanotubes, this means that there can be light emission. Further, because the relative amounts of electron and hole injection depend on the band-bending at the two contacts, i.e., on V_{GS}, then the position along the tube of the maximum in the $n(x)p(x)$-product will be determined by the gate bias. Thus, we have a **light-emitting transistor** with voltage control of the site of emission. Surely there is an application for this novel phenomenon?

Exercises

18.1 Expressions for the mean, unidirectional velocity of electrons in 1-D and 3-D equilibrium distributions appear in the text.
   (a) Write down these expressions, and from them infer the expression for the mean, unidirectional velocity of electrons in a 2-D equilibrium distribution.
   (b) Confirm your intuition by deriving an expression for the 2-D v_R.
18.2 It is instructive to calculate the current in a 1-D nanosystem using (18.11).
   To do this, consider an asymmetrical rectangular barrier, of the type shown in Fig. 5.8, and imagine this to represent a 1-D transistor, with Region 1 being the source, Region 2 being the channel, and Region 3 being the drain.

For a particular example, take the energies in eV to be 0, 0.1 and -0.5 for U_1, U_2 and U_3, respectively. Make the channel 10 nm long, take $m^* = 0.1m_0$ in all regions, and evaluate the transmission probability $T(E)$ for energies up to about $10k_B T$ above U_1.

(a) You should find that $T(E)$ peaks at values close to 1 at energies of 137, 251 and 443 meV.

(b) Now use (18.11) to get the spectral current density. You should find it peaking at about 131 meV.

(c) Finally, integrate the spectral current density to find that the current in this example is about 19 nA.

18.3 In Exercise 14.11 the small-signal y-parameters for a 2-port network were introduced, and used to examine an intrinsic MOSFET. Adding the parasitic resistances of the source, drain and gate to the circuit allows the extrinsic performance of the device to be examined. It is easier to add series resistances to an impedance network than it is to add them to an admittance network, so it is usual to employ z-parameters for the extrinsic case [8]. The resulting *extrinsic z-parameters* are:

$$z_{22e} = y_{33}/Y + R_{sd}$$
$$z_{23e} = -y_{23}/Y + R_s$$
$$z_{32e} = -y_{32}/Y + R_s$$
$$z_{33e} = y_{22}/Y + R_{sg}$$
$$Y = y_{33}y_{22} - y_{32}y_{23} , \qquad (18.28)$$

where $R_{sg} = R_s + R_g$ and $R_{sd} = R_s + R_d$.

Actual measurements on high-frequency devices usually employ s-parameters, as mentioned in Section 14.8. Conversions between all 2-port parameters are well-documented [9].

Here, we wish to examine the model parameters of a nanowire field-effect transistor. The measured s-parameters are given in the table below. The source and load impedances were 50 kΩ.

(a) Use the conversion formulae to obtain the z-parameters and the y-parameters at the given frequencies.

(b) Plot the appropriate functions of the z- or y-parameters on a graph from which it is possible to estimate f_T.

(c) Plot Mason's Unilateral Gain as a function of frequency and extrapolate to find f_{max}.

(d) Given that the resistances of the source, drain and gate are all 1 kΩ,[2] estimate the following device parameters: $C_{gs}, C_{gd}, C_{sd}, C_m, g_m, g_{dd}$. The smallness of the capacitances will give you further appreciation of the scale of nanodevice properties.

---

[2] Yes, these are very large. That's because it's a nanowire transistor. It's also why the source and load impedances for the s-parameter measurement are so high.

(e) Use your 'reverse-engineered' parameter values from the previous question to evaluate f_T from (14.40). Hopefully there is excellent agreement with the value computed above from part (b).

f (GHz)	s_{33}	s_{32}
1	$0.99985150454822 - 0.01634532553216i$	$0.00001727511869 + 0.00187960147356i$
4	$0.99762701570972 - 0.06530039195981i$	$0.00027605948194 + 0.00750899184418i$
10	$0.98527101475528 - 0.16212743411278i$	$0.00171348446647 + 0.01864175295064i$
40	$0.79015015095037 - 0.57753956483546i$	$0.02441224807920 + 0.06630950138903i$
100	$0.18708292391116 - 0.89547971380160i$	$0.09455821595567 + 0.10197010357623i$

f (GHz)	s_{23}	s_{22}
1	$-2.64657736369830 + 0.02617069629805i$	$-0.32343549257613 - 0.003079266667866i$
4	$-2.64300928913059 + 0.10455300242331i$	$-0.32385070665862 - 0.01230196618653i$
10	$-2.62319029818602 + 0.25958029893259i$	$-0.32615703010880 - 0.03054522445715i$
40	$-2.31021796439278 + 0.92448253847297i$	$-0.36257812914205 - 0.10893550076219i$
100	$-1.34292424363059 + 1.43160307403116i$	$-0.47515899209335 - 0.16999464828556i$

References

[1] Kinneret Keren, Rotem S. Berman, Evgeny Buchstab, Uri Sivan and Erez Braun, DNA-templated Carbon Nanotube Field-effect Transistor, *Science*, vol. 302, 1380–1382, 2003.

[2] K.K. Likharev, Single-electron Devices and Their Applications, *Proc. IEEE*, vol. 87, 606–632, 1999.

[3] J. Chen, M. A. Reed, A. M. Rawlett and J. M. Tour, Large On-Off Ratios and Negative Differential Resistance in a Molecular Electronic Device, *Science*, vol. 286, 1550–1552, 1999.

[4] S. Datta and B. Das, Electronic Analog of the Electro-optic Modulator, *Appl. Phys. Lett.*, vol. 56, 665–667, 1990.

[5] D.L. John, L.C. Castro, P.J.S. Pereira and D.L. Pulfrey, A Schrödinger-Poisson Solver for Modeling Carbon Nanotube FETs, *Tech. Proc. NSTI Nanotechnology Conf. and Trade Show*, vol. 3, 65–68, 2004.

[6] L.C. Castro, D.L. John, D.L. Pulfrey, M. Pourfath, A. Gehring and H. Kosina, Method for Predicting f_T for CNFETs, *IEEE Trans. Nanotechnology*, vol. 4, 699–704, 2005.

[7] L.C. Castro, D.L. John and D.L. Pulfrey, An Evaluation of CNFET DC Performance, *Device and Process Technologies for MEMS, Microelectronics, and Photonics III, Proc. SPIE*, vol. 5276, 1–10, 2004.

[8] W. Liu, *Fundamentals of III-V Devices: HBTs, MESFETs, and HFETs/HEMTs*, p. 440, John Wiley & Sons Inc., 1999.

[9] D.A. Frickey, Conversions between S, Z, Y, H, ABCD, and T Parameters which are Valid for Complex Source and Load Impedances, *IEEE Trans. Microwave Theory and Techniques*, vol. 42, 205–211, 1994.

19   Appendices

19.1   Appendix A: Physical constants

Constant	Symbol	Value	Units
Boltzmann's constant	k_B	1.38×10^{-23}	J/K
Dirac's constant	$\hbar$	1.05×10^{-34}	J s
Elementary charge	q	1.6×10^{-19}	C
Electron rest mass	m_0	9.1×10^{-31}	kg
Electron volt	eV	1.6×10^{-19}	J
Permittivity of free space	ϵ_0	8.85×10^{-12}	F/m
Planck's constant	h	6.63×10^{-34}	J s
Speed of light in vacuum	c	3×10^8	m/s
Thermal energy at 300 K	$k_B T$	0.0259	eV
Thermal voltage at 300 K	V_{th} or $k_B T/q$	0.0259	V

19.2   Appendix B: Selected material properties

This appendix indicates where to find information on the numerical values of those material properties that are used in the figures and examples in this book, or are required to complete some of the exercises.

The symbols for the properties are also listed, thereby providing a glossary of selected terms.

Those values that are listed here, and which do not appear elsewhere in the book, come mostly from *Semiconductors on NSM*; available online: [http://www.ioffe.rssi.ru/SVA/NSM/Semicond/].

The polarization properties of AlGaN come from Ambacher *et al.*, *J. Appl. Phys.*, vol. 85, 3222–3233, 1999.

For evaluating Fermi-Dirac integrals, please see: R. Kim and M.S. Lundstrom, *Notes on Fermi-Dirac Integrals*, 3rd Edn., posted 23 September, 2008. Online [http://nanohub.org/resources/5475/].

Diffusion length	Si	compute from $\sqrt{D\tau}$
L_e, L_h	GaAs	see Fig. 9.3
	AlGaAs & InGaP	take to be same as GaAs
Diffusivity	Si	see Fig. 5.3
D_e, D_h	GaAs	see Fig. 5.3
	AlGaAs & InGaP	take to be same as GaAs
Drift velocity	Si	see Fig. 5.2
v_{de}, v_{dh}	GaAs	see Fig. 5.2
	InGaAs & Ge	see Fig. 11.1
Effective density of states	Si	see Table 4.1
N_C, N_V	GaAs	see Table 4.1
	AlGaAs & InGaP	take to be same as GaAs
Effective mass	Si	see Section 5.4.2 and Table 4.2
(conductivity)	GaAs	see Section 5.4.2 and Table 4.2
$m^*_{e,\mathrm{CON}}$, $m^*_{h,\mathrm{CON}}$	InGaP	take to be same as GaAs
Effective mass	Si	see Table 3.2
(density of states)	GaAs	see Table 3.2
$m^*_{e,\mathrm{DOS}}$, $m^*_{h,\mathrm{DOS}}$	InGaP	take to be same as GaAs
Effective mass	Si	see Table 2.1
(band structure)	SiO$_2$	0.3
m^*_e/m_0, m^*_h/m_0	GaAs	see Table 2.1
	InGaP	take to be same as GaAs
	Al$_x$Ga$_{1-x}$As	$m^*_e = 0.067 + 0.083x$, $m^*_h = 0.48 + 0.31x$
Electron affinity	Si	4.01 eV
χ	SiO$_2$	0.9 eV
	GaAs	4.07 eV
	In$_{0.49}$Ga$_{0.51}$P	4.07 eV
	Al$_x$Ga$_{1-x}$As	$4.07 - 0.64 \times 1.247x$ eV
	Al$_x$Ga$_{1-x}$N	$4.1 - 1.87x$ eV
Energy bandgap	Si	see Table 2.1
E_g	SiO$_2$	8 eV
	GaAs	see Table 2.1
	In$_{0.49}$Ga$_{0.51}$P	1.89 eV
	Al$_x$Ga$_{1-x}$As	$1.424 + 1.247x$ eV
	Al$_x$Ga$_{1-x}$N	$x^2 + 1.7x + 3.42$ eV
Intrinsic carrier	Si	see Table 4.1
concentration	GaAs	see Table 4.1
$n_i = p_i$	AlGaAs & InGaP	compute from (4.19)
Lattice constant	Si	0.543 nm
a	SiGe	see Fig. 8.3
	GaAs	0.565 nm
	Al$_x$Ga$_{1-x}$As	0.565 nm for $x \leq 0.4$
	In$_{0.49}$Ga$_{0.51}$P	0.565 nm
	Al$_x$Ga$_{1-x}$N	$0.3189 - 0.0077x$ nm

(cont.)

	Si	see Table 4.2
Mean unidirectional		
thermal velocity	GaAs	see Table 4.2
$v_{R,e}$	InGaP	take to be same as GaAs
	$Al_xGa_{1-x}As$	use $m_e^*(x)$ in (4.30)
Mobility	Si	see (5.30) and Fig. 5.3
μ_e, μ_h	GaAs	see (5.31) and Fig. 5.3
	AlGaAs & InGaP	take to be same as GaAs
Recombination-Generation lifetime	Si	see (3.21)
$\tau_{e,RG}$, $\tau_{h,RG}$	GaAs	see (3.22)
	AlGaAs & InGaP	take to be same as GaAs
Recombination parameters	Si	see Table 3.1
A_e, A_h, B, C, D	GaAs	see Table 3.1
	AlGaAs & InGaP	take to be same as GaAs
Relative permittivity	Si	11.9
(static value)	SiO_2	3.9
ϵ_r or ϵ/ϵ_0	GaAs	12.9
	$Al_xGa_{1-x}As$	$12.9 - 2.84x$
	$In_{0.49}Ga_{0.51}P$	11.8
	$Al_xGa_{1-x}N$	$9.5 - 0.5x$
Polarization – piezoelectric	$Al_{0.3}Ga_{0.7}N$	-0.0113 C/m^2
P_{Pz}		
Polarization – spontaneous	$Al_xGa_{1-x}N$	$-0.029 - 0.052x$ C/m^2
P_{Sp}		
Saturation velocity	Si	see Fig. 5.2 and Table 16.1
(electrons)	GaAs	see Fig. 5.2 and Table 16.1
v_{sat}	GaN	see Table 16.1

19.3   Appendix C: N-MOSFET parameters

The parameters listed below are used in many of the examples, figures, and exercises in this book. They are intended to be representative of Si NFETs at the 90-nm and 65-nm technology nodes. Data for the 45-nm technology node is not widely available, but perhaps will become so in the near future, in which case you can fill-in the blanks.

	Units	CMOS90	CMOS65	CMOS45
L	nm	90	65	45
t_{ox}	nm	2.3	1.7	
ϵ_{ox}	ϵ_0	3.9	3.9	
N_A	cm^{-3}	8.3×10^{17}	2.6×10^{18}	
μ_{eff}	cm^2(Vs)$^{-1}$	230	600	
v_{sat}	cm s^{-1}	7×10^6	9×10^6	
V_{DD}	V	1.0	1.0	

Index